ACS SYMPOSIUM SERIES **303**

Petroleum-Derived Carbons

John D. Bacha, EDITOR
Chevron Research Company

John W. Newman, EDITOR
Ashland Petroleum Company

J. L. White, EDITOR
The Aerospace Corporation

American Chemical Society, Washington, DC 1986

Library of Congress Cataloging-in-Publication Data

Petroleum-derived carbons.
  (ACS symposium series; 303)

  Includes bibliographies and indexes.

  1. Carbon—Congresses. 2. Petroleum coke—Congresses. 3. Carbon black—Congresses.

  I. Bacha, John D., 1921–   . II. Newman, John W., 1937–   . III. White, J. L., 1925–   . IV. American Chemical Society. V. Series.

TP245.C4P43   1986   661'.0681   86-7894
ISBN 0-8412-0964-2

Copyright © 1986

American Chemical Society

All Rights Reserved. The appearance of the code at the bottom of the first page of each chapter in this volume indicates the copyright owner's consent that reprographic copies of the chapter may be made for personal or internal use or for the personal or internal use of specific clients. This consent is given on the condition, however, that the copier pay the stated per copy fee through the Copyright Clearance Center, Inc., 27 Congress Street, Salem, MA 01970, for copying beyond that permitted by Sections 107 or 108 of the U.S. Copyright Law. This consent does not extend to copying or transmission by any means—graphic or electronic—for any other purpose, such as for general distribution, for advertising or promotional purposes, for creating a new collective work, for resale, or for information storage and retrieval systems. The copying fee for each chapter is indicated in the code at the bottom of the first page of the chapter.

The citation of trade names and/or names of manufacturers in this publication is not to be construed as an endorsement or as approval by ACS of the commercial products or services referenced herein; nor should the mere reference herein to any drawing, specification, chemical process, or other data be regarded as a license or as a conveyance of any right or permission, to the holder, reader, or any other person or corporation, to manufacture, reproduce, use, or sell any patented invention or copyrighted work that may in any way be related thereto. Registered names, trademarks, etc., used in this publication, even without specific indication thereof, are not to be considered unprotected by law.

PRINTED IN THE UNITED STATES OF AMERICA

ACS Symposium Series

M. Joan Comstock, *Series Editor*

Advisory Board

Harvey W. Blanch
University of California—Berkeley

Alan Elzerman
Clemson University

John W. Finley
Nabisco Brands, Inc.

Marye Anne Fox
The University of Texas—Austin

Martin L. Gorbaty
Exxon Research and Engineering Co.

Roland F. Hirsch
U.S. Department of Energy

Rudolph J. Marcus
Consultant, Computers &
 Chemistry Research

Vincent D. McGinniss
Battelle Columbus Laboratories

Donald E. Moreland
USDA, Agricultural Research Service

W. H. Norton
J. T. Baker Chemical Company

James C. Randall
Exxon Chemical Company

W. D. Shults
Oak Ridge National Laboratory

Geoffrey K. Smith
Rohm & Haas Co.

Charles S. Tuesday
General Motors Research Laboratory

Douglas B. Walters
National Institute of
 Environmental Health

C. Grant Willson
IBM Research Department

FOREWORD

The ACS SYMPOSIUM SERIES was founded in 1974 to provide a medium for publishing symposia quickly in book form. The format of the Series parallels that of the continuing ADVANCES IN CHEMISTRY SERIES except that, in order to save time, the papers are not typeset but are reproduced as they are submitted by the authors in camera-ready form. Papers are reviewed under the supervision of the Editors with the assistance of the Series Advisory Board and are selected to maintain the integrity of the symposia; however, verbatim reproductions of previously published papers are not accepted. Both reviews and reports of research are acceptable, because symposia may embrace both types of presentation.

CONTENTS

Preface vii

1. **The Chemistry of Mesophase Formation** 1
   Harry Marsh and Carolyn S. Latham

2. **Chemical Characterization and Preparation of the Carbonaceous Mesophase** 29
   Isao Mochida and Yozo Korai

3. **The Pitch-Mesophase-Coke Transformation As Studied by Thermal Analytical and Rheological Techniques** 45
   B. Rand

4. **Microstructure Formation in Mesophase Carbon Fibers and Other Graphitic Materials** 62
   J. L. White and M. Buechler

5. **Electron Microscopic Observations on Carbonization and Graphitization** 85
   A. Oberlin, S. Bonnamy, X. Bourrat, M. Monthioux, and J. N. Rouzaud

6. **Residual Oil Processing: Predicting Slurry Oil and Coke Yields** 99
   W. P. Hettinger, Jr., D. P. Wesley, and R. H. Wombles

7. **Synthetic Aromatic Pitch: Aromatic Pitches from the Asphaltene-Free Distillate Fraction of Catalytic Cracker Bottoms** 118
   G. Dickakian

8. **Synthetic Aromatic Pitch: Aromatic Pitches from the Distillate Fraction of Catalytic Cracker Bottoms and Residue Fractions** 126
   G. Dickakian

9. **Synthetic Aromatic Pitch: Aromatic Pitch Production Using Steam-Cracker Tar** 134
   G. Dickakian

10. **Petroleum-Coke Overview** 144
    James H. Waller, Gary W. Grimes, and John A. Matson

11. **Delayed-Coking Process Update** 155
    Robert DeBiase, John D. Elliott, and Thomas E. Hartnett

12. **Petroleum-Coke Calcining Technology** 172
    H. H. Brandt

13. **New Calcining Technology of Petroleum Coke** 179
    M. Kakuta, H. Yamasaki, H. Tanaka, J. Sato, and K. Noguchi

14. **Petroleum-Coke Desulfurization: An Improved Thermal-Chemical Process** 193
    H. H. Brandt and R. S. Kapner

15. Granular Graphitic Carbon...200
    W. M. Goldberger, P. R. Carney, R. F. Markel, and F. J. Deutschle

16. Carbonization and Coke Characterization............................215
    Harald Tillmanns

17. Anode-Carbon Usage in the Aluminum Industry......................234
    Samuel S. Jones

18. Utilization of Petroleum Coke in Metallurgical Coke Making............251
    Kenji Matsubara, Hidetoshi Morotomi, and Takashi Miyazu

19. Mechanism of Carbon-Black Formation in Relation
    to Compounded-Rubber Properties...................................269
    James E. Lewis

20. High-Surface-Area Active Carbon....................................302
    T. M. O'Grady and A. N. Wennerberg

21. Dispersion of Metallic Derivates on Carbon Supports..................310
    Pierre Ehrburger and Jacques Lahaye

22. Progress of Pitch-Based Carbon Fiber in Japan.......................323
    Sugio Ōtani and Asao Ōya

23. Growth of Carbon Fibers in Stainless Steel Tubes by Natural Gas
    Pyrolysis...335
    G. G. Tibbetts

24. Carbon-Fiber-Reinforced Carbon Composites Fabricated by Liquid
    Impregnation...346
    Erich Fitzer and Antonios Gkogkidis

25. Carbon–Carbon Composites: Matrix Microstructure and Its Possible
    Influence on Physical Properties....................................380
    R. A. Meyer and S. R. Gyetvay

Author Index...395

Subject Index..395

PREFACE

THE ECONOMIC CRISIS of the early 1970s upset petroleum supply and price patterns throughout the world and led to an uncertain outlook for carbon precursor development. Nevertheless, promising concepts were emerging for a wider range of carbon products, including the idea that the "bottom-of-the-barrel" petroleum residues should not be viewed as a disposal problem but as a valuable source of heavy molecular species.

In 1975 the American Chemical Society (ACS) sponsored the first symposium on petroleum-derived carbons as part of the 169th National Meeting. This first symposium was organized by M. L. Deviney and T. M. O'Grady; the strong favorable response of the participants led these cochairmen to edit the papers for publication in the ACS Symposium Series. This volume (1) has become a standard reference work for carbon scientists and organizations with vested interests in the refining of petroleum precursors, the fabrication of carbon and graphitic materials, and the use of carbon and graphitic materials in our increasingly high-technological environment.

Since 1975, the increased intensity of research on carbon materials and precursors has been evidenced by the growth of various conferences on carbon, including those conducted biennially by the American Carbon Society (2-6) as well as the international conferences in Europe (7-11) and Japan (12). The number of published abstracts on carbon materials nearly doubled from the 12th American Conference (1975) to the 16th Conference (1983). Furthermore, new patterns of petroleum supply have formed a more stable basis for carbon precursor development and thus foster an optimistic outlook for the development of high-technological carbon products.

A second symposium on petroleum-derived carbons was held as part of the 187th National ACS Meeting. The primary objectives were to discuss research progress since the first symposium and to appreciate the potential of the higher added-value carbon products that are or could be based on petroleum-derived precursors. By inviting selected speakers, we sought to cover most topics of interest to the petroleum industry, and the papers included in this volume fall into five general categories:

- Chemistry and industrial processing of carbon precursors
- Chemical and plastic behavior of the carbonaceous mesophase, which is the liquid crystalline phase where the microstructure of coke and many other carbon products is established
- Petroleum coke and coking processes

- Carbon products ranging from conventional materials, such as electrodes for the aluminum and steel industries, to new high-technological applications, such as biomedical implants
- Carbon fibers and carbon–carbon composite materials

The topic of graphitic intercalation compounds was intentionally omitted because intercalation science has become a major field, encompassing disciplines beyond those conventionally associated with petroleum chemistry.

Thirty papers were presented at the 1984 symposium. Speakers were free to develop their topics, but encouraged to focus on their own work and to provide full reference lists for readers who wish to pursue particular topics in further depth. The twenty-five papers appearing in this volume were completed after the symposium so that authors could write with the benefits of the questions and discussions at the symposium. One paper entitled "Feedstocks for Carbon Black, Needle Coke, and Electrode Pitch," was published elsewhere, and readers are referred to that publication (*13*) for an analysis of market trends and an outlook for heavy aromatic oil supplies. Two other papers were also published elsewhere, and these appeared substantially in the same form as in the symposium preprints (*14, 15*).

We believe that three factors contributed to the success these symposia have achieved. The first factor is the growing recognition that the high-carbon "bottom-of-the-barrel" residues offer enormous potential for higher added-value carbon products. The second factor is the general appreciation that success in the development of carbon products depends on improved understanding of carbonization chemistry. These views are certainly well justified by the development of such products as high-modulus carbon fibers spun from refined petroleum (or coal-tar) pitches.

The third factor was the attendance of nine well-known carbon scientists from Europe and Japan in the 1984 symposium. Support for their attendance was obtained by a grant from The Petroleum Research Fund supplemented by contributions from the following industrial sponsors: Aluminum Company of America, Ashland Petroleum Company, Arco Petroleum Products Company, Exxon Research & Development Laboratories, GA Technologies Inc., Gulf Canada, Ltd., Gulf Research & Development Company, Koppers Company, Inc., Mobil Oil Corporation, The Standard Oil Company (Ohio), and UOP, Inc.

The active participation by European and Japanese scientists contributed to the scientific quality of the symposium and provided an international perspective that will be increasingly significant to future carbon technology. We were particularly impressed by the vigor of Japanese research and development on carbon materials in the absence of stimulus by large aerospace and defense industries. Reflecting the dedicated activity of many workers, the Japanese effort also seems to result from clear recognition by government and industry of the important role of carbon products in future technology.

In addition to the foregoing industrial sponsors, we also wish to thank the reviewers of the papers. Although they must remain anonymous, we wish to join a number of authors in expressing gratitude for their contributions by careful review, constructive criticism, and good questioning. Finally we thank our respective organizations for the opportunity to undertake the duties of symposium cochairmen as well as coeditors of the present volume.

JOHN D. BACHA
Chevron Research Company
Richmond, CA 94802

JOHN W. NEWMAN
Ashland Petroleum Company
Ashland, KY 41114

J. L. WHITE
The Aerospace Corporation
Los Angeles, CA 90009

Literature Cited

1. *Petroleum-Derived Carbons;* Deviney, M. L.; O'Grady, T. M., Eds.; ACS Symposium Series 21; American Chemical Society: Washington, DC, 1976.

2. Extended Abstracts, 12th Conference on Carbon, Pittsburgh, PA; American Carbon Society, 1975.

3. Extended Abstracts, 13th Conference on Carbon, Irvine, CA; American Carbon Society, 1977.

4. Extended Abstracts, 14th Conference on Carbon, University Park, PA; American Carbon Society, 1979.

5. Extended Abstracts, 15th Conference on Carbon, Philadelphia, PA; American Carbon Society, 1981.

6. Extended Abstracts, 16th Conference on Carbon, San Diego, CA; American Carbon Society, 1983.

7. Preprints, Carbon '76, Baden-Baden, Federal Republic of Germany; Deutsche Keramische Gesellschaft, 1976.

8. Proceedings, 5th London International Carbon and Graphite Conference; London, England, 1978.

9. Preprints, Carbon '80, Baden-Baden, Federal Republic of Germany; Deutsche Keramische Gesellschaft, 1980.

10. Proceedings, 6th London International Carbon and Graphite Conference; London, England, 1982.

11. Extended Abstracts, Carbone '84, Bordeaux, France, 1984.
12. Extended Abstracts, International Symposium on Carbon, Toyohashi University; Carbon Society of Japan, 1982.
13. Stokes, C. A.; Guercio, V. J. *Erdöl und Kohle* **1985**, *38*, 31.
14. Riggs, D. M. In *Polymers for Fibers and Elastomers;* Arthur, Jett C., Jr., Ed.; ACS Symposium Series 260; American Chemical Society: Washington, DC, 1984; pp 245–262.
15. Janssen, H. R.; Leaman, Gordon L. *Oil & Gas J.* **1984,** June 25, 79–83.

August 1, 1985

The Chemistry of Mesophase Formation

Harry Marsh and Carolyn S. Latham

Northern Carbon Research Laboratories, University of Newcastle upon Tyne, Newcastle upon Tyne NE1 7RU, England

> The origins and development of the concept of carbonaceous mesophase, as derived from discotic aromatic nematic liquid crystals, enable applications to be made to industrial processes. The world availability of pitch materials is such that there is an abundance of pitch which produces cokes of little commercial value. A major incentive for research into the chemistry of mesophase formation is the commercial up-grading of such pitches and the development of specialized pitches. Structure in cokes is described in terms of optical texture. The importance of viscosity of pyrolyzing pitch in controlling size of optical texture is stressed. Pitch viscosity itself is related to chemical composition of the parent material. Those rich in oxygen and sulphur tend to produce cokes with small sized optical texture; those rich in hydrogen tend to produce cokes with large sized optical texture related to needle coke formation. Generally, cokes with small sized optical texture tend to have higher strength, higher reactivity and higher CTE values. Cokes with larger sized optical texture, e.g. needle cokes, have lower reactivity and CTE values. The important role of transferable hydrogen in carbonizing systems is stressed. Attention is drawn to the fact that inerts or quinoline-insoluble material in pitch may not be inert, but can be chemically and physically active in the carbonization system. Future research and development is outlined.

The carbonaceous mesophase is the intermediate material formed during carbonization of parent pitch and leading to the resultant coke. The discovery and development of the concept of mesophase over the last twenty years must represent one of the most significant advances in carbon science. Mesophase is a term borrowed from the science of conventional liquid crystals and means "intermediate phase". The term "carbonaceous mesophase" is distinguished from the term 'nematic liquid crystal' the former being a polymerized liquid crystal system. Mesophase is composed of lamellar molecules the structures of which are based on the hexagonal network of carbon atoms of the graphite lattice. Currently, carbonaceous mesophases can be produced with a range of solubilities, in e.g. quinoline. The term 'mesophase pitch' with thermotropic properties has been introduced recently. Here the carbonaceous mesophase is formed by cooling the fluid isotropic pitch. Prior to the recognition of mesophase, the formation of anisotropic coke from a fluid phase was described in terms of a coking principle (1).

As early as 1944, Blayden, Gibson and Riley (2) looked for significant changes in structure between non-graphitizing and graphitizing carbons during the carbonization process, using X-ray diffraction techniques. Diffraction patterns were interpreted in terms of the concept that lamellar constituent molecules within the carbonizing system formed stacked units, called crystallites. Although the X-ray diffraction approach showed that essential differences existed between the mechanisms of formation of non-graphitizing and graphitizing carbons (definitions adopted from the study of Rosalind Franklin (3)), the method was not specific enough to be able to identify clearly "the coking principle". Wandless (4), in 1971, was anticipating the future quite clearly when he wrote that the basic strength of coke is determined in the plastic phase and the very early stages of resolidification.

Although Taylor (5) discussed the development of optical properties of anisotropic material formed during carbonization in 1961, and together with Brooks (6) reviewed the concept of liquid crystals as an intermediate to coke formation in 1968, it was not until well into the 1970's that the potential of this knowledge was realised.

The work of Brooks and Taylor led to a resurgence of research activity into the carbonization process with the advancement of knowledge leading to changes and developments within the carbon and graphite manufacturing industries. The tantalising prospect of cheap carbon fibres by spinning mesophase was one possibility, see Figure 1 (7). A window was opened into the mysteries of the delayed coker (8). The microstructure of carbon-carbon fibre composites using pitch carbon as the matrix material became better understood, see Figure 2 (9). Improved isotropic graphites of high density could be prepared by isostatic pressurised carbonizing techniques. The theory of coal co-carbonization processes was advanced considerably, and the use of pitch additives to up-grade effective coal rank was understood (10). Aspects of coke strength and physical properties e.g. the thermal expansivities of graphites were more fully understood. The role of quinoline insolubles (QI), both primary and secondary, in pitch carbonizations was explained, leading to the manufacture of prime needle cokes from coal-tar pitch (11, 12) by filtration of the QI material. A knowledge of the conditions of formation of mesophase from coal liquefaction products enabled a more precise control of operating conditions thus preventing, in this case, the retrogressive formation of mesophase and coke in pipe-work leading to plant closure (13). Improvements became possible in the manufacture of baked anodes for aluminium production as a knowledge of the structure of binder coke became available and the relationships between pitch properties and resultant cokes were further clarified. This in turn enabled pitch producers to begin to 'tailor-make' pitch materials relative to

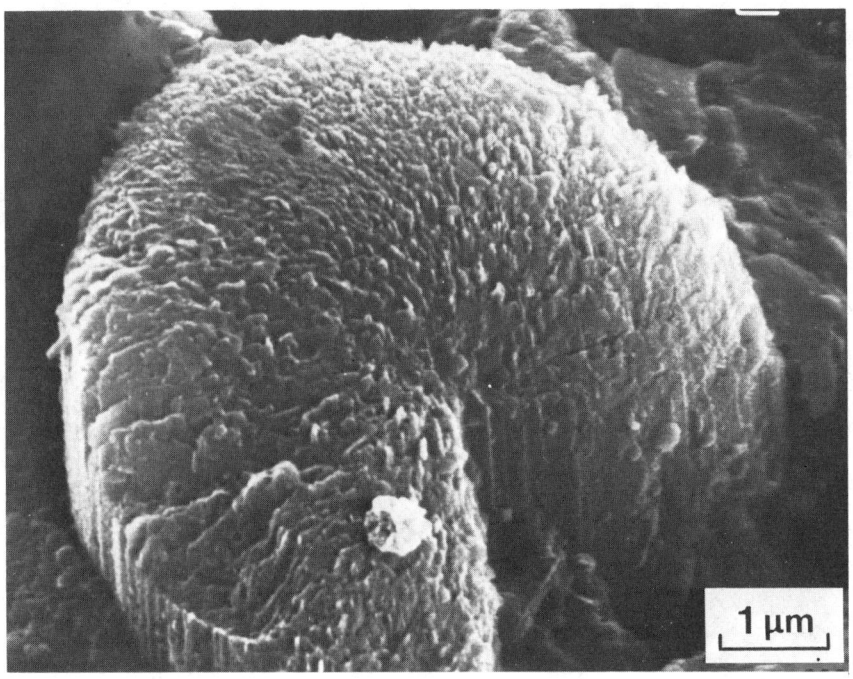

Figure 1. Scanning electron micrograph of fracture surface of a mesophase pitch carbon fiber etched with chromic acid to reveal the radial arrangement of constituent lamellar molecules.

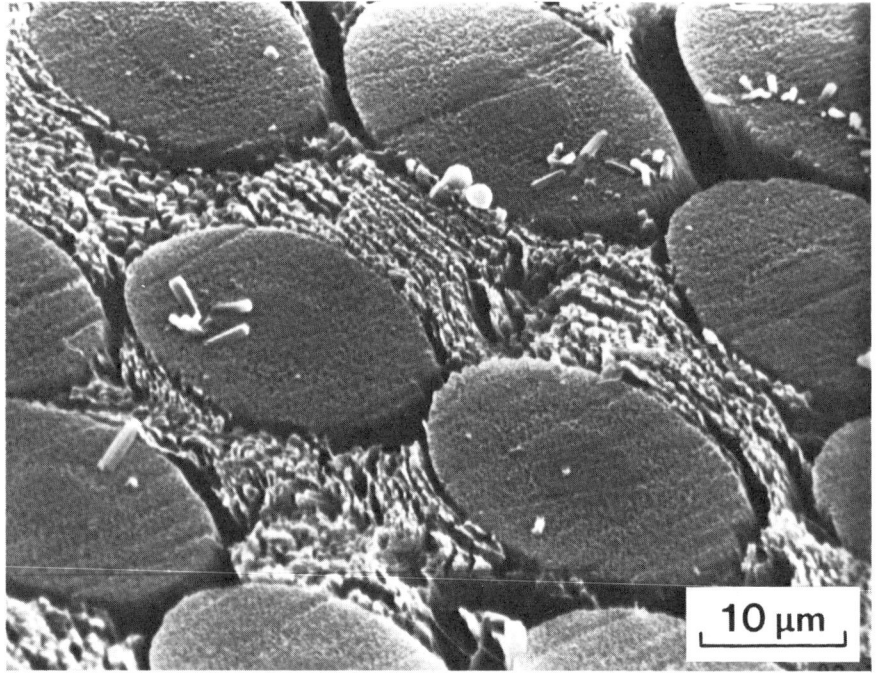

Figure 2. Scanning electron micrograph of a polished surface subsequently etched with chromic acid, of a carbon fiber-pitch carbon composite showing the orientation of a constituent lamellar molecules of pitch carbon parallel to surfaces of the carbon fiber.

a desired type of carbon product (14). Thus, the decade 1974-1984 has been one of intense interest and development of knowledge of many aspects of the conversion of pitch materials, with their various origins, to the structures of the many grades of cokes.

During this period, useful reviews were published. White (15) and Zimmer and White (16) describe the disclinations present in mesophase microstructures and their relation to coke properties. Lewis and Singer (17) overview the role and importance of stable free radicals in carbonization processes. Fitzer et al (18) provide a comprehensive description of the chemistry involved in the conversion of specific organic compounds to carbon. Marsh (19-23) and Forrest and Marsh (24) relate the chemistry of mesophase formation to its properties and applications. The journal CARBON published an issue devoted specially to studies of mesophase and its applications (25). The world availability of pitch materials is such that there is an abundance of pitch which produces cokes of little industrial value. A major incentive into research of mesophase is the commercial upgrading of such pitches.

Mesophase and Coke Structure

Petroleum and coal are the dominant parent sources of carbons, cokes and graphites. Apart from such exceptions as the carbonization of sucrose and related materials, the carbonization systems leading to graphitizable carbons (for definitions see Ref. 26) all pass through a fluid phase and all produce the liquid crystal/mesophase material. The cokes produced from such materials, e.g. filler coke, needle coke, shot coke, metallurgical coke and fluid coke, may not be distinguishable structurally in a meaningful way by X-ray diffraction. A powerful analytical tool in studies of coke structure is the polarized light optical microscope (16, 21) which, whether operating with cross-polarized light or making use of reflection interference colours (27, 28), very definitely categorizes carbons according to their optical texture. This optical texture is a measure of the size and coalescence behaviour of mesophase developing from the isotropic parent pitch and enables structure and orientation of lamellae within the mesophase to be established. A nomenclature (24) used to describe optical texture in cokes derived from petroleum and coal is in Table 1. The microscopic appearance of optical texture can be quantified in an arbitrary way (29) by making use of an optical texture index (OTI) calculated using the formula:-

$$OTI = \Sigma f_i \times (OTI)_i$$

where: f_i = fraction of component of optical texture in the overall appearance of polished surfaces of the coke

$(OTI)_i$ = an arbitrary (29) factor for each recognizable component of optical texture related to the relative sizes of the component as in Table 1. For example, for an isotropic carbon OTI = 0; for an anisotropic carbon composed entirely of domains, OTI = 30. All other anisotropic carbons have OTI values in the range 0-30 being summation of $(OTI)_i$ values. Examples of optical textures are the micrographs of Figures 3-6.

There is a limitation to the use of optical microscopy in terms of its resolution. The correlation of anisotropy (optical activity) with graphitizability and of isotropy (zero optical activity) with

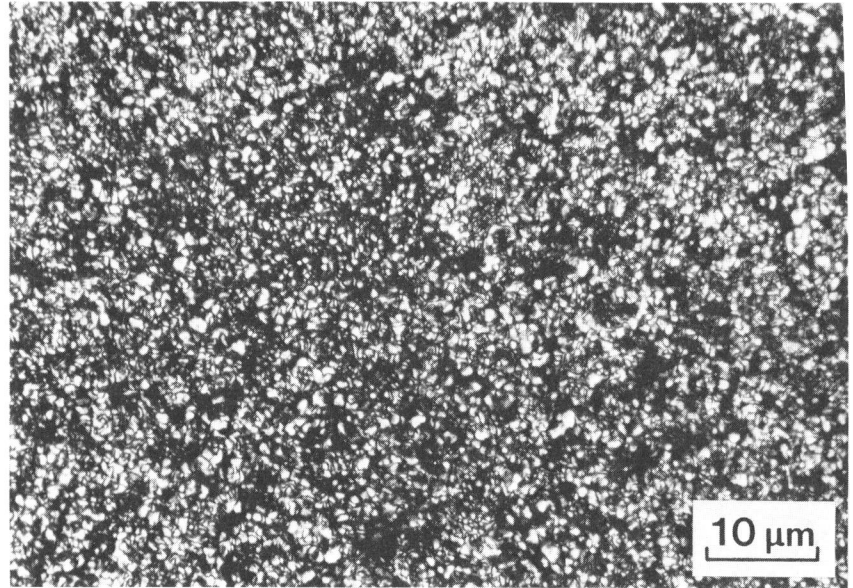

Figure 3. Optical micrograph of a coke surface showing an optical texture of fine-grained mosaics <1.5 μm diameter, OTI = 1.

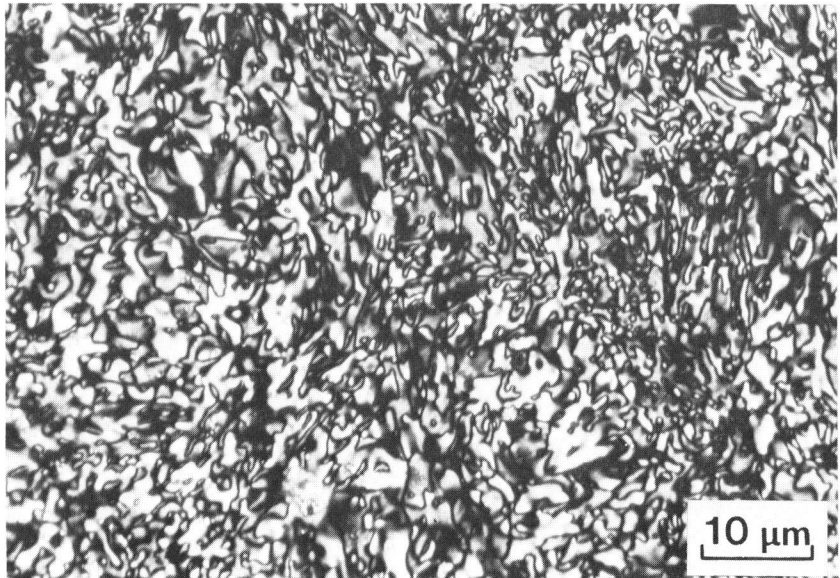

Figure 4. Optical micrograph of a coke surface showing an optical texture of medium-and coarse-grained mosaics, 1.5 - 10 μm diameter, OTI = 5.

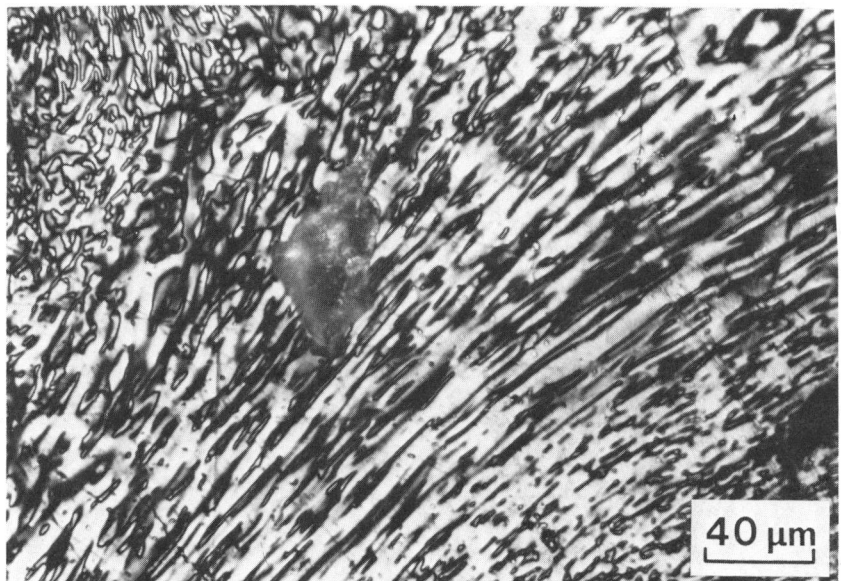

Figure 5. Optical micrograph of a coke surface showing an optical texture of coarse-flow anisotropy 30-60 μm length, 5-10 μm width, OTI = 20.

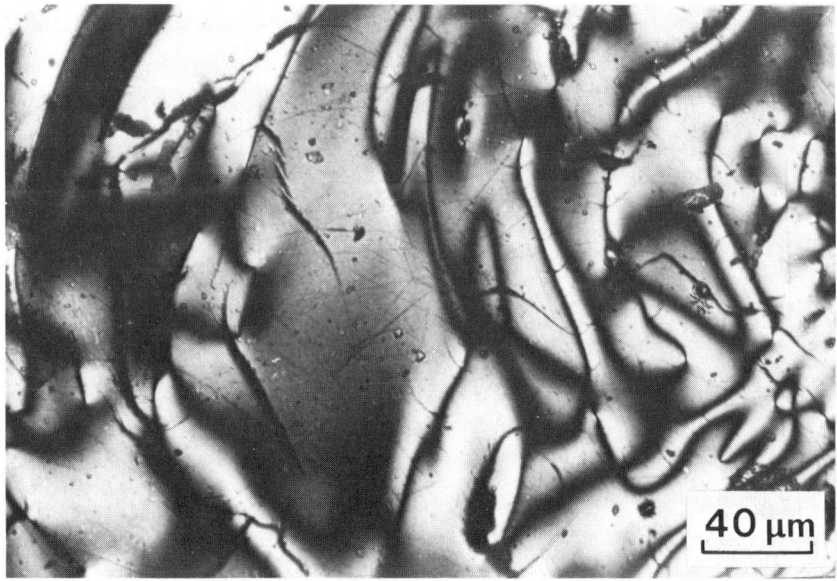

Figure 6. Optical micrograph of a coke surface showing an optical texture of domains, >60 μm, OTI = 30.

Table I. Nomenclature to Describe Optical Texture in Cokes

Component of Optical Texture seen in Microscopy	Abbreviation	Size	Optical Texture Index (OTI)
Isotropic	I	No optical activity	0
Fine-grained mosaics	Mf	<1.5 μm	1
Medium-grained mosaics	Mm	1.5–5.0 μm	3
Coarse-grained mosaics	Mc	5.0–10.0 μm	7
Supra-mosaics	Ms	Aligned mosaics	10
Small domains	SD	10.0–60.0 μm	20
Domains	D	>60 μm	30
Medium-flow anisotropy	MFA	<30 μm length <5 μm width	7
Coarse-flow anisotropy	CFA	30–60 μm length 5–10 μm width	20
Flow domain anisotropy	FD	>60 μm length >10 μm width	30

non-graphitizability is extremely useful but fails e.g. when examining cokes from coals of National Coal Board rank 600–400. It must be emphasised that mesophase can also exist when it is not visible in the optical microscope. Units of mesophase can exist, in pitch materials, of size e.g. <0.5 μm diameter. Such a pitch would appear to be isotropic to the optical microscope. However, examination of a fracture surface of this coke by scanning electron microscopy (SEM) clearly shows the interlocked (not coalesced) mode of attachment of these units of mesophase (30). This is illustrated in Figure 7 where the growth units within a coke are 0.1 μm diameter. Such a coke can be described as isotropic (to the polarized light microscope) but with limited graphitizability.

The growth units of mesophase can be detected of size <0.1 μm diameter in the pitch using transmission electron microscopy (TEM) (31). This theme was developed further by Monthioux et al. (32) and Auguie et al. (33) who examined by TEM and selected area electron diffraction (SAD) the carbonization of heavy petroleum products and report on the initial stacking of molecules which must represent the beginnings of mesophase. It is the subsequent growth and coalescence of these associations of planar aromatic structures which dictate coke properties. If the polymerisation of the pitch constituent molecules occurs to form non-lamellar molecules then random relative orientation will develop and the resultant coke becomes truly non-graphitizable (isotropic) (33). Hence all forms of carbon can exist between non-graphitizable, graphitizable and graphitized. As reported (32), the concept of two distinct forms of carbons can be discarded because of the known continuity of structure between the two extremes.

Figure 8, is an interesting SEM scanning electron micrograph of units of mesophase formed by co-carbonization of anthracene and phenanthrene (3:7) to 823 K at 300 MPa pressure (19). The effect of enhanced pressure is to increase the viscosity of the mesophase and

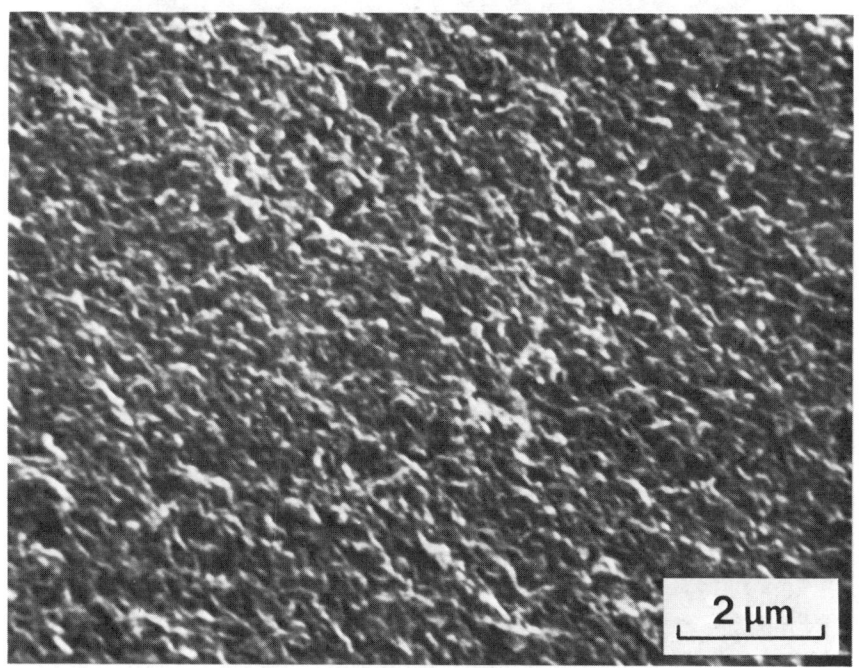

Figure 7. Scanning electron micrograph of fracture surface of coke from Chinese Shuang Ya coal, CR 502 (See Ref. 30).

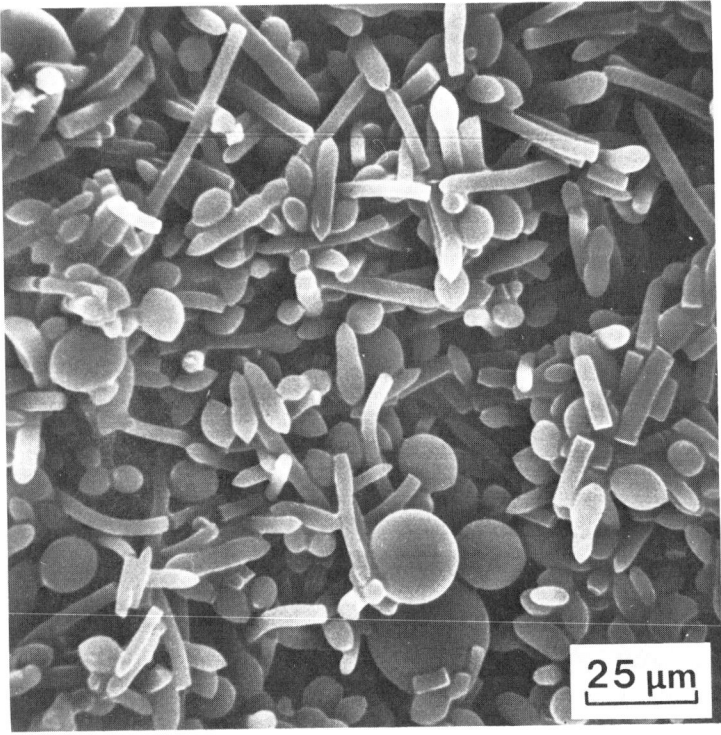

Figure 8. Scanning electron micrograph of growth units of mesophase from anthracene-phenanthrene (3:7), 823 K, 300 MPa pressure (See Ref. 19).

to prevent coalescence. Figure 8 shows mesophase which is cylindrical (narrow), ovoid and spherical in shape. The stacking of the lamellar molecules is parallel to a transverse section of the cylinder (as in a stack of coins). Figure 8 illustrates how some of the cylinders have converted into non-coalesced spherical shapes (of lower surface energy), perhaps because of small differences in temperature within the sample critically affecting viscosity.

The resultant optical texture of coke is directly dependent upon the viscosity of the preceding mesophase. Direct measurements of viscosity of pitch/mesophase systems have been reported (34-37). The use of hot-stage optical microscopy is particularly revealing. Viscosity measurements as pointed out by Nazem (36) can be misleading and have to be interpreted with caution. This is because rheological properties often reflect the two-phase composition of mesophase pitches and/or their liquid crystalline nature. Consequently, they are somewhat viscoelastic and cannot always be regarded as Newtonian liquids. In hot-stage microscopy, although quantification of data is not possible, direct physical presentation of phenomena occurs (38-43). The carbonization of a selected range of parent pitch materials using hot-stage optical microscopy demonstrates the relationship between viscosity and optical texture. As discussed by Yokono and Marsh (44) the physical property of viscosity of mesophase has a dominant role in determining optical texture of cokes. The diagram of Figure 9 summarises essential considerations. For pitch systems (relatively rich in oxygen and sulphur) which possess (relatively) the highest viscosities and the shortest range of temperature at minimum viscosity, the resultant coke possesses the smallest size of optical texture. Conversely, for pitch systems (relatively rich in hydrogen) which show the lowest viscosities and have the widest range of temperature at minimum viscosity, the resultant coke possesses the largest size of optical texture.

Chemical Aspects of Formation of Mesophase

Although the great majority of petroleum and coal-based pitch materials, as well as model compounds such as polyvinyl chloride, acenaphthylene, decacyclene and polynuclear aromatic hydrocarbons, form anisotropic graphitizable carbons, it is an almost impossible task to predict the type of optical texture of a coke from an elemental analysis of the pitch. The size, shape and reactivity of pericondensed polynuclear aromatic molecules in the products of pyrolysis of a pitch play a more important role in determining optical texture.

A major stumbling block to an understanding in depth of the chemistry of conversion of pitch to coke is the extreme chemical complexity of pitch composition. Even with a detailed knowledge of pitch composition based on e.g. chromatographic analyses, nuclear magnetic resonance (NMR) thermal analysis, mass spectrometry and electron spin resonance (ESR), the infinite number of possible reaction steps leading to mesophase growth will always be elusive (22). Rather the 'general' properties of pitch composition rather than specific molecular analysis and characterization provide a more usable base.

It is interesting to observe that the elaborate, detailed and extensive studies of Lewis and Singer (17) are essentially restricted to what (relatively) could be called simple systems (e.g. carbonization

of anthracene) - although the conclusions of these studies show they
are far from being simple. Cokes of largest size of optical texture,
produced from mesophase of minimum viscosity and maximum thermal
stability, are produced from polynuclear aromatic hydrocarbons. Lewis
(45) in a study of such model compounds showed the importance of
molecular rearrangements and dehydrogenative polymerizations in the
overall carbonization process. In the carbonization of acenaphthylene
(I) (Figure 10) rather complex rearrangements occur to give zethrene
(II). In the carbonization of anthracene (III), molecular size
increases by dehydrogenative polymerization without molecular rearrangements to form dibenzoperylene (IV) (46, 47).

Lewis and Singer (17) studied the radical chemistry of pyrolysis
of a wide range of model organic compounds. Most ESR studies are of
systems which have been heated to temperatures above which mesophase
is formed. An initial study by Marsh et al. (48) reported that in a
mixed mesophase-pitch system, the free radicals observable by ESR were
predominantly within the mesophase and later studies reviewed by
Lewis and Singer (17) support this. The ESR approach therefore does
not examine significantly the short-lived transient reactive radicals
leading to the larger discotic molecules which pass into the mesophase
(from the liquid crystal state). Rather the technique examines the free
radical chemistry of stable free radicals within mesophase, stabilized
by resonance within the extensive aromatic network of carbon atoms.
Yokono et al. (49), using high-pressure high temperature ESR in the
early stages of carbonization of pitch, monitor transient radical
removal by hydrogen. Lewis and Kovac (50) state that size of molecule
or radical is more important to mesophase growth than the properties
assigned to radicals. These authors noted that additions of radicals
derived from benzanthrene and naphthanthrene did not induce the formation of mesophase when melted with other polynuclear hydrocarbons.

An important clue to the relevance of radical chemistry prior to
mesophase formation comes from the study of Singer and Lewis (51) who
observed that, in the carbonization of ethylene tar pitch (which leads
to a smaller size of optical texture in resultant coke), both the
concentration and rate of build-up of radicals are larger than for
a petroleum-derived pitch. As discussed below, lower concentrations
of radicals lead to larger sized optical textures. The relationships
between transient and stable radicals have yet to be established.

As the pyrolysis of model, polynuclear hydrocarbon compounds
represents, possibly, the ultimate in ability to form the largest and
most stable of mesophase molecules leading to the domains (Table 1) of
optical texture in cokes, then smaller sizes of optical texture can
be explained by processes which restrict or inhibit polymerization to
larger molecule sizes. Conversely, it may be possible to increase the
size of the optical texture of coke by suitable ameliorative treatments
to a pitch.

The following is a discussion of factors which can be used to
control or influence the size of optical texture of cokes by manipu-
lation of the parent pitch composition and carbonization treatment.

<u>Rates of Carbonization</u>. It is reported (52) that decreased rates of
carbonization enhance the size of resultant optical textures. This
effect is concerned with the thermolysis of parent coal or pitch and
rates of evolution of volatile matter, some of which, if retained,
promote the fluidity (decrease the viscosity) of the system and hence

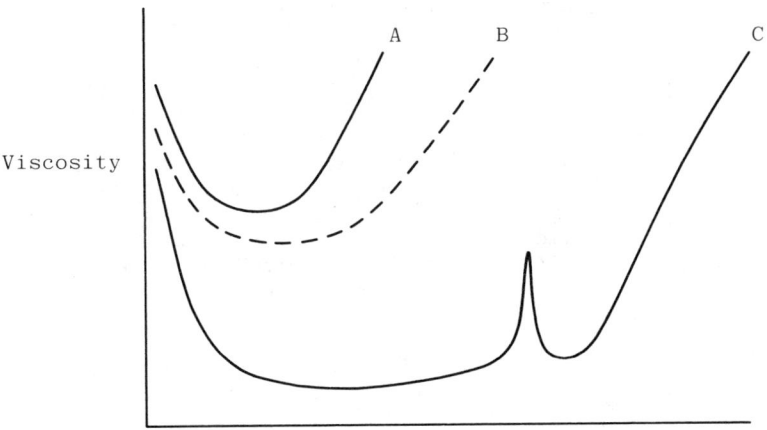

Figure 9. Diagram of possible variation of viscosity of pyrolysing systems with temperature of pyrolysis. System A: pyrolysis of coal to give coke of fine-grained mosaics, ~1 μm diameter; System B: pyrolysis of coal-pitch blend with good modification of coal by pitch to give coke with coarse-grained mosaics, <10 μm diameter; System C: pyrolysis of pitch to give coke with optical texture of domains >60 μm diameter, (See Ref. 44).

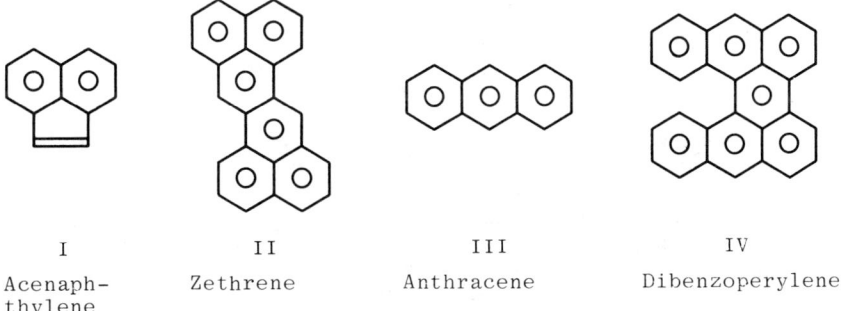

| I | II | III | IV |
| Acenaph-thylene | Zethrene | Anthracene | Dibenzoperylene |

Figure 10. Compounds used in carbonization systems.

promote the size of optical texture. This study (52) is of changes within one system only.

Heat Treatment Temperature and Soak Time. A study of Mochida and Marsh (53) indicates, unlike classical kinetics, that time and temperature for mesophase formation are not interdependent. The reason for this is the controlling influence of viscosity (not found for reactions in the gas or solution phase). Maximum size of optical texture and coalescence results if the mesophase is formed under conditions which provide a minimum viscosity as quickly as possible. Probably, rate controlling processes for mesophase growth are not the dehydrogenated polymerization reactions. Therefore, the attainment, relatively quickly, of temperature of ~400°C has provided the necessary size of molecule and consequently the resultant mesophase shows minimum viscosity because it is at a high temperature (~400°C). Mesophase formed at lower (relative) temperatures can have a higher viscosity and coalescence behaviour can be restricted.

Pressure of Carbonization. The effect of a pressurized carbonization is to create a closed system preventing loss of volatile materials. Hence, carbon yields increase. Further, the material normally lost as volatiles in open systems is now retained and the effect of this, by reducing turbulence and bubble formation, is to enhance the size of resultant optical textures. Hüttinger and Rosenblatt (54) report such effects when gas pressures up to 15 MPa pressure (150 bar) were applied to the carbonization of a coal-tar pitch. If higher pressures are used, the pressure being applied hydraulically to the carbonization system, then the effect of pressures at, say, 300 MPa, is to enhance the viscosity of the total system and this prevents coalescence of the mesophase. The resultant appearance of the carbon has been described as 'botryoidal' (55, 56) and an example is Figure 8.

Composition of Parent Pitch. Once the chemical composition of the carbonizing system moves away from the comparative simplicity of polynuclear aromatic hydrocarbons to that of industrial pitches, then the pyrolysis chemistry incorporates effects caused by the presence of heteroatoms (O, N and S) and alkyl and naphthenic groups. In general terms, the system becomes more 'reactive' creating higher concentrations of radicals detectable by ESR. This in turn, leads to enhanced cross-linkages and polymerization of molecular constituents of any mesophase which is formed, and this causes enhanced viscosity and a reduction in size of optical texture.

The review of Marsh and Walker (22) places emphasis upon the relationships between molecular structure and size of resultant optical texture. Certainly, the presence of reactive groups attached to aromatic nuclei, e.g. phenolic, carboxylic and the presence of heteroatoms, all lead to decreased size of optical texture. A principal conclusion of such studies is the difficulty of precise prediction of optical texture for a given carbonizing system. This is because it is difficult to quantify, precisely, for such many and diverse systems being carbonized, the critical balance which has to be maintained between the rate of carbonization, controlled by the reactivity of components of carbonizing systems (to increase the average size of constituent molecules), and the viscosity of the resultant pitch necessary for the movement and orientation of these

larger molecules to adopt the structure of the discotic nematic
liquid crystal leading to mesophase. Too high a reactivity or
viscosity both could result in the formation of the non-graphitizable
isotropic coke, the system having solidified before any mesophase
could be formed.

Analyses of Pitch. Modern analytical facilities of high-pressure
liquid chromatography, gel permeation chromatography, ^{13}C and ^{1}H
nuclear magnetic resonance and mass spectrometry, associated with IR
and UV spectroscopy enable a total molecular constituent analysis of
pitch composition to be obtained. The use of such information could
then possibly be the route to prediction of pitch quality on carbonization. It would appear that such an approach would not be successful
(ignoring the cost factor for such detailed analysis). The pitch
cannot be considered as an assembly of molecules which pyrolyse
independently of each other. The pitch carbonizes as a multi-phase
system and experience today would indicate the impossibility of
predicting all interactions, physical and chemical.

Irrespective of the chemistry of dehydrogenative polymerization
necessary to form the discotic molecules leading to liquid crystals,
the pyrolyzing pitch system has to maintain the necessary fluidity
(low viscosity). That is, part of the pitch, during the initial
stages of mesophase growth, has to play the role of solvent. This
permits the necessary movement or diffusion of larger sized molecules
in the growth process of the mesophase forming molecules. There also
exists the possibility that part of the pitch system may be soluble
in the mesophase and this acts as an internal plasticizer within the
mesophase so reducing its viscosity and enhancing coalescence and a
larger size of optical texture (57).

Thus, in general terms, pitch material may be considered as being
made up of three types of components: (a) the low molecular weight
fraction which acts as a solvent during the period of the carbonization
process and which, towards the completion of the formation of mesophase, either becomes incorporated into the mesophase by pyrolyses to
the discotic, mesophase-forming molecule or vaporizes; (b) a higher
molecular weight fraction which is central to formation of mesophase
from the pitch, i.e., the mesomorphic pitch fraction discussed by
Chen, Eilenberg and Diefendorf (58), (c) a fraction which is insoluble
in benzene and tetrahydrofuran (59) which, when carbonized independently, gives an isotropic, non-graphitizable carbon. However, it can
be assimilated when in the pitch by mesophase without apparent detriment. Within a single pitch system, considerations of solute/solvent
apply. With some thermally processed pitches an isotropic quinoline-
insoluble fraction can be isolated (60). The parent pitch does not
possess this fraction.

Thus, one approach to understanding the chemistry of pyrolysis
of pitch leading to mesophase is not to make a complete molecular
analysis but to solvent fractionate the pitch using solvents of
increasing solubility parameters (58). An early study of fractionation and NMR analysis of fractions is that of Smith et al. (61).
Bacha et al. (62) used cyclohexane, acetone, toluene and tetrahydrofuran to fractionate petroleum pitches and related the properties of
the fractions to optical texture of resultant cokes. For Ashland A240
pitch the toluene extracted fraction showed a higher temperature of
mesophase formation than the tetrahydrofuran fraction. The carboniza-

tion behaviour of pitches is discussed in terms of composition as measured by solvents of known solubility parameters (63).

Marsh et al. (60, 64) examined optical textures of cokes from carbonizations of soluble and insoluble fractions of coal-extract solutions as well as conditions of extraction. Generally, the low-molecular-weight fractions of the coal-extract solution produced cokes with larger sized optical textures than the coke from the parent coal-extract solution. The higher-molecular-weight fractions produced cokes with smaller sized optical texture. Isotropic coke was produced from material which was insoluble in benzene and tetrahydrofuran. Within the parent coal-extract solution the minor component of smaller molecules extends a "dominant partner effect" (21) by providing the necessary physical fluidity of the system and possibly some chemical stability (see below, on Hydrogen Transfer Reactions).

Kakuta et al. (65) fractioned petroleum feedstocks and noted that the saturate fraction which had a low aromaticity gave a coke with a needle-coke structure. A relationship was obtained between the chemical structural parameters of the fractions and X-ray peak intensity of the graphitized cokes derived therefrom. Coke quality related to the number of condensed aromatic rings and substituted side chain groups of the raw materials. Seshadri et al. (66) compared the aromatic and asphaltene fractions of a decant oil and of an ethylene pyrolysis tar by an NMR technique. The average molecule in these fractions differed in certain structural units and this may account for their different behaviour on pyrolysis. This study indicates that comparable fractionation procedures do not necessarily separate comparable molecular components and the need for further analysis still exists. For example, aromaticity may indicate good coking ability, but only if contained within certain limits of molecular size. Weinberg and Yen (67) analyzed coal-liquid solvent fractions and examined microscopically the resultant cokes. When different fractions had comparable aromaticities, determined by NMR (68), then mesophase growth was suppressed in the material with more oxygen.

Chen et al. (58) combine their results into a master plot depicting optical texture of coke against molecular weight of extract, carbonization temperature and time. This master plot indicates not only how size of resultant optical texture is developed but also its growth characteristics. The toluene insoluble fraction of significantly broader molecular weight distribution formed coalesced mesophase at 375°C in <0.5 h; lower molecular weight fractions formed spheres of mesophase. Higher molecular weight fractions formed high percentages of very small spheres of mesophase <1.0 μm diameter. Chen et al. (58) consider temperature coefficients of domain growth (coalesced mesophase) to be the same as for viscosity (69).

The world availability of pitch materials is such that there is an abundance of pitch which produces cokes of little industrial value. The economic pressures within the industry are to upgrade the commercial value of these pitches. Three ways of doing this appear as possibilities:- (i) from fractionation separation of pitch systems to select molecular species (a process of "molecular cropping") most amenable to formation of mesophase. (ii) to add, by blending, materials which have been proven to up-grade the quality of resultant cokes. This is equivalent to a dilution of unsuitable contents (as (i) above) but also there may be significant changes induced in the pyrolysis (see discussion of the dominant partner effect) (Ref. 21); (iii) to modify chemically by alkylation, hydrogenation, etc. the chemical composition of the pitch such that the resultant pyrolysis chemistry leads to improved coke quality.

Industrial practice (usually of a proprietary nature) may combine all three of these approaches (14).

Studies of blending of industrially based pitch materials are limited. Yokono et al. (70) in a preliminary publication, studying carbonization behaviour by proton relaxation measurements using Fourier-Transform NMR, observed that spin-lattice relaxation times and free spin concentrations of blends of pitch did not follow the additive rule (Figure 11). The spin-lattice relaxation time passed through a maximum and the spin concentration through a minimum. The positions of these maximum and minimum effects related to an increased size of optical texture. Such studies as these are of major industrial relevance as they are examining the central mechanisms leading to formation of mesophase.

Chemical Modifications to Pitch. The earlier attempts to improve the commercial value of pitch residues must have been essentially exploratory research. Sanada et al. (71) in 1973 methylated the hydroxyl groups of 3,5-dimethyl phenol formaldehyde resin and noted, on carbonization, the formation of spheres of mesophase, the original resin giving an optical texture of mosaics in resultant carbons. Mochida et al. (72) carbonized naphthalene, anthracene and pyrene with aluminium chloride, sodium and potassium and examined the structure of the resultant carbons by optical microscopy and high resolution, fringe-imaging transmission electron microscopy (TEM). The catalytic action of polymerization by the alkali metals was so severe that isotropic carbon was formed. That is, the increase in molecular weight occurred too soon in the carbonization process so precluding an existence of a fluid phase necessary for the formation of mesophase. The aluminium chloride was less severe and anisotropic carbon resulted, the size of optical texture being smaller for pyrene (mosaic) than anthracene carbons (flow). The TEM fringes, descriptive of molecular ordering within the carbons, were smaller for pyrene carbon than the anthracene carbon, a result in agreement with the study of Marsh et al. (21, 73) indicating smaller, less ordered molecules within the smaller sized optical textures compared with molecular stacking in cokes of larger sized optical texture. Mochida et al. (72) then went on to examine the co-carbonization of the heterocyclic compounds carbazole, phenazine and acridine with the Lewis acid-aluminium chloride. Whereas in the absence of the catalyst these compounds would have vaporised from the system, in the presence of the catalyst carbons of needle-coke appearance were formed from carbazole and phenazine, with fine-grained mosaics (Table 1) being observed in the coke from the acridine.

Sulphur, or sulphur-containing heterocyclics induce the formation of cokes of small sized optical texture (22); Mochida et al. (75) examined the carbonization behaviour, in the presence of aluminium chlorides, of diphenyl sulphide, thioxanthene and thianthrene, Figure 12. The catalyst up-grades the size of optical texture. Mochida et al. (75) discuss polymerization mechanisms and rates leading to mesophase as well as sulphur elimination.

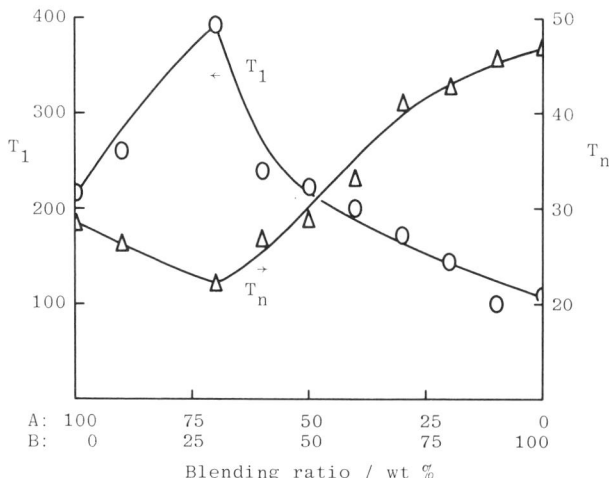

Figure 11. Variation of spin lattice relaxation time and spin concentration for carbonization of blends of two petroleum pitches (See Ref. 70).

–O—O– T_1 = spin lattice relaxation time in msec.
–△—△– T_n = spin concentration x $10^{18} g^{-1}$.

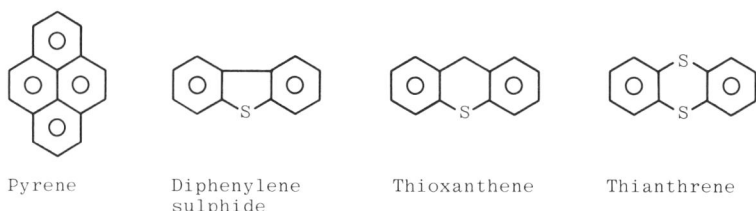

Pyrene   Diphenylene sulphide   Thioxanthene   Thianthrene

Figure 12. Model compounds used in carbonizations (75).

Mochida et al. (76) report that in carbonizations of several heterocyclic compounds containing nitrogen, sulphur and oxygen, the presence of the heteroatoms markedly affects optical texture of resultant cokes. When co-carbonized with aluminium chloride the heteroatoms exert a less marked effect where heteroatom evolution becomes important. Again, Mochida et al. (76) emphasise that the rate of the carbonization process, by influencing viscosity, is dominant in determining optical textures of cokes and related physical properties, i.e. CTE.

Mochida et al. (77) studied blending by co-carbonizing heterocyclic compounds with anthracene, and 9,10-dihydroanthracene catalyzed by aluminium chloride. Small additions of anthracene (~20 wt %) significantly up-graded the size of optical textures confirming the Dominant Partner Effect postulated by Marsh et al. (21) in which minor constituents of a carbonizing system significantly up-grade coke quality.

In studies running parallel to carbonization with catalysts, Mochida et al. (78, 81) studied the separation of parent feedstocks into benzene-soluble (BS) and benzene-insoluble fractions (BI) and attempted co-carbonization of blends of these fractions. In addition, certain fractions were hydrogenated or alkylated and these alkylated fractions themselves hydrogenated. Mochida et al. (78-81) were looking for 'compatibility' between fractions, that is the ability of a blend of fractions to carbonize to a needle-coke. The commercial application of good compatibility could involve the use of low percentage additions of an active fraction (component) to produce a good needle-coke.

This approach to pitch blending studies appeared to be successful. Careful and extended studies are obviously essential.
The conclusions available are:-
(i) Hydrogenation improves significantly the co-carbonization compatibility of solvent refined coal with benzene insoluble fractions which give isotropic coke.
(ii) Good compatibility can be attributed in part to naphthenic structures.
(iii) Alkylation destroys the co-carbonization compatibility of BS fractions suggesting poor compatibility in pitches could be associated with alkylation in constituent pitch molecules.
(iv) Hydrogenation of the alkylated material restored the compatibility.

Effects upon optical texture of cokes of other additives to pitch have been reported. Evans and Marsh (82) added anhydrides, Ōtani and Ōya (83) added ferric chloride and Murty et al. (84) added metals.

The Role of Transferable Hydrogen. In attempts to up-grade effective coal rank (i.e. increase the size of optical texture of resultant coke) by pitch addition based on the Dominant Partner Effect (21), Mochida et al. (85, 86) found it impossible to predict the ability of the industrial pitches to modify coal carbonizations (up-grade size of optical texture) using elemental analyses and fractionation techniques. Those pitches with excellent modifying ability must possess effective molecular structures which were difficult to recognise. In a separate study, Marsh and Neavel (87) discussed the common stage in mechanisms of coal liquefaction and of co-carbonization of coal blends. The role of hydrogen in liquefaction is one of transfer to reactive radicals, so stabilizing the radical/molecule and preventing the retrogressive reactions of coke formation (13). The suggestion was made (87) of the hydrogen-donor facility of a pitch additive to coal being the controlling factor in the modifying ability of the pitch additive to increase the effective rank of the coal (10, 88). This, in turn, leads to the suggestion that cokes with optical textures of domain were formed from pyrolyzing systems in which hydrogen transfer reactions have a significant role in stabilizing reactive species, thereby facilitating the formation of the polynuclear aromatic molecules leading to mesophase (Figure 13).

The most powerful analytical methods available to confirm this theory are NMR and ESR. Yokono et al. (89, 90) used both NMR and ESR to elucidate coal structure and the composition of coal extracts. Spin-lattice and spin-spin relaxation times for protons were measured as well as carbon aromaticity using high resolution gated decoupling ^{13}C spectra (90). Miyazawa et al. (91) examined the early stages of carbonization with a high temperature, in situ, probe of a pulsed FT-NMR spectrometer. These authors report an excellent relationship between the temperature dependence of line-width at half-height of proton spectral bands ($\Delta H_{\frac{1}{2}}$) and the optical texture of the mesophase. The line-width is considered to be influenced principally by the proton dipole-dipole interactions which reflect the degree of fluidity (mobility) of the pyrolyzing system.

Miyazawa et al. (92) related rates of decrease of aliphatic hydrogen protons during pyrolysis of ethylene tar pitch to formation of mesophase. Yokono et al. (93) used the model compound anthracene to monitor the availability of transferable hydrogen. Co-carbonizations of pitches with anthracene suggested that extents of formation of 9,10-dihydroanthracene could be correlated with size of optical texture. The method was then applied to the carbonization behaviour of hydrogenated ethylene tar pitch (94). This pitch, hydrogenated at 573 K, had a pronounced proton donor ability and produced, on carbonization, a coke of flow-type anisotropy compared with the coarse-grained mosaics (<10 µm dia) of coke from untreated pitch.

Obara et al. (95) co-carbonized a petroleum pitch which gave a coke of mosaic size of optical texture with the strong Lewis acid catalyst, aluminium chloride, which promoted the size of the optical texture and extents of hydrogen transfer to added anthracene. A correlation was established between size of optical texture of the resultant cokes and extents of formation of 9,10-dihydroanthracene plus evolved hydrogen gas.

The above concepts have been applied with considerable success to coal-based systems as distinct from petroleum-derived carbons. The subject area is extensive and beyond the scope of this paper. However, several references are available which provide a lead into this subject area (96-101). In both petroleum pitch and coal-tar pitch the presence of hydro-aromaticity and methylene linkages may provide the hydrogen for the transfer reactions. The removal of hydrogen from such linkages does not in itself produce stable radicals.

The Role of Inerts in Pitch

Industrial pitch often contains components which are insoluble in quinoline, i.e. QI material. This QI material can conveniently be

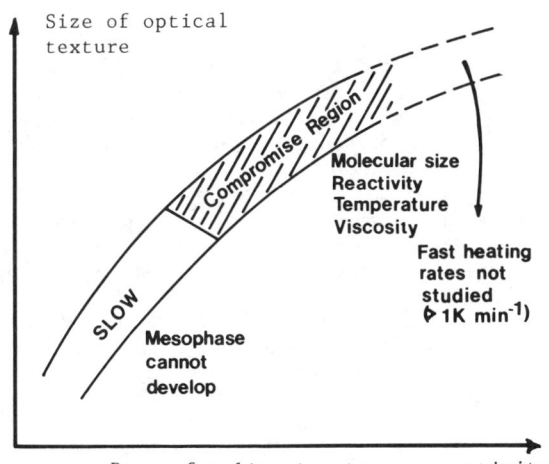

Figure 13. A diagrammatic representation of how size of optical texture of coke is related to rates of polymerization or reactivity of constituent molecules.

placed into several types. Primary QI is usually found in coal-tar pitches, is a product of the carbonization/distillation process, is particulate, usually <1 μm diameter, resulting from cracking processes in the vapour phase and resembles carbon black. Petroleum pitches usually have low values of primary QI. Secondary QI occurs in pitches following thermal treatment and is essentially made up of spheres of mesophase, without coalescence, <10 μm diameter. Also in pitch there are to be found specimens of pyrolytic carbon, of coke dust and 'dirt' particles. A tertiary type of QI, representing molecular components soluble in the pitch but not in quinoline, is described by Marsh et al. (60). These QI materials are usually referred to as the 'inert' contents of pitches.

This is a serious misnomer as these inert constituents of pitch are certainly not inert during the carbonization processes. It is well-established that the size of the optical texture of a coke can be reduced by the presence within the pitch of primary QI material (102-105). The QI material within the pitch becomes adsorbed on the surfaces of the growth units of mesophase. This thereby prohibits coalescence of these growth units into the larger sized optical textures. When this process is viewed by hot-stage optical microscopy (106) this lack of coalescence is seen to reduce markedly the flow characteristics of the mesophase - it becomes almost static. This influences the structural features of the mesophase which remains more disordered, a point made by Cranmer et al. (43). Stadelhofer (107) found that the presence of QI did not change rates of formation of mesophase. Romovacek et al. (108) consider that pyrolytic particles in pitch (primary QI) retard the development of mesophase and suppress coalescence. Decrease in size of optical texture, as brought about by 'mechanical' modification as distinct from 'chemical' modification of pitch properties can increase both the strength and reactivity to oxidising gases of the resultant coke, as recently put forward by Markovic et al. (109).

The presence of inerts can override chemical factors within pitches undergoing pyrolysis. In fact, Marsh et al. (110) postulate that the chemistry of polymerization within mesophase may be dependent upon the flow properties of the mesophase. Hence, inerts may influence pyrolysis chemistry because they can restrict movement of mesophase.

Ōbara et al. (111) investigated the influence of silica gel (<325 mesh) as an inert additive on the carbonization reactions of A240 pitch by optical microscopy, measurements of transferable hydrogen, high-temperature ESR and high temperature ^{1}H-NMR. Additions of silica gel have the effect of reducing the size of the optical texture of mesophase which appears in the early stage of carbonization. The higher the concentration of silica gel in pitch, the less is the amount of transferable hydrogen (H_{DHA}) for A240 pitch. The spin concentration of pitch increases with increasing silica gel content. Measured values of $\Delta H_{\frac{1}{2}}$, using high temperature ^{1}H-NMR, suggest that the molecular motions in the pitch become rigid with the addition of silica gel. There is thus an effect of silica gel on the physical properties and chemical reactivity of carbonizing pitch.

Future Research and Development

The last twenty years have seen an appreciation of the origins of anisotropy or graphitizability in carbons and the beginnings of the understandings of its formation.

The complexity of formation of mesophase must not be underestimated. With the exception of a few model compounds, it is the industrial pitch which is the source of mesophase. Such materials contain thousands of reactive molecules and there is an interdependence in the carbonization system which currently is known to us but not analyzed in depth. This is an area for further research. Formation of mesophase is further complicated because it involves chemistry within a fluid/plastic system of increasing viscosity. And in the delayed coker, volatile release and liquid turbulence are yet additional factors in influencing final structure in mesophase.

Summary of Known Methods of Preparation of Mesophase. At the time of writing, it appears that a major objective of research into the chemistry of mesophase formation is to obtain mesophase at as low a temperature as possible and to have a soluble mesophase, e.g. in 1,2,4 trichlorobenzene (112). The applications of such mesophase(s) must be in the production of graphitizable pitch fibres. However, this direction of research could have applications in many other areas of carbon/graphite production. Therefore it is appropriate, now, to make a list of known methods of preparation of mesophase (113).

Bulk processes of carbonization can be "stopped" at an appropriate stage and mesophase separated from the remaining pitch by solvent extraction. Spheres of mesophase can then be isolated. However, from the viewpoint of specialised industrial usage, e.g. spinning/moulding (not traditional coke-making) such mesophase can be said to be "overcooked". It is not readily soluble in generally available solvents; it does not represent a minimum in its visco-elastic properties.

Crystallization of Mesophase. As mentioned above, there exists an approach which selects solvent fractions from pitch materials as being suitable for manufacture of carbon artifacts. This is 'trial and error' molecular engineering. If the selected cropped feedstock is suitable then it is possible:-
  (i) to heat the feedstock to a temperature below that at which spheres are normally seen.
 (ii) to cool this isotropic pitch, of pre-determined heat treatment temperature, such that spheres of mesophase crystallize out.

Because temperatures have been kept low (relative to traditional systems), this mesophase is formed utilizing the established thermotropic properties of nematic liquid crystals. Such 'spheres of crystallization' can therefore be dissolved in the pitch (by re-heating) or in other solvents. They tend to have a lower viscosity than traditionally made mesophase.

Precipitation of Mesophase. If the modified, cropped feedstock is suitable then it is possible:-
  (i) to heat feedstocks to a temperature below that at which spheres are normally seen,
 (ii) to cool this isotropic pitch, heat-treated appropriately, to crystallize out spheres of material,
(iii) to separate these spheres from residual pitch,
 (iv) to dissolve the spheres in a solvent of high solution power (solubility parameter), e.g. quinoline,

(v) to gradually remove the solvent in a distillation process causing the <u>preciptation</u> of pitch-based liquid crystal material. This material has the properties of good solubility and of low viscosity.

These methods, described above, are to produce material (still loosely called mesophase) which essentially is a feedstock for other process developments. This "mesophase" prepared at temperatures below normal carbonization temperatures can be called low temperature mesophase pitch (LTMP). The term mesophase pitch has crept into the vocabulary of this subject and is thought to refer mainly to mesophase as a feedstock. Its anisotropy can be detectable by polarized light optical microscopy.

<u>Lyotropic Mesophase</u>

It is possible to develop further the specificity of 'molecular engineering' as follows:-
  (i) use the low temperature mesophase pitch (LTMP) as described in paragraphs above,
 (ii) selectively extract molecular species from the LTMP by use of e.g. benzene to give benzene-insoluble (BI),
(iii) prepare new systems by blending BI with BS in ratios different from that in the original LTMP. In this way some further optimization of feedstock properties/suitability of LTMP can be achieved.

<u>Current Mesophase Research</u>. With a view to production of carbon fibres and of graphite/carbon artifacts of specialized use, the research programmes appear to be orientated towards:-
(1). Optimization of cropping or molecular engineering of the molecular constituents of pitches available industrially. This essentially is the creation of new parent feedstocks by separation/blending techniques.
(2). Upgrading the properties of parent feedstocks by chemical methods:-
  (i) hydrogenations
 (ii) alkylations
(iii) polymerizations
 (iv) aromatizations, etc.
(3) the formation of mesophase pitch or low temperature mesophase pitch (LTMP) according to details given above.
(4) Analyses of chemical compositions of parent pitches to optimize a pitch origin so leading to optimization of molecular engineering techniques. This involves considerable analytical work of separation and analysis by GC/MS, NMR, IR, etc..
(5) The establishment of relationships between molecular sizes/shape, and structure of compounds in pitches and solubilities in solvents of known solubility parameters.
(6) the establishment of relationships between the viscosity of mesophase and its structure with the surface energy associated with the containing fluid phase.

<u>Conclusions</u>

Industrially available pitch materials are extremely complex chemical mixtures. Detailed chemical analyses in terms of characterization of molecular species may not be rewarding in terms of effort required.

Pitch materials may best be characterized in terms of compositions of molecular weight fractions or solubility fractions.

Polynuclear hydrocarbons, e.g. anthracene, form mesophase by dehydrogenative polymerization reactions. Functional groups and heteroatoms hinder mesophase growth.

Pitch materials can be improved with regard to coke quality by:
(a) solvent or distillation separation techniques;
(b) chemical modification of the pitch, e.g. hydrogenation;
(c) judicious selection of blending procedures;
(d) relating pitch properties to optical texture and quality of resultant cokes.

A major factor to consider in pitch carbonizations is the facility for hydrogen transfer reactions. These stabilize otherwise reactive radical species and permit the growth of mesophase in a fluid of low viscosity. Hydroaromaticity and methylene linkages are able to transfer hydrogen without creating radicals.

Inert material in pitch, the QI material, is not an inert constituent and its behaviour should be noted seriously, both with regard to the improvement and to the detriment of coke properties.

Acknowledgments

C.S.L. is grateful to British Alcan Lynemouth Ltd. for support to enable her to study in The Northern Carbon Research Laboratories. The authors appreciate the support of Miss B.A. Clow, Mrs. M. Poad and Mrs. P.M. Wooster in the preparation of the manuscript.

Literature Cited

1. Marsh H. Fuel, 1973, 52, 205.
2. Blayden, H.E.; Gibson, J.; Riley, H.L. Proc. Conf. Ultrafine Structure of Coals and Cokes, BCURA (London) 1944, p.176.
3. Franklin, R. Proc. Roy. Soc. 1951, A209, 196.
4. Wandless, A.M. J. Inst. Fuel, 1971, 44, 531.
5. Taylor, G.H. Fuel 1961, 40, 465.
6. Brooks, J.D.; Taylor, G.H. 'Chemistry and Physics of Carbon'. Ed. P.L. Walker Jr., Marcel Dekker, N.Y., 1968, 4, p. 243.
7. Singer, L.S. Fuel 1981, 60, 839.
8. Markovic, V.; Ragan, S.; Marsh, H. J. Mat. Sci. 1984, 19, 3287.
9. Forrest, M.A.; Marsh, H. J. Mat. Sci. 1983, 18, 973.
10. Grint, A.; Marsh, H. Fuel 1981, 60, 513.
11. Morris, E.G.; Tucker, K.W.; Joo, L.A. Ext. Abs. 16th Conf. on Carbon, The American Carbon Society, San Diego, U.S.A., 1983, p. 595.
12. Stadelhofer, J.; Marrett, R.; Gemmeke, W. Fuel 1981, 60, 877.
13. Wakeley, L.D.; Davis, A.; Jenkins, R.G.; Mitchell, G.D.; Walker, P.L. Jr. Fuel 1978, 58, 379.
14. Newman, J.W. 'Petroleum Derived Carbons', Eds. M.L. Deviney and T.M. O'Grady, American Chemical Society, Washington D.C., U.S.A., Symposium Series 1976, No. 21, p. 52.
15. White, J.L. Ref. 14, p. 282.
16. Zimmer, J.E.; White, J.L. Adv. in Liq. Cryst. 1982, 5, 157.
17. Lewis, I.C.; Singer, L.S. 'Chemistry and Physics of Carbon', Ed.P.L. Walker Jr. and P.A. Thrower, Marcel Dekker, N.Y. 1981, 17, p. 1.

18. Fitzer, E.; Mueller, K.; Schaefer, W. 'Chemistry and Physics of Carbon', Ed. P.L. Walker Jr. and P.A. Thrower, Marcel Dekker, N.Y. 1971, 7, p. 237.
19. Marsh, H. Proc. 4th Intern. Conf. on Industrial Carbon and Graphite, 1974, Society of Chemical Industry, London, 1976, pp. 2-38.
20. Marsh, H.; Cornford, C. Ref. 14, p. 266.
21. Marsh, H.; Smith, J. Analytical Methods for Coal and Coal Products, Ed. Clarence Karr Jr., Academic Press, N.Y. 1978, p. 371.
22. Marsh, H.; Walker, P.L. Jr. 'Chemistry and Physics of Carbon', Eds. P.L. Walker Jr. and P.A. Thrower, Marcel Dekker, N.Y. 1979, 15, p. 229.
23. Marsh, H., Ironmaking Proceedings of the Iron and Steel Society of AIME, 1982, 41, 2.
24. Forrest, M.A.; Marsh, H. 'Coal and Coal Products: Analytical Characterization Techniques', Ed. E.L. Fuller Jr., American Chemical Society, Washington, D.C., U.S.A., Symposium Series, 1982, 205, p. 1.
25. Carbon 1978, 16, Number 6, pp. 408-505.
26. International Committee for Characterization and Terminology of Carbon:- Carbon 1982, 20, 445; Carbon 1983, 21, 517.
27. Forrest, R.A.; Marsh, H., Carbon 1977, 15, 348.
28. Forrest, R.A.; Marsh, H.; Cornford C.; Kelly B.T. 'Chemistry and Physics of Carbon' Ed. Peter A. Thrower, Marcel Dekker, N.Y. 1984, 19, p. 211.
29. Ōya, A.; Qian, Z.; Marsh, H. Fuel 1983, 62, 274.
30. Qian, Z.; Clarke, D.E.; Marsh, H. Fuel 1983, 62, 1355.
31. Ihnatowicz, M.; Chiche, P.; Deduit, J.; Pregermain, S.; Tournant, R. Carbon (1966), 4, 41.
32. Monthioux, M.; Oberlin, M.; Oberlin, A.; Bourrat, X. Carbon 1982, 20, 167.
33. Auguie, D.; Oberlin, M.; Oberlin, A.; Hyvernat, P. Carbon 1980, 18, 337.
34. Collett, G.W.; Rand, B. Carbon 1978, 16, 477.
35. Briggs, D.K.H. Fuel 1980, 59, 201.
36. Nazem, F.F. Fuel (1980), 59, 851.
37. Balduhn, R.; Fitzer, E. Carbon (1980), 18, 155.
38. Davis, A.; Hoover, D.S.; Wakeley, L.D.; Mitchell, G.D. J. of Microscopy 1983, 132, 315.
39. Fitzer, E.; Holley, C. 'International Symposium on Carbon: New Processing and New Applications, Carbon Society of Japan, Toyohashi, 1982, p. 144.
40. Mochida, I.; Tamaru, K.; Fujitsu, H.; Korai, Y. Ref. 37, p. 153.
41. Buechler, M.; Ng, C.B.; White, J.L. Carbon 1983, 21, 603.
42. Atkinson, C.J.; Marsh, H. Ext. Abs. 6th London Intern. Carbon and Graphite Conf., Society of Chemical Industry, London, 1982, p. 198.
43. Cranmer, J.H.; Plotzker, I.G.; Peebles, L.H.; Uhlmann, D.R. Carbon 1983, 21, 201.
44. Yokono, T.; Marsh, H. 'Coal Liquefaction Products', Ed. H.D. Schultz, John Wiley & Sons, N.Y. 1983, 1, p. 125.
45. Lewis, I.C. Carbon 1982, 20, 519.
46. Evans, S.; Marsh, H. Carbon 1971, 9, 733; 747.
47. Lewis, I.C. Carbon 1980, 18, 191.

48. Marsh, H.; Akitt, J.W.; Hurley, J.M.; Melvin, J.; Warburton, A.P. J. Appl. Chem. Biotechnol. 1971, 21, 251.
49. Yokono, T.; Iyama, S.; Sanada, Y,; Making, K. Carbon 1985, 22, 624.
50. Lewis, I.C.; Kovac, C.A. Carbon 1978, 16, 425.
51. Singer, L.S.; Lewis, I.C. Carbon 1978, 16, 417.
52. Marsh, H.; Gerus-Piasecka, I.; Grint, A. Fuel 1980, 59, 343.
53. Mochida, I.; Marsh, H. Fuel 1979, 58, 809.
54. Hüttinger, K.J.; Rosenblatt, U. Proc. of 4th London Intern. Carbon and Graphite Conf. 1974, Society of Chemical Industry, London, 1976, p. 50.
55. Marsh, H.; Dachille, F.; Melvin, J.; Walker, P.L. Jr. Carbon 1971, 9, 159.
56. Forrest, M.A.; Marsh, H. J. Mat. Sci. 1983, 18, 991.
57. Park, Y.D.; Ōya, A.; Otani, S. Fuel 1983, 62, 700; 1499.
58. Chen, S.H.; Eilenberg, S.L.; Diefendorf, R.J. Ref. 39, p. 42.
59. Marsh, H.; Mochida, I.; Macefield, I.; Scott, E. Fuel 1980, 59, 520.
60. Marsh, H.; Latham, C.S.; Gray, E. To be published in Carbon, 1985.
61. Smith, W.E.; Napier, B.; Horne, O.J. 'Petroleum Derived Carbons', Eds. M.L. Deviney and T.M. O'Grady, American Chemical Society, Washington D.C., U.S.A. 1976, No. 21, p. 63.
62. Perrotta, A.J.; Henry, P.M.; Bacha, J.D.; Albaugh, E.W. Preprints, Carbon '80, Deutschen Keramischen Gesellschaft, Baden-Baden, 1980, 350.
63. Riggs, D.M.; Diefendorf, R.J. Carbon '80, Proc. 3rd International Carbon Conf., Deutschen-Keramischen Gesellschaft, Baden-Baden, 1980, 326.
64. Marsh, H.; Kimber, G.M.; Rantell, T.; Scott, E., Fuel 1980, 59, 520.
65. Kakuta, M.; Kohriki, M.; Sanada, Y. J. Mat. Sci. 1980, 15, 1671.
66. Seshadri, K.S.; Albaugh, E.W.; Bacha, J.P. Fuel 1982, 61, 336; 1095.
67. Weinberg, V.A.; Yen, T.F. Carbon 1983, 21, 39.
68. Brown, J.K.; Ladner, W.R. Fuel 1960, 39, 87.
69. Sakai, M.; Inagaki, M. Carbon 1981, 19, 37.
70. Yokono, T.; Nomura, Y., Ōzawa, H.; Sanada, Y. Ref. 39 p. 176.
71. Sanada, Y.; Furuta, T.; Kimura, H.; Honda, H. Fuel 1973, 52, 143.
72. Mochida, I.; Nakamura, E.; Maeda, K.; Takeshita, K. Carbon 1976, 14, 341.
73. Crawford, D.; Marsh, H. Preprints Carbon '76, Deutschen Keramischen Gesellschaft, Baden-Baden, 1976, 231.
74. Mochida, I.; Ando, T.; Maeda, K.; Takeshita, K. Carbon 1978, 16, 453.
75. Mochida, I.; Ando, T.; Maeda, K.; Fujitsu, H.; Takeshita, K. Carbon 1980, 18, 131.
76. Mochida, I.; Inoue, S-I.; Maeda, K.; Takeshita K. Carbon 1977, 15, 9.
77. Mochida, I.; Ando T.; Maeda K.; Fujitsu H.; Takeshita K. Carbon 1980, 18, 319.
78. Mochida, I.; Kudo, K.; Takeshita, K.; Takehashi, R.; Suetsugu Y.; Furumi J. Fuel 1974, 53, 253.
79. Mochida, I.; Amamoto, K.; Maeda, K.; Takeshita, K. Fuel 1978, 57, 225.
80. Mochida, I.; Matsuoka, H.; Fujitsu, H.; Korai, Y.; Takeshita, K. Carbon 1981, 19, 213.

81. Mochida, I.; Amamoto, K.; Maeda, K.; Takeshita, K.; Marsh, H. Fuel 1979, 58, 482.
82. Evans, S.; Marsh, H. Carbon 1971, 9, 747.
83. Otani, S.; Ōya, A. Bull. Jap. Chem. Soc. 1972, 45, 623.
84. Murty, H.N.; Biederman, D.L.; Heintz, E.A. Carbon 1973, 11, 163.
85. Mochida, I.; Marsh, H. Fuel 1979, 58, 797.
86. Mochida, I.; Marsh, H.; Grint, A. Fuel 1979, 58, 803.
87. Marsh, H.; Neavel, R.C. Fuel 1980, 59, 511.
88. Mochida, I.; Marsh, H. Fuel 1979, 58, 790.
89. Yokono, T.; Sanada, Y. Fuel 1978, 57, 334.
90. Yokono, T.; Miyazawa, K.; Sanada, Y. Fuel 1978, 57, 555.
91. Miyazawa, K.; Yokono, T.; Sanada, Y. Carbon 1979, 17, 223.
92. Miyazawa, K.; Yokono, T.; Sanada, Y.; Yamada, E.; Shimokawa, S. Carbon 1981, 19, 143.
93. Yokono, T.; Marsh, H.; Yokono, M. Fuel 1981, 60, 607.
94. Ōbara, T.; Yokono, T.; Miyazawa, K.; Sanada, Y. Carbon 1981, 19, 263.
95. Ōbara, T.; Yokono, T.; Kohno, T.; Sanada, Y. Ref. 39, p. 172.
96. Grint, A.; Marsh, H. Fuel 1981, 60, 1115.
97. Marsh, H.; Clarke, D.E. Submitted to Edöl und Köhle (1985).
98. Qian, Z.; Marsh, H., Fuel 1984, 63, 1588.
99. Petrakis, L.; Jones, G.L.; Granby, D.W.; King, A.B. Fuel 1983, 62, 681.
100. Sprecher, R.F.; Retcofsky, H.L. Fuel 1983, 62, 473.
101. Shibaoka, M. Fuel 1982, 61, 303.
102. Bradford, D.J.; Greenhalgh, E.; Kingshott, R.; Senior, A.; Bailey, P.A. Proc. of 3rd Conf. Ind. Carbon and Graphite, Society of Chemical Industry, London, 1971, p. 520.
103. Marsh, H.; Dachille, F.; Iley, M.; Walker, P.L. Jr.; Whang, P.W. Fuel 1973, 52, 253.
104. Tillmanns, H.; Pietzka, G.; Pauls, H. Fuel 1978, 57, 171.
105. Forrest, M.; Marsh, H. Fuel 1983, 62, 612.
106. Marsh, H.; Atkinson C. Unpublished results.
107. Stadelhofer, J.W. Fuel 1980, 59, 369.
108. Romovacek, G.R.; McCullough, J.P.; Perrotta, A.J. Fuel 1983, 62, 1236.
109. Markovic, V.; Lander, J.R.; Marsh, H.; Taylor D. Ext. Abs. of 15th Biennial Conf. on Carbon, The American Carbon Society, Philadelphia, U.S.A., June, 1981, p. 415.
110. Marsh, H.; Calvert, C.; Latham, C.S.; Markovic, V.; Lander, J.R.; Watson, R. Unpublished results (1984).
111. Ōbara, T.; Yokono, T.; Sanada, Y.; Marsh, H. Submitted to Fuel (1985).
112. Diefendorf, R.J. U.S. Patent 4 443 324, 1984.
113. Mochida, I.; Marsh, H. Submitted to Fuel (1985).

RECEIVED July 11, 1985

Chemical Characterization and Preparation of the Carbonaceous Mesophase

Isao Mochida and Yozo Korai

Research Institute of Industrial Science, Kyushu University 86, Kasuga, Fukuda, 816, Japan

> The development of mesophase carbon fiber provides a major incentive for both basic and practical investigations of the chemistry of the carbonaceous mesophase. Recent work on the properties and structural chemistry of the mesophase is surveyed, with emphasis on the chemical structures formed in the pyrolysis of pure compounds, interactions between molecules in the carbonization of practical pitch materials, and the preparation of mesophase pitches. Naphthenic intermediates play useful roles in improving anisotropy and solubility and in reducing viscosity. The growing knowledge of such mechanisms as hydrogenation, oxidation, dealkylation, and solvent fractionation opens increasingly precise approaches to control the molecular order, viscosity, and reactivity of mesophase pitch. An example is the use of hydrogenation to improve the fusibility and solubility of mesophase pitch, and under certain conditions, to prepare "dormant mesophase", an isotropic pitch that becomes anisotropic under stress.

The optical anisotropy observed in most carbon materials reflects the ordered stacking of graphite-like microcrystalline units that has been recognized to be essential in determining their properties. Pitch-based carbon fiber, electrode and metallurgical cokes, and carbons for nuclear reactors are characterized by their anisotropic texture since this structural factor is fundamentally related to their mechanical, thermal, electronic, and chemical properties (1-5).

The optical anisotropy has been shown by Brooks and Taylor (5) to be built in through the carbonaceous mesophase, the liquid crystalline state formed during the liquid-phase carbonization of those organic materials that can be pyrolyzed and heat-treated to the graphitic state. The mesophase is the critical intermediate state in which the quality of carbon products is determined. The chemistry of its characterization, preparation, and control is most relevant to modern carbonization technology.

The technology of mesophase-pitch-based carbon fiber has stimulated the rapid development of the chemistry of mesophase behavior and preparation. The carbonization schemes and mechanisms leading to optical anisotropy via the mesophase, the control of carbonization with emphasis on the preparation of spinnable mesophase, and the mesophase transition and reactivity in relation to the structure of its constituent molecules are summarized in this paper.

Carbonization Schemes and Mechanisms

When aromatic materials (including coal, coal tar, petroleum residues, and bitumens) are heated in inert atmosphere under ambient pressure, some of their components undergo pyrolytic reactions to produce radical fragments, although many constituents may distill. Some radical fragments decompose to produce lighter components which are also evolved, but others recombine with reactive molecules to yield larger molecules. Such reactions are repeated until finally solid coke appears in the reactor. The viscosity of the reacting liquid phase increases because of the increase in molecular weight.

The large aromatic molecules interact through $\pi - \pi$ electron forces and segregate from the matrix of smaller molecules to produce anisotropic spheres. The anisotropic spheres grow in diameter, coalesce to form broader isochromatic regions in spite of the increasing viscosity, and finally solidify into anisotropic coke. This series of steps leading to optical anisotropy is observed in the quenched mesophase specimens shown in Figure 1, although some effects due to quenching may be included. Hot-stage micrography shows similar in situ progress during carbonization (6-8). This carbonization mechanism may be defined as a "sphere mechanism".

Since the optical anisotropy reflects the ordered stacking of aromatic molecules, the anisotropic spheres and the coalesced regions (which can be quite viscous but deformable) are in the liquid crystalline state during the carbonization process. Their quenched state can be described as "liquid crystal glass" according to Diefendorf (9).

The present authors found another pyrolysis mechanism leading to optical anisotropy (10), in which no definite liquid phase was observable during the carbonization. In some carbonaceous substances, such as semi-anthracite coal, optical anisotropy over broad regions develops promptly at certain temperatures from the highly viscous stage. The component layers appear to be stacked, and are rearranged by heat-treatment to show optical anisotropy. Such a mechanism can be called the "preordered layer-transformation mechanism".

The reaction schemes of carbonization have also been investigated (11-15). The molecular structure of the carbonization intermediates can influence strongly the optical anisotropy of the resultant coke. The carbonization intermediates have been reported for the pyrolysis of acenaphthylene, which provides a rare example of atmospheric carbonization of a pure organic chemical (11). The carbonization scheme is illustrated in Figure 2. The intermediates II, III, and VI are proposed based on

2. MOCHIDA AND KORAI *Characterization and Preparation of the Mesophase* 31

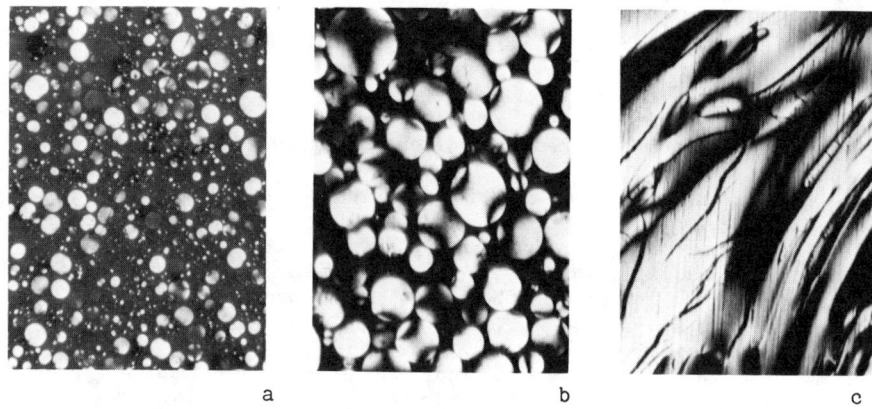

Figure 1. Three steps in the development of optical anisotropy via the mesophase. (a) Nucleation of mesophase spherules. (b) Growth and coalescence of mesophase. (c) Flow anisotropy.

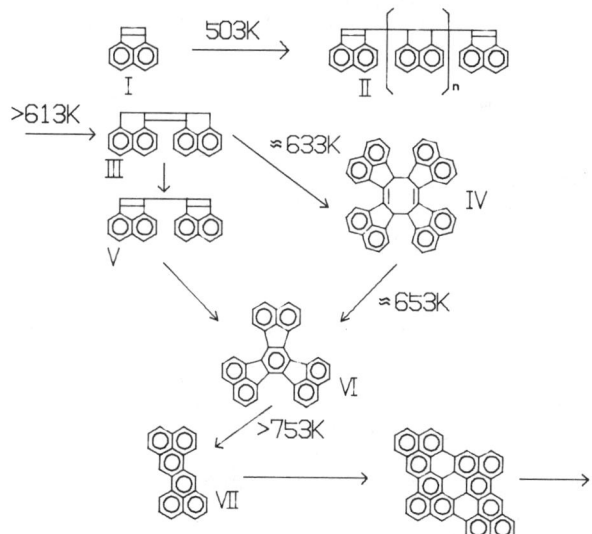

Figure 2. Carbonization scheme for acenaphthylene
Reproduced with permission from reference 12.
Copyright 1979 IPC Business Press, Ltd.

ultraviolet spectroscopy (UV), nuclear magnetic resonance (NMR), and vapor phase osmometry (VPO). Comparison of the carbonization reactivity of acenaphthylene (I) and decacyclene (VI) indicates the complexity of such reaction schemes. Decacyclene always forms as an intermediate in the pyrolysis of acenaphthylene. In fact, decacyclene is stable at 500°C, whereas acenaphthylene develops optical anisotropy at lower pyrolysis temperatures (12). These results suggest a role for reactive intermediates which are not included in the major steps of the reaction scheme and may also imply that a minor component can often affect the final results of carbonization.

Figure 3 describes reaction schemes for naphthalene carbonization catalyzed by metallic potassium or by aluminum chloride (13,14); these catalysts produce contrasting isotropic and anisotropic carbons, respectively. The intermediate structures are similar except for more naphthenic structure induced in the $AlCl_3$-catalyzed carbonization. The role of naphthenic structures leading to optical anisotropy has been recognized in many examples, and their introduction can improve the anisotropic development, as described later. Higher fusibility, lower melting temperature, and higher solubility of the intermediate molecules may be obtained by the formation of partially naphthenic structures (15).

Since the viscosity increase in the later stages of carbonization influences mesophase development, any factor influencing the viscosity of the carbonizing system may affect mesophase development. The rate (16,17) and atmosphere (18) of carbonization are important factors. Coexisting substances are also influential even if they are not carbonized (19). The carbonaceous sources are often mixtures of complex components so that their mutual interaction is important for understanding the carbonization behavior of pitch materials. The authors introduced "compatibility" (20,21) among the carbonizing components to describe this situation. The components perform the interactions of liquid-liquid mixing, solvation, and solvolysis and influence the carbonization by modifying the carbonization rate, the intermediate structures and composition, and the viscosity of the system (22). The concept of compatibility is not only useful in understanding carbonization reactions in practical materials, but also to improve the quality of a product by proper additives. Carbonization reactions are also influenced strongly by the starting structure and structural distribution since they govern the carbonization reactivity and the carbonization phase (23-25). A complex mixture is more properly described by its structural distribution than by its average structure.

Control of Carbonization to Mesophase

An important goal is to be able to produce a carbon material with well-defined properties from a given carbonaceous precursor. Based on carbonization mechanisms, some concepts concerning the control of anisotropic development from a given starting material can be developed. There are two principal approaches:
  1. Design of carbonization conditions.

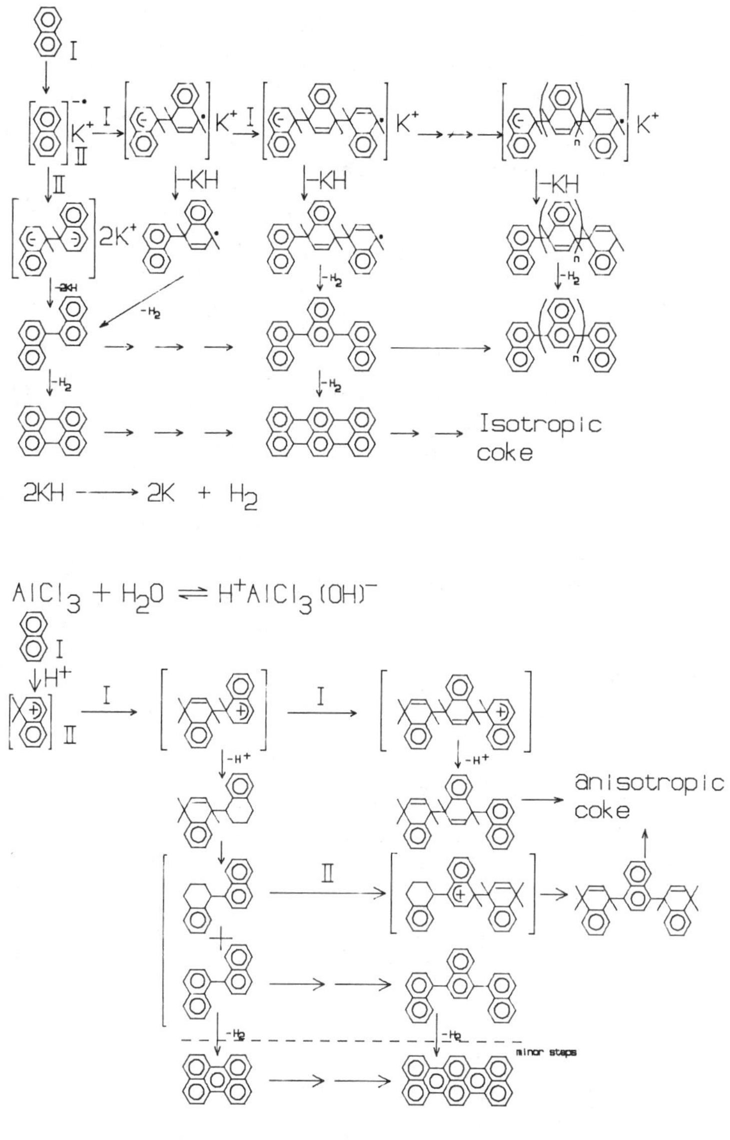

Figure 3. Sequences of condensation reactions leading to carbon from naphthalene (I) using potassium or aluminum chloride catalysts. Reproduced with permission from reference 14. Copyright 1976 Pergamon Press, Inc.

2. Design of starting material including structural modification.

The first approach includes catalytic (13-17) and pressurized carbonization (23-25). Catalysts may be used to create naphthenic structure in the condensation reaction. Heating rate is another important factor (26). The authors have also emphasized the importance of cocarbonization; carbonization reactions tend to be governed by minor components (Figure 4). Marsh et al. (27) defined such a situation as the "dominant partner effect". By a suitable additive, the carbonization reaction can be modified to produce a desired optical texture. The aromaticity and hydrogen-donating ability of the additive are recognized to be important for the modifying ability of the additive (28-30).

The extent of modification in the cocarbonization process is also influenced by the principal carbonizing substances (28-31). The authors introduced the term "cocarbonization susceptibility" to describe this situation. The presence of oxygen-containing functional groups, which is often observed in coal-derived pitch, is one of the major factors which influence the susceptibility (31). The molecular-size distribution of aromatic and paraffinic components in the feedstock is another factor (28). The combined effect of modifying ability and susceptibility on the resultant optical texture is illustrated schematically in Figure 4.

The second approach includes separation of undesirable components and chemical modification of the structure. Anti-solvent techniques may be applied to remove quinoline-insoluble particles (QI) from coal tar pitch prior to delayed coking (32). Various kinds of chemical modification, thermal and catalytic treatments, and their combination are possible (33-36). These treatments may perform dealkylation, cracking, condensation, hydrogenation, and hydrocracking. The authors have reported that partial hydrogenation can enhance carbonization yield as well as anisotropic development. Partially hydrogenated pyrene (produced using Li-ethylenediamine) carbonized under atmospheric pressure while pyrene did not (37). Some typical hydrogenated pyrenes are illustrated in Figure 5. Among these hydropyrenes, 1,6- and 1,8-dihydropyrene yield carbons by thermal oligomerization at a coke yield of about 20%. Oxidative pretreatment at 150°C was effective in increasing the coke yield to as high as 35% without decreasing the size of optical texture, while some naphthenic partial structure survived even after oxidative or oxygenative oligomerization as shown in Figure 6 (33).

Based on the creation of naphthenic structures in the condensation reaction, the modification by aluminum chloride increased carbon yield and improved the potential for anisotropic development. Oxidative pretreatment usually impairs anisotropic development although it increases carbon yield. The oxidized pitch is also modified by aluminum chloride (34); this may be used to prepare additives (38) and mesophase pitches (39). It should be noted that these processes allow the catalyst to be readily separated (in contrast to catalytic carbonization) since the modified product is still either soluble in some solvents or fusible.

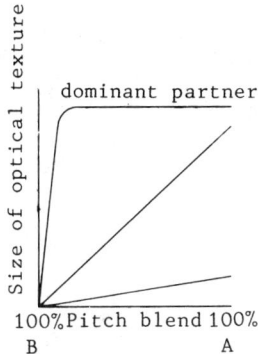

Figure 4. Schematic diagram to illustrate three possible effects of cocarbonization on the development of optical anisotropy. A, additive; B, principal carbonizing substance.

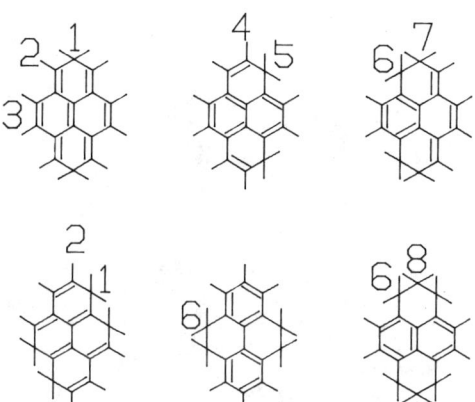

Figure 5. Some typical structures of hydrogenated pyrenes (prepared using Li and ethylenediamine). ^{1}H-NMR identification: 1, 3.2 ppm; 2, 6.5 - 7.0 ppm; 3, 7.0 - 8.0 ppm; 4, 5.5 - 6.0 ppm; 5, 4.0 ppm; 6, 3.0 ppm; 7, 2.5 ppm; 8, 2.0 ppm.
Reproduced with permission from reference 33.
Copyright 1982 Pergamon Press, Inc.

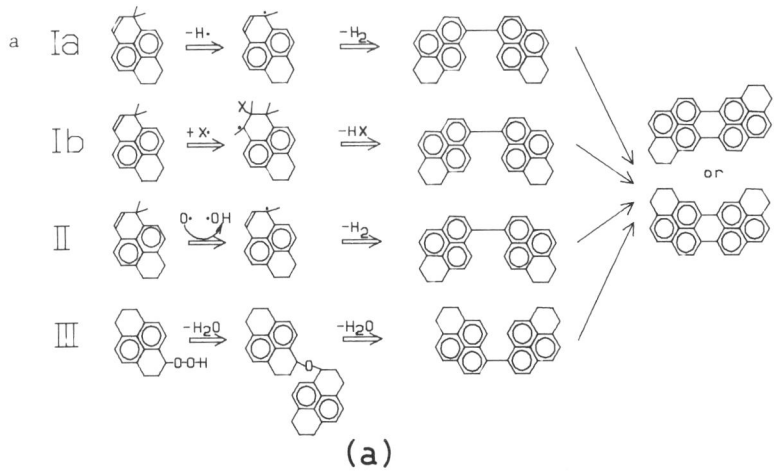

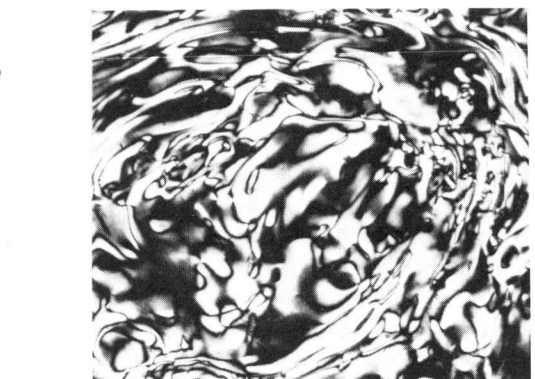

Figure 6. (a) Oligomerization mechanisms of hydrogenated pyrene: Ia, radical-coupling oligomerization; Ib, olefinic oligomerization; II, oxidative oligomerization; III, oxygenative oligomerization (x = oligomerization initiator). (b) Micrograph of coke from hydropyrene.
Reproduced with permission from reference 33.
Copyright 1982 Pergamon Press, Inc.

Preparation of Mesophase Pitch

Mesophase pitch, a precursor for high performance carbon fiber, must be spinnable at moderate temperatures, highly oriented, reactive for oxidation, and of high coking value, although these factors are sometimes conflicting. The general approaches to prepare such a mesophase pitch can be classified into three categories:
1. Preparation from appropriate starting substances.
2. Extraction or separation of the appropriate fraction during or after the preparation.
3. Modification of ordinary mesophase.

Acenaphthylene and tetrabenzophenazine (40) are known to be appropriate starting materials; hydrogenated or catalytically condensed aromatic hydrocarbons and pitches are also suitable (41). In thermal treatment of pitch, it is still a question whether short residence times at high temperatures or long residence times at low temperatures are preferable. In any case, the reaction should be homogeneous. The solvent may thus be important.

Examples of the second approach are described in patents from Union Carbide (42) and Exxon (43), in which the separation is made during or after the preparative carbonization. Aromatic hydrocarbons large enough for anisotropic development, but not to exceed a threshold size, are prepared by the removal of lighter fractions. By the solvent approach, heavy fractions can also be eliminated if they are produced.

A last approach is the hydrogenation of mesophase. Hydrogenation using Li-ethylenediamine (44) and hydrogen transfer from tetrahydroquinoline (32) have been reported. More conventional hydrogenation is also possible (45,46). The authors have reported that the combination of alkylation and hydrogenation of mesophase increases the yield of solubilized mesophase (47). Hydrogen transfer at high temperature for short contact time can be effective (48). "Dormant mesophase", in which the isotropic precursor displays anisotropy after stressing, has been prepared by the hydrogenation of mesophase (44); this transformation may provide another example of the preordered layer transformation mechanism. Physical mixing (blending) can also increase the fusibility of mesophase, thus indicating the importance of structural distribution.

Structure and Phase Transition of Mesophase

The molecular structure and constitution of the anisotropic mesophase sphere have been analyzed (49,50). NMR, UV, IR, and other modern techniques are applied after the mesophase solubility is enhanced by non-destructive hydrogenation or reductive alkylation. A model structure is illustrated in Figure 7 (49,50) where aromatic sheets 6 to 15 Å in extent are connected directly or through methylene bonds to form large molecules with molecular weights in the range of 400 to 4000. This wide molecular weight distribution is important since it may be responsible for the ordered stacking of smaller molecules through $\pi - \pi$ interaction, the extent of insolubility in quinoline, and the high viscosity at elevated temperatures.

The optical texture of mesophase and resultant carbons is observed readily by means of a reflected polarized light microscope and may be classified according to the shape and size of the isochromatic units. Such a classification is useful to evaluate the properties of mesophase and carbons such as needle cokes. The mesophase has been defined as the intermediate state which shows optical anisotropy and is quinoline-insoluble at room temperature (5,51) (liquid crystal glass), although it is a viscous liquid crystal during the carbonization process (6).

Recent advances in the technology of carbon fiber have revised the definition of mesophase. Its nature as a liquid crystal has been emphasized. Diefendorf proposed a phase diagram of carbonaceous mesophase analogous to that of conventional nematic liquid crystals (9). The schematic phase diagram of Figure 8 summarizes the thermal behavior of the mesophase. At present, the abscissa cannot be precisely defined, although Diefendorf has used the content of meso-species and non-meso-species according to conventional liquid crystal theory (9). The real mesophase consists of complex molecules that may interact; hence the abscissa may need to take the structural distribution into account. The phase diagram in Figure 8 consists of isotropic liquid, anisotropic liquid crystal, and liquid crystal glass. The phase boundaries correspond to the melting and glass transition temperatures. The anisotropic liquid crystal will not often exhibit the isotropic liquid state when the temperature rises since the pyrolysis reactions usually form non-fusible coke, as usually observed in needle coke production. The liquid crystal and liquid crystal glass may display the various morphologies summarized in Table I of the paper by Marsh and Latham (52). Interconversion among the morphologies of the glass is allowed via isotropic liquid and liquid crystal states as shown by the following scheme:

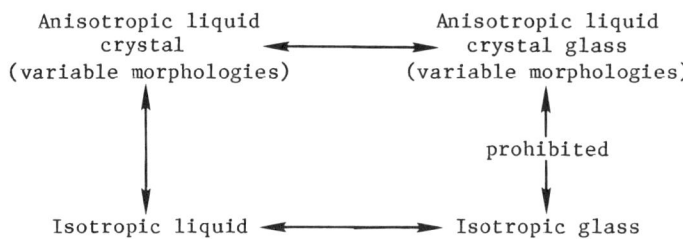

The interconversion includes recrystallization and rearrangement by annealing. The interconversion processes are rate-dependent, and various morphologies (even isotropy from the potentially anisotropic components) may result, depending on the recrystallization rate. In general, slow rates tend to enlarge the unit size. This also suggests that the isotropic pitch may be spun to anisotropic fiber if the cooling and extension permit anisotropic development.

Substances which show optical anisotropy and are soluble in quinoline or even in tetrahydrofuran (THF) have been prepared (53). Thus, solubility and optical anisotropy are now considered distinct phenomena. This is important since the carbon fiber

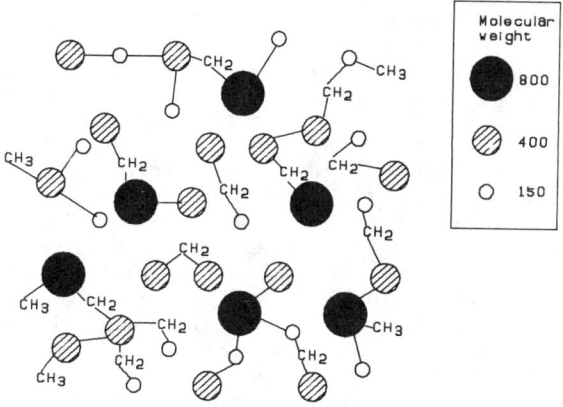

Figure 7. Models for the constituent molecules of mesophase spheres. Reproduced with permission from reference 20. Copyright 1977 IPC Business Press, Ltd.

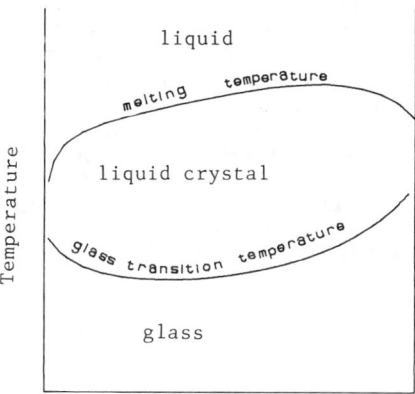

Figure 8. A schematic phase diagram for the carbonaceous mesophase.

precursor should be highly oriented and at the same time spinnable (plastic) without decomposition at the appropriate temperature prior to the carbonization. The molecules of high solubility or fusibility may differ considerably in their sizes and functional groups from those illustrated in Figure 7.

The authors have prepared soluble mesophase pitches from petroleum pitch (A240) and coal tar pitches (CTP), applying essentially the method reported by the Union Carbide group (42), and have analyzed the chemical structures of the constituent molecules. A240 heat-treated at 380°C for 30 hr. and CTP heat-treated at 380°C for 25 hr. developed anisotropic contents of more than 90 vol%. According to the material balance a considerable portion of both the THFS (THF-soluble) and THFI-PS (THF-insoluble pyridine-soluble) fractions, in addition to the PI (pyridine-insoluble) fraction from both A240 and CTP, exhibited optical anisotropy. However, only the THFI-PS fraction from A240 exhibited anisotropy as recrystallized (Figure 9), although both PI fractions were anisotropic. Some structural parameters for the THFS and THFI-PS fractions from the mesophase pitches, analyzed according the Brown-Ladner method (54), are summarized in Table I. The aromaticity values (fa) for the THFS and THFI-PS fractions from A240 were significantly smaller than those from CTP, while the average molecular weights of the A240 were definitely larger than those of the CTP fractions. Both Rnus (naphthenic ring number per unit structure) and σ (number of substitution groups per unit structure) of the A240 fractions were also larger than those of the CTP fractions. The naphthenic and alkyl groups enhance the solubility and fusibility through more intimate interaction with solvent molecules and somewhat looser intermolecular interaction within the mesophase. The higher molecular weight favors molecular association of layered stacking to exhibit anisotropy. Thus, the degree of anisotropy of the THFI-PS fractions recrystallized from A240 and CTP may reflect the structure of their components. Smaller molecules from CTP, molecules which do not develop anisotropy by themselves, were located in layers of the insoluble fraction, which displayed anisotropy before the fractionation.

Table I. Structural Indices of THFS and THFI-PS Fractions

	f_a	Rnus	σ	AMW
A240: 380°C, 30 h.				
THFS	0.84	0.1	0.19	540
THFl-PS	0.88	5.7	0.19	1800
CTP: 380°C, 25 h.				
THFS	0.86	0.0	0.14	380
THFI-PS	0.96	0.0	0.12	860

f_a: aromaticity
Rnus: naphthenic ring number per unit structure
σ: number of substitution groups per unit structure
AMW: average molecular weight

Figure 9. Optical micrographs of recrystallized THFI-PS fractions from a) A240- and b) CTP-derived mesophase pitches.

Mesophase Reactivity

Mesophase is susceptible to chemical reactions other than those induced by pyrolysis. Modifications to enhance fusibility or solubility for easier spinning (see Preparation of Mesophase Pitch) and to induce thermosetting for carbonization without deformation are both practical steps in carbon fiber manufacture.
  The reactivity for hydrogenation is governed both by the reactivity of hydrogenation sources and the susceptibility of the mesophase. Generally, mesophase prepared at higher temperature is more difficult to hydrogenate, probably due to the extensively condensed structure of the constituents (55). Nevertheless selective hydrogenation is required for effective modification. Although the heterogeneous catalyst system has seldom been applied for the hydrogenation of mesophase, it may be applicable if a proper solvent is selected.
  Alkylation enhances the solubility of mesophase, but the alkyl groups tend to be thermally eliminated (56) and thus contribute little to the fusibility. Alkylation prior to the hydrogenation is thus very effective to increase the yield of modified mesophase, as described above (47).
  Oxidative treatments can be applied to remove hydrogen from or to introduce oxygen into the mesophase (34,57,58), thus leading to further condensation of mesophase constituent molecules (oxidative condensation) sufficient for thermosetting. The reaction should be performed at temperatures below the softening point and thus proceeds at a slow rate. Mesophase reactivity may originate from the constituent molecules. Naphthenic hydrogen is known to be much more reactive than aromatic hydrogen, as indicated by hydrogenated pyrene (33). The benzyl group may be also susceptible. Thus the introduction of a sufficient number of such groups for the induction of thermosetting properties may be useful to shorten the infusibilization procedure.
  Oxygen may be incorporated into the mesophase (57,58). Incorporated oxygen is lost as carbon monoxide or dioxide in the calcination step so that the oxygenation reaction should be minimized. The precise control of such mesophase modification reactions may be a future problem.

Literature Cited

1. Marsh, H.; Walker, P. L., Jr. Chem. and Phys. of Carbon 1979, 5, 229.
2. Lewis, L. S. Fuel 1981, 60, 839.
3. Bradshaw, W.; Mamone, V. In "Petroleum Derived Carbons"; Deviney, M. L.; O'Grady, T. M. Eds.; ACS SYMPOSIUM SERIES No. 21, American Chemical Society: Washington, DC, 1976.
4. Patrick, J. W.; Wilkinson, H. C. In "Analytical Methods for Coal and Coal Products", Karr, C., Jr., Ed.; Academic: New York, 1978; Vol. II, p. 339.
5. Brooks, J. D.; Taylor, G. H. Chem. and Phys. of Carbon 1968, 4, 243.
6. Hoover, K. S.; Davis, A.; Perrotta, A. J.; Spackman, W. Ext. Abstr. 14th Conf. Carbon, 1979, p. 393.

7. Buechler, M.; Ng, C. B.; White, J. L. Ext. Abstr. 15th Conf. Carbon, 1981, p. 182.
8. Mochida, I.; Matsuoka, H.; Fujitsu, H.; Korai, Y.; Takeshita, K. Carbon 1981, 19, 213.
9. Diefendorf, R. J. Ext. Abstr. 16th Conf. Carbon, 1983, p. 26.
10. Mochida, I.; Korai, Y.; Fujitsu, H.; Takeshita, K.; Komatsubara, Y.; Koba, Y. Fuel 1981, 60, 1083.
11. Fitzer, E.; Mueller, K.; Schaefer, W. Chem. and Phys. of Carbon 1971, 7, 237.
12. Mochida, I.; Marsh, H. Fuel 1979, 58 626.
13. Mochida, I.; Nakamura E.; Maeda, K.; Takeshita, K. Carbon 1975, 13, 489.
14. Mochida, I.; Nakamura, E.; Maeda, K.; Takeshita, K. Carbon 1976, 14, 123.
15. Mochida, I.; Maeda, K.; Takeshita, K. High Temp. High Press. 1977, 9, 123.
16. Mochida, I.; Ando, T.; Maeda, K.; Takeshita, K. Carbon 1978, 16, 453.
17. Mochida, I.; Ando, T.; Maeda, K.; Fujitsu, H.; Takeshita, K. Carbon 1980, 8, 319.
18. Weintraub, A.; Walker, P. L., Jr. Proc. 3rd Int. Conf. on Carbon and Graphites 1971, p. 136.
19. Korai, Y.; Fujitsu, H.; Takeshita, K.; Mochida, I. Fuel 1981, 60, 1106.
20. Mochida, I.; Amamoto, K.; Maeda, K.; Takeshita, K. Fuel 1977, 56, 1977.
21. Mochida, I.; Amamoto, K.; Maeda, K.; Takeshita, K. Fuel 1978, 57, 225.
22. Mochida, I.; Marsh, H.; Grint, A. Fuel 1979, 58, 633.
23. Otani, S.; Okamoto, Y.; Oshima, T.; Oya, A. Tanso 1977, No. 8, 9.
24. Mochida, I.; Sakata, K.; Maeda, K.; Fujitsu, H.; Takeshita, K. Fuel Process Tech. 1980, 3, 207.
25. Marsh, H.; Akitt, J. W.; Hurley, J. M.; Melvin, J.; Warburton, A. P. J. Appl. Chem. 1971, 21, 251.
26. Evans, S.; Marsh, H. Carbon 1971, 9, 733.
27. Marsh, H.; Macefield, I.; Smith, J. Ext. Abstr., 13th Conf. Carbon 1977, p. 21.
28. Mochida, I.; Korai, Y.; Fujitsu, H.; Takeshita, K.; Mukai, K.; Migitaka, W.; Suetsugu, F. Fuel 1981, 60, 405.
29. Mochida, I.; Matsuoka, H.; Korai, Y.; Fujitsu, H. Takeshita, K. Fuel 1982, 61, 595.
30. Mochida, I.; Matsuoka, H.; Korai Y.; Fujitsu, H.; Takeshita, K. Fuel 1982, 61, 595.
31. Korai, Y.; Mochida, I. Fuel 1983, 62, 893.
32. Yamada, Y.; Honda, H. Japanese patent 58-18421, 1983.
33. Mochida, I.; Tamaru, K.; Korai, Y.; Fujitsu, H.; Takeshita, K. Carbon 1982, 20, 231.
34. Mochida, I.; Inaba, T.; Korai, Y.; Fujitsu, H.; Takeshita, K. Carbon 1983, 21. 535.
35. Mochida, I.; Takeshita, Y.; Korai, Y.; Fujitsu, H.; Takeshita, K. Ind. Eng. Chem. Prod. Res. Dev. 1982, 21, 505.
36. Mochida, I.; Oishi, T.; Korai, Y.; Fujitsu, H.; Takeshita, K. Fuel Process Tech. 1983, 7, 109.

37. Mochida, I.; Matsuoka, H.; Fujitsu, H.; Korai, Y.; Takeshita, K. Carbon 1981, 19, 213.
38. Mochida, I.; Iwamoto, K.; Korai, Y.; Takeshita, K. J. Japan Petrol. Inst. 1983, 26, 201.
39. Mochida, I.; Inaba, T.; Korai, Y.; Takeshita, K. Carbon, in press.
40. Ōtani, S. Mol Cryst. Liq. Cryst. 1981, 63, 249.
41. Mochida, I.; Sone, Y.; Matsuoka, H.; Korai, Y. Abstracts, Int. Symp. on Carbon (Toyohashi, Japan) 1982, p. 157.
42. Chwastiak, S. U.K. Patent 2005 2984, 1979.
43. Diefendorf, R. J.; Riggs, D. M. U.K. Patent GB 2002 024A, 1979.
44. Ōtani, S. Japanese Patent 57-100186, 1982.
45. Mochida, I.; Marsh, H. Fuel 1979, 58, 797.
46. Mochida, I.; Maeda K.; Korai, Y.; Fujitsu, H.; Takeshita, K. Fuel 1981, 60, 747.
47. Mochida, I.; Maeda, K.; Korai, Y. Ext. Abstr. 16th Conf. Carbon 1983, p. 32.
48. Mochida, I.; Moriguchi, Y.; Shimohara, T.; Korai, Y.; Fujitsu, H.; Takeshita, K. Fuel 1982, 61, 1015.
49. Mochida, I.; Maeda, K.; Takeshita, K. Carbon 1977, 15, 17.
50. Mochida, I.; Maeda, K.; Takeshita, K. Carbon 1978, 16, 459.
51. Yamada, Y.; Imamura, T.; Kakiyama, H.; Honda, H.; Oi, S.; Fukuda, K. Carbon 1974, 12, 307.
52. Marsh, H.; Latham, C. S. This volume.
53. Chwastiak, S.; Lewis, I. C. Carbon 1978, 16, 156.
54. Brown, J. K.; Ladner, W. R. Fuel 1954, 33, 79.
55. Mochida, I.; Maeda, K. J. Materials Sci. 1983, 18, 3736.
56. Greinke, R. A.; Lewis, I. C. Ext. Abstr. 16th Conf. Carbon 1983, p. 7.
57. Ōtani, S. Carbon 1965, 3, 31.
58. Barr, J. B.; Lewis, I. C. Carbon 1978, 16, 439.

RECEIVED November 27, 1985

3

The Pitch-Mesophase-Coke Transformation As Studied by Thermal Analytical and Rheological Techniques

B. Rand

Department of Ceramics, Glasses, and Polymers, University of Sheffield, Sheffield S10 2TZ, United Kingdom

> Some recent "in situ" continuous shear rheological measurements are discussed; the complexity of the behaviour is outlined and the factors that may influence the apparent viscosity are considered. During transformation to coke the mesophase pitch is regarded as an emulsion with each phase continuously changing in composition and in rheological behaviour and in particular in the temperature dependence of viscosity. Finally a transformation diagram is outlined that describes the transformation to coke and shows the regions of chemically stable liquids. This diagram also enables changes in the properties of the pyrolysis residue to be related to the weight loss at each stage of pyrolysis.

The pyrolysis of pitch occurs with considerable loss of volatile matter and produces marked changes in the physical properties of the pyrolysis residue. Thermogravimetric analysis is widely used to follow the evolution of the volatiles whilst the changes in the structure of the pyrolysis residue have largely been studied by optical microscopy, as outlined by Marsh and Cornford([1]) and White([2]) in the first symposium on "Petroleum Derived Carbons". However, as a result of developments in processing mesophase pitch into high-grade carbon products, there is now increasing interest in characterising the changes in other physical properties of the pyrolysis residue as the pitch is transformed through the mesophase state to coke. Perhaps the most profound change is in the rheological behaviour of the residue. In general, on heating, a pitch passes from the state of an organic glass, through its glass transformation zone to form a liquid phase, which passes through the mesophase transformation before hardening to an "elastic" coke. Attempts to study these changes in rheological behaviour have been few, and, as yet, our understanding of the rheological characteristics of the pyrolysis products is far from complete. This article discusses certain aspects of the rheology of mesophase pitch that have led to a method whereby one simple rheological parameter can be combined with thermogravimetric data to produce a schematic representation of the pyrolysis procedure, a so-called 'Transformation Diagram'.

Thermogravimetry of Pitch Pyrolysis

Fitzer et al.(3) discuss in some detail the application of thermo-gravimetry to the pyrolysis of pitch. The characteristic parameters that can be obtained, such ; temperature of decomposition (i.e., onset of weight loss), temperatu.e of maximum rate of weight loss, rate of weight loss as a function of extent of weight loss, and total weight loss (i.e., coke yield), are all affected by the experimental conditions employed, which must be standardised to enable comparison between different pitches.

Variations in experimental conditions also affect kinetic parameters that can be obtained by application of the non-isothermal kinetic techniques, as shown by Hüttinger(4,5). Collett and Rand(6) showed, using four different methods of analysis, that the apparent activation energy E_a for the volatilization from pitches of different types could increase from about 60 to 200 kJ mole^{-1} as the heating rate or extent of pyrolysis increased and Hayward et al(7,8)came to similar conclusions. These kinetic parameters are of doubtful significance and merely reflect the fact that the processes leading to volatilization of species from a molten pitch become more energetic as the temperature increases, probably because the balance between straight evaporation of molecules and cracking reactions followed by evaporation of volatile fragments is altered in favour of the latter. The increase in E_a implies that when pitch samples from the same source, previously heat treated to various extents at the same heating rate, atmosphere, etc., are reheated under the same conditions to complete the carbonization the later stages of the pyrolysis process are different. This seems unlikely and is thus an artefact of the kinetic method of analysis.

One of the problems of this kind of kinetic analysis is that the thermogram is fitted to an equation of the form

$$f(\alpha) = kt = k^1 T \tag{1}$$

where α is the fraction of total weight loss that has occurred at time t and temperature T. This procedure obscures the fact that the total weight loss is a function of the previous extent of pyrolysis of the pitch material prior to its thermogravimetric analysis, see Figure 1a. In Figure 1b the weight loss data from Figure 1a are replotted in terms of the volatile content γ, which is the fraction of the pyrolysis residue that comprises potentially volatile species (defined as $\gamma = (W_T - W_F)/W_T$, where W_T is the sample weight at temperature T and W_F is the weight at some final temperature, e.g., 800°C). The points are brought together at high temperature (Figure 1b) indicating that the previous heat treatment has not substantially changed the later stages of pyrolysis. However, although this type of plot is more meaningful than that shown in Figure 1a, it is less amenable to linearization and determination of a kinetic parameter. The volatile content correlates well with toluene and quinoline insolubles for many samples from Ashland A240 pitch, pyrolysed either at different heating rates or isothermally at different temperatures(9,10) as shown in Figure 2.

The above discussion illustrates some problems in detailed kinetic analysis of non-isothermal thermogravimetry of pitch and

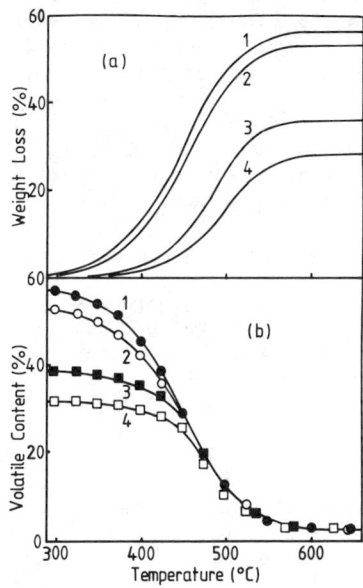

Figure 1. Thermogravimetric data for pyrolysis products of A240 petroleum pitch. (Heating rate = 1 K min^{-1})
1 – as received; 2 – HTT, 365°C; 3 – HTT, 400°C; 4 – HTT, 460°C.

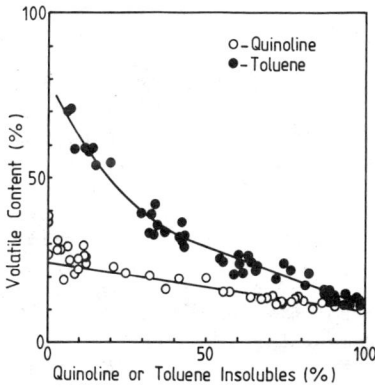

Figure 2. Relationship between volatile content and insoluble contents for A240 petroleum pitch samples produced by isothermal or dynamic heating conditions.

emphasises the importance of standardization in experimental conditions for comparative work. The volatile content is suggested to be a useful measure of the extent of pyrolysis, and this, although rather simplified, will be utilized later. There is a need to be able to define such a parameter in order to compare pyrolysis products and to enable their properties to be related to the extent of volatilization. The advantage of this particular one is that it can be extended into the mesophase range where the pyrolysis products are not entirely soluble in solvents.

Rheology of Pitch and Mesophase Pitch

As outlined earlier, the rheological properties of pitch and mesophase pitch are important in the processing of these materials to carbon products. The rheology of isotropic pitch will be considered briefly and then the effects of pyrolysis to mesophase will be described.

Isotropic Pitch. At low temperatures, pitch is an organic glass (11), and its glass transition temperature T_g can be determined by standard techniques such as differential thermal analysis, e.g., DTA or DSC (12-14), or by the change in thermal expansion coefficient. Above the glass transition temperature the viscosity decreases rapidly with increasing temperature, and hence the rheology of pitch-coke mixes is strongly dependent upon the relative values of T_g and ambient temperature. The glass transition temperature is related to molecular composition, and, in general, as the volatile content decreases and the concentration of aromatic high-molecular-weight species increases, the T_g will increase. However, T_g will also be strongly affected by small amounts of low molecular weight oils, etc., that may be blended into the pitch. The T_g can be regarded as the temperature at which the viscosity is 10^{12} Pa.s and it is usually of the order of 60 degrees lower than the ring and ball softening point where the viscosity is of the order of 10^3 Pa.s (15-17).

At temperatures above the softening point, isotropic pitch often displays Newtonian flow characteristics (18,19), but this may well depend upon the concentration of any insoluble particles (i.e., primary QI in the case of coal tar based materials) present within the pitch. A high concentration of QI could lead to non-Newtonian character as a result of the particle-particle attractive forces. Figure 3 shows η-T curves for a variety of pitch materials and their pyrolysis products. Pyrolysis increases the T_g of the system and shifts the viscosity-temperature curve to higher temperatures.

Although a number of equations have been used to describe the temperature dependence of viscosity of pitch systems (15), the Williams, Landel, Ferry equation (WLF) has received relatively little attention.

$$\log \eta = \log \eta_g - \frac{C_1(T-T_g)}{C_2 + T-T_g} \qquad (2)$$

(η and η_g are respectively the viscosities at temperatures T and T_g, whilst C_1 and C_2 are constants). Collett et al. (15) showed that this equation, which is widely used to account for the temperature dependence of viscosity and viscoelastic parameters of polymer systems, can be applied to the viscosities of pitch and mesophase pitch. It has also been used to describe the viscoelastic constants of bitumens (20,22). The equation is useful in the characterization of pitch in that, in addition to its ability to describe η-T data over a wider range of temperature than other equations, it contains the glass transition temperature as a useful temperature shifting parameter amenable to independent experimental determination. Also the constants C_1 and C_2 do not vary widely from pitch to pitch provided that the pitch has not received extensive thermal treatment (i.e., into the mesophase regime). T_g is seen to be a fundamental parameter and is an essential characteristic of mesophase pitch also, as discussed below.

Mesophase Pitch. Mesophase pitch is a generic term for those products of pyrolysis that contain mesophase, or are entirely mesophase, but which on reheating still pass through a fluid phase prior to the formation of semi-coke. The following are areas of scientific or industrial interest where the rheological properties of mesophase pitch are of interest.

1. The total transformation of pitch to coke, as already outlined, involves a profound change in rheological behaviour. Rheological studies form an essential part of any detailed understanding of this process.

2. Sedimentation of the mesophase from the isotropic pitch due to its slightly higher density is dependent, at least partly, on the viscosity of the isotropic phase. (Under the conditions of normal pyrolysis other factors, such as turbulence or gas evolution, may also play major roles.)

3. Evolution of gas bubbles is determined also by the viscosity of the media through which the bubbles rise, in this case both of the mesophase and of the isotropic phase.

4. Coalescence of mesophase is often said to be determined by the mesophase viscosity. This aspect requires much further investigation. However, it is clear that, amongst other factors, the rheological behaviour (including viscoelastic effects) of each phase is important in mesophase growth and coalescence. Diffusion of molecular species through the isotropic pitch to the mesophase spheres is likely to be related to the viscosity of the isotropic medium.

5. Coke microstructure is determined by the conditions in the coalesced mesophase prior to solidification. It is determined by the combined effects of convection currents, bubble percolation, and imposed shear stresses, all of which tend to deform the mesophase. The extent to which the mesophase pitch is deformed, as well as the relaxation after deformation, depends on the basic rheological behaviour.

6. <u>Fibre manufacture</u> - The conditions for spinning of fibres from mesophase pitch are determined by the pitch rheology which may also influence the degree of preferred orientation. The rheological properties of the spun fibres are important in defining the conditions for thermosetting.

7. <u>Other manufacturing routes for carbon products</u> based on mesophase are almost certain to be devised in the future and will depend upon an understanding of the rheological behaviour of the mesophase pitch.

8. <u>Binderless carbons</u> are one form of material manufactured from "green-cokes". The rheology of the green-coke is important in determining the pressing characteristics and interparticle bonding that is developed during firing.

<u>Rheological studies during pyrolysis</u>. The general changes in apparent viscosity of pitches during pyrolysis are now quite well documented (<u>18</u>,<u>19</u>) as a result of a number of "in-situ" investigations, and they are illustrated in Figure 4 by the solid line. The viscosity first decreases with increasing temperature, then levels out before rising steeply as the mesophase appears, increases in volume fraction, and coalesces. A maximum has often been reported in the apparent viscosity-temperature curve at the inversion point for the two-phase emulsion; the apparent viscosity then rises sharply due to rapid molecular growth in the continuous mesophase. These effects are dependent upon the rate of shear prevailing during the measurement of apparent viscosity, indicating that the fluid in this region is shear-thinning (i.e., pseudoplastic). There have also been reports of systems displaying a yield stress(<u>12</u>).

Detailed interpretation of such "in-situ" results is difficult because of the complexity of the system and the various competing effects that influence the measured shear stress and hence the value of apparent viscosity. These are summarised below:

1. The average molecular weight of each phase and hence its viscosity at any fixed rate of shear and temperature will increase as pyrolysis proceeds.

2. The volume fraction of the dispersed phase changes. It increases up to the phase inversion point when mesophase is the dispersed phase and decreases above this point when isotropic liquid is the dispersed phase.

3. Increasing temperature decreases the viscosity at any stage.

4. Volatile species emanate from the fluid during pyrolysis. If the concentration of bubbles at any stage is appreciable then we have a 3-phase colloidal system and the gaseous bubbles will also tend to raise the viscosity of the system.

5. Coagulation and coalescence of the dispersed phases will change the apparent viscosity at any rate of shear. Coagulation without coalescence will tend to increase it whilst coalescence of spheres by

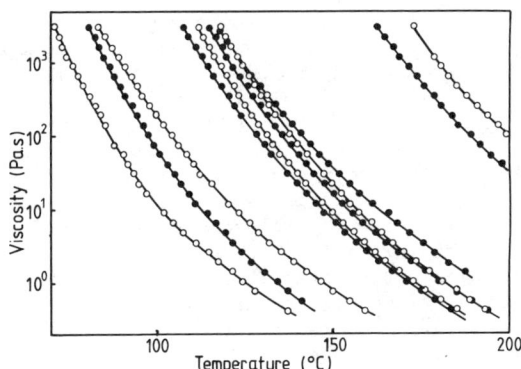

Figure 3. Typical viscosity-temperature curves for a variety of isotropic pitch samples and partially pyrolysed pitches.

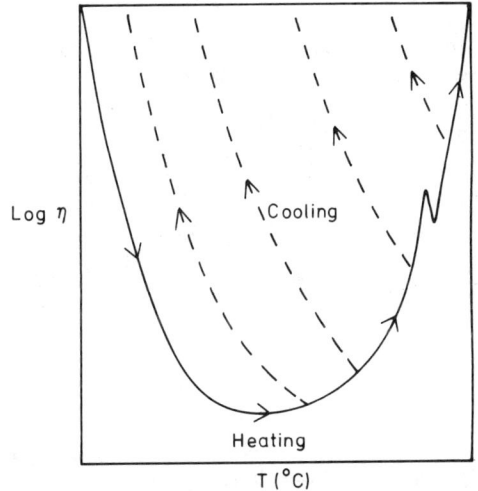

Figure 4. Schematic diagram depicting the change in apparent viscosity during 'in-situ' rheological studies of pitch pyrolysed under fixed conditions of shear and heating.

increasing their average size may tend to decrease this property until the phase inversion region is reached when a connected network of mesophase regions begins to form.

6. It is now known that the rate of shear prevailing in such "in situ" studies can influence the rate of formation of mesophase (23) and this has recently been explained by Sorensen and Diefendorf (24) as being due to the release of volatiles being facilitated during flow of the hot pitch.

Despite these complicating factors, "in situ" measurements are useful in comparing the pyrolysis behaviour of pitches, and, just as the thermogravimetric curve determined under standardized conditions can give a kind of fingerprint of the pitch and its pyrolysis behaviour, particularly the rate of release of volatile matter, so the in situ apparent viscosity-temperature curve at a fixed heating rate and rate of shear is characteristic of the changes in the pyrolysis residue. This has been very clearly demonstrated by Fitzer and co-workers (19,23,25). It would, however, be desirable to be able to relate the extent of volatilization to the changes in rheological behaviour of the residue.

Effect of the dispersed phase on the apparent viscosity-temperature curve. Maxima do not always appear in the apparent viscosity-temperature curves. A recent report (25) shows that the appearance of the maximum in the η - HTT curve could be an artifact produced by experimental procedure. Whilst this is undoubtedly the case in some systems, whether it is always the case is not yet clear. At this stage it will suffice to review factors that may be important in determining the form of these "in-situ" results.

Consider first the effect of a dispersed phase, of volume fraction ϕ, on the viscosity η of a suspension or emulsion with a Newtonian continuous phase of viscosity η_o and dispersed particles (droplets) which do not attract. At low volume fractions the Einstein equation should apply to a suspension of solid particles at constant temperature,

$$\eta_{rel} = \eta/\eta_o = 1 + k\phi \qquad (3)$$

(k = 2.5 for spherical particles). Taylor (26) extended Einstein's hydro-dynamic treatment to dilute emulsions in which the interfacial film posed no impediment to the transmission of tangential and normal stresses from the continuous phase to the dispersed phase and there was no interfacial slippage. These stresses cause internal fluid circulation within the droplets of the dispersed phase which reduces the distortion of flow patterns around the droplets and hence the relative viscosity

$$\eta_{rel} = \eta/\eta_o = 1 + k \left[\frac{\eta_d + 0.4\, \eta_o}{\eta_d + \eta_o}\right] \phi \qquad (4)$$

where η_d is the viscosity of the dispersed phase. The relative viscosity is always increased by the presence of the dispersed phase by an amount dependent upon its viscosity. However, the interfacial film is important also.

These equations are not valid at high concentrations of dispersed phase. A number of expressions have been discussed in detail by Sherman ([27](#)), Goodwin([28](#)) and Frisch and Simha([29](#)). One such equation, for a solid dispersed phase, is due to Mooney

$$\eta_{rel} = \exp\left[\frac{k_1\phi}{1-k_2\phi}\right] \quad (5)$$

which is included here to show the exponential nature of the viscosity increase. This is summarised in Figure 5 which shows schematically for two mixed immiscible liquid phases, A and B, how the viscosity changes with volume fraction at constant temperature. Phase B is assumed to have the higher viscosity. It is clear that because the volume fraction of dispersed phase decreases above the phase inversion point, there is a decrease in viscosity, abcd. The value of the maximum depends upon the relative viscosities of the two phases and the effects of volume fraction of each dispersed phase in raising the viscosity of the continuous medium. In practice there will be additional effects, due to coagulation of droplets and coalescence, to take into account in the region near the inversion point. Figure 5 also shows the η-ϕ curves resulting from different viscosities of the continuous phases, a'b'c'd'. In a mesophase-pitch system, pyrolysis at constant temperature could result in the viscosity of each phase increasing at the same time as the volume fraction of the mesophase increases. In Figure 5 the phase B can be identified with the mesophase and phase A with the isotropic phase. Thus, the variation in viscosity with extent of transformation may follow the path aef in which the maximum is almost absent or if η_B increases very rapidly, path aeg could be traversed, in which there is no maximum. Thus, the form of the curve depends on the rate of rise in the viscosities of the separate phases and the degree of interaction at phase inversion.

The behaviour depicted above is, however, further complicated in dynamic heating experiments, such as were described earlier. Firstly, increasing the temperature will tend to reduce the viscosity of each phase, at any particular composition; although it will also have the effect of enhancing the rate of increase of viscosity of each phase due to chemical reaction. Secondly, the continuous phase above the phase inversion point is the mesophase which may show shear thinning character due to the preferred orientation of the lamellar molecules in the shear field. Thus, if there is a maximum it may be larger the lower the rate of shear because higher rates of shear impose orientation at lower mesophase contents. It is clear from the above discussion that information about the change in the properties of both phases in a mesophase pitch throughout the transformation is required before rheological behaviour can be well understood. Specifically, the following information about each phase at each stage in the pitch-mesophase-carbon transformation is desirable:

1. The molecular composition

2. The optical texture

3. The viscosity-temperature relationship and its dependence on previous shear history

4. Non-Newtonian behaviour including viscoelasticity

5. The effect of shear on the optical texture

This requires a detailed study of the systems when they are cooled from the reaction temperature and are chemically stable.

Rheological behaviour of chemically stable mesophase-pitch.
Nazem(30,31) has carried out an excellent study of mesophase pitches in the temperature range where the chemical reactions are quenched. These pitches were entirely or predominantly mesophase. A variety of techniques were used to investigate the behaviour over a wide range of shear rate. Non-Newtonian behaviour was observed in some cases but not in all. Some pitches showed a linear relationship between rate of shear and shear stress, implying Newtonian behaviour, but also exhibited viscoelastic characteristics such as normal stresses and die swell on extrusion. It was not clear from these experiments what was the pyrolysis history of the pitches or whether measurements at low shear rates were carried out before or after those at high shear rates when there would be extensive orientation of molecules in the shear field. In some cases the temperature dependence followed the Arrhenius equation, but in other cases this equation did not adequately describe the data. This is not surprising. The Arrhenius equation is usually used to describe the temperature dependence of viscosity at temperatures well above the glass transition temperature. As explained earlier, the WLF equation is likely to give a better representation of the viscosity- temperature data over a wide temperature range.

The Transformation Diagram

The form of the $\eta - T$ curve for "in situ" pyrolysis studies shown in Figure 4 has been discussed. Figure 4 also shows the η-T curves that might be obtained on cooling the pyrolysis residue at any stage of the process. (Strictly, this $\eta = f(T)$ should be depicted as being a function of the rate of shear in the mesophase zone, where non-Newtonian character is observed.) The effect of pyrolysis is to shift the $\eta - T$ cooling curve to higher temperatures, and this represents an increase in the glass transition temperature of the mesophase pitch. In fact each phase of the two phase emulsion will have its own glass transition temperature T_g, and as pyrolysis proceeds, the T_g of each phase should increase. Since the "in situ" $\eta - T$ curve is complicated by the change in temperature during the measurement, a better representation of the changes in the system brought about by pyrolysis would be the change in some temperature-independent parameter, and it is suggested that one suitable parameter is the glass transition temperature. It will now be shown that by combining glass transition temperatures with thermogravimetric data to form so-called Transformation Diagrams, as already outlined by Whitehouse and Rand(32-33), a good representation of the pyrolysis process can be obtained.

Riggs and Diefendorf (34) first suggested a schematic phase diagram for pitch to account for the role of the mesogenic molecules in the separation of mesophase from the isotropic liquid. This diagram showed clearly the effect on the formation of mesophase brought about by reducing the concentration of 'disordering' low molecular weight species by selective dissolution. The schematic diagram as outlined here accounts for the changes in the residue as a result of thermal treatment.

The transformation diagram is depicted schematically in Figure 6 ; it comprises characteristic temperatures plotted against some measure of the extent of pyrolysis. To extend the range of the diagram, a parameter defining the extent of pyrolysis is required that can be applied into the so-called semi-coke or even coke regions. Two measures have been used: (a) the volatile content, as defined earlier, which enables thermogravimetric data to be used in the compilation of the diagram ; (b) the atomic ratio C/H.

The characteristic temperatures used are:-

(i) the glass transition temperature of the pyrolysis residue at the particular extent of pyrolysis, and

(ii) the decomposition temperature, T_o.

These points respectively define the temperature at which, on reheating, a cooled or quenched pyrolysis residue begins to soften and develop fluidity and the temperature at which the liquid residue begins to change its chemical composition, i.e., when pyrolysis recommences. This latter point is more ambiguous since it is determined from a thermogravimetric experiment which is subject to the effects of experimental procedure, as discussed earlier. Therefore, it is essential to use a fixed heating rate, etc., in the determination of the decomposition temperature. The volatile content can be determined in the same experiment.

Before presenting some experimental data, it will first be appropriate to consider the significance of the transformation diagram proposed. In Figure 6 consider first point A, which represents an isotropic pitch typical of a coking precursor, i.e., with a low C/H ratio and a low glass transition temperature T_{gi}. Its composition is such that, on heating, the first pyrolysis process is loss of low-molecular-weight species by distillation; this begins at temperature T_o. If the temperature is raised to T_1, then the composition of the liquid pitch will move towards point B whilst that of the vapour might be given by point V. When the material of composition B is cooled, volatilization is arrested, and the liquid residue increases in viscosity until it attains its T_g value, which is greater than that for composition A, reflecting the change in average molecular weight, C/H and volatile content. Further heating causes the composition to move to the right with corresponding

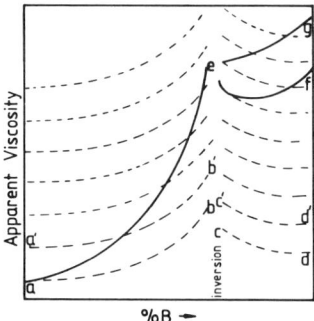

Figure 5. Schematic diagram showing the effect of changing the volume fraction of second phase on the apparent viscosity at a fixed rate of shear of a two-phase emulsion. The different dotted lines refer to different viscosities of the pure phases A and B. The solid line suggests the viscosity that may be displayed by a system in which both the viscosities of the pure phases and the relative proportions of phases are changing continuously, as in a pyrolysis run.

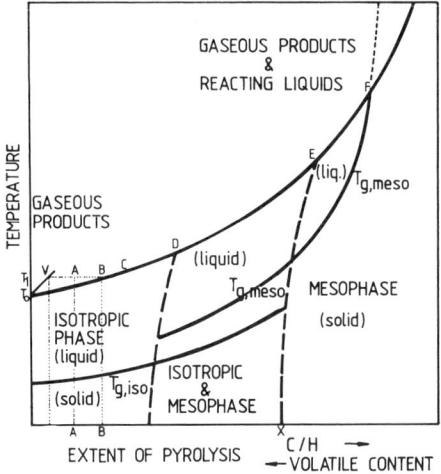

Figure 6. Schematic diagram describing the pitch-mesophase-coke transformation.

increases in the characteristic temperatures. Eventually, when cracking reactions take place, there will no longer be a simple relationship between liquid and vapour phases. At point D, the mesophase range is attained, and two phases now exist in the pyrolysis residue, each with its own characteristic temperatures. The characteristic temperatures for the mesophase should be the larger as a result of its higher average molecular weight. However, in the diagram only the lowest decomposition temperature is shown, being the one that defines the limit of chemical stability of the two phase system as a whole, whilst both T_g values are shown.

Boundaries are also shown in Figure 6 defining the upper and lower compositions of the two phase region. The shape of these phase boundaries depends upon the particular precursor and the manner of its thermal transformation. However, in Figure 6 they have been drawn to allow the existence of mesophase above the line CF when the system is undergoing further pyrolysis and also to allow the possibility of additional mesophase separating out on cooling and redissolving on reheating prior to further pyrolysis. At an extent of transformation greater than depicted by line EX on the diagram, only one phase exists, the mesophase, and, on further pyrolysis, its rheological properties change rapidly towards those of coke. This is depicted by a rapidly increasing T_g which exceeds the decomposition temperature, T_o. The point where T_g and T_o become equal is significant. It defines the upper limit for the existence of a stable liquid mesophase. Although a pitch with T_g greater than this value can be heated through its glass transition to form a fluid phase, this fluid will be evolving gaseous species and changing its composition continuously with time.

Figure 7 presents an experimentally constructed transformation diagram for Ashland A240 pitch. Ashland A240 (low S) pitch was pyrolysed in a stirred vessel at a heating rate of 0.1 K min^{-1}. The values of T_o and T_g of residues extracted from the pyrolysis vessel at different times were determined as outlined elsewhere(32,33,35). T_o is the temperature at which weight loss could be detected. T_g was determined by a penetrometry technique that did not allow the separate determination of the values for each phase in the two-phase region; consequently only one T_g value for the total pitch is shown in the experimental diagram. This value is characteristic of that phase which forms the continuous matrix in the sample. (Recently, it has been possible (35,36) to separate each phase by a hot centrifugation technique, and T_g's for the separate phases have been determined).

In the diagram, phase boundaries have been indicated by vertical lines, since their dependence upon temperature during reheating of a quenched sample is not known.

The experimental diagram resembles the schematic one except that the decomposition temperature is relatively insensitive to the degree of pyrolysis until the volatile content is reduced below about 10%. This point corresponds approximately to that at which the residue is entirely mesophase, which is also the point at which the T_g of the mesophase equals the decomposition temperature. This means that it is not possible to produce, under these conditions of pyrolysis, a chemically stable single-phase mesophase pitch in the fluid state (although this may be possible by other techniques).

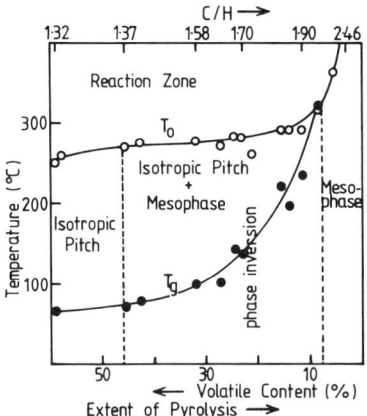

Figure 7. Experimental transformation diagram for A240 (low S) pitch. (Samples produced at 0.1 K min^{-1} and 1 atm pressure).

Use of the Transformation Diagram. Firstly, it is useful in relating the weight changes during pyrolysis, i.e., the extent of removal of volatile species, to the changes in physical properties of the residue. In this respect it would be desirable if the mesophase pitch properties could be predicted from a given time temperature treatment. This can be achieved as follows. Figure 8 shows the thermogravimetric data (heating rate 3 K min^{-1}) for further pyrolysis of each of the residues used to construct Figure 7. If T_o and T_g data were available for the pyrolysis of similar sized specimens at the same heating rate, then these data could be combined to produce a transformation diagram of the form of Figure 9, enabling the properties of the residues to be related to the weight loss curve. If the iso-viscous temperatures within the fluid region are known, then they can be included, as shown.

Secondly, the diagram, partial diagrams, or at least knowledge of T_g and T_o values can be of significance in processing mesophase pitch to carbon products. At any stage of pyrolysis the relative magnitudes of T_g and T_o determine the temperature range over which a stable fluid exists; this range decreases with increasing extent of pyrolysis and the minimum viscosity that can be attained in this stable state increases. Such information is relevant to the production of precursors to be spun into mesophase-pitch fibres.

The relative magnitudes of the T_g values in the two-phase region should be related to the development of microstructure in mesophase pitch, as discussed earlier. T_g will also be a useful parameter in defining the maximum temperature at which fibres can be oxidized without molecular motion causing some decrease in extent of preferred orientation.

However, the precise location of the decomposition and glass transition lines may depend upon pyrolysis conditions because kinetic factors such as gas flow rate, stirring conditions, heating rate, etc., all affect the relative extents to which evaporation and cracking reactions contribute to the change in composition.

The diagram enables us to make a definition of the term 'coke'. After extensive pyrolysis a residue will be obtained which on cooling and reheating will not pass through the glass transition temperature at all, i.e., it changes its composition within the solid state. This material can be defined as coke. A simple definition of semi-coke, however, is more difficult to provide, although it will probably be based on rheological behaviour. For example, the diagram shows the rapidly increasing T_g in the later stages of pyrolysis. At some stage it will be possible only for the residue, on reheating, to pass through a reacting viscoelastic state of high apparent viscosity (perhaps in excess of 10^6-10^7 Pa. s) and not attain a very fluid stage because of its continually changing composition which increases its T_g and therefore the magnitude of its viscoelastic parameters. Perhaps such a material is 'semi-coke'.

Acknowledgments

The author expresses thanks to his co-workers, G.W. Collett, P.M. Shepherd, S. Whitehouse and M. Benn who have contributed to the concepts described in this article. Thanks are also due to the National Coal Board and Science and Engineering Research Council for supporting these studies.

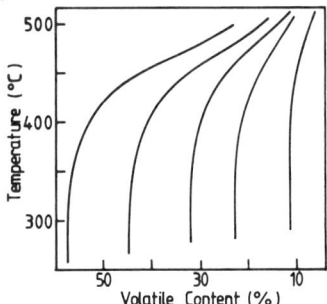

Figure 8. Further decomposition at 3 K min^{-1} of A240 (low S) pitch pyrolysis products used in compiling fig. 7.

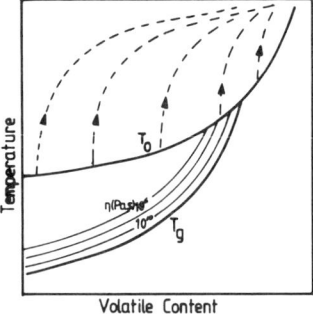

Figure 9. Schematic Predictive Transformation Diagram. Isoviscous temperatures are shown in the stable liquid region. Above the decomposition temperature the variation in volatile content with temperature is shown enabling extent of weight loss to be related to properties of the pyrolysis liquid.

Literature Cited

1. Marsh, H.; Cornford, C. In "Petroleum Derived Carbons"; Deviney, M.L.; O'Grady, T.M., Eds.; ACS SYMPOSIUM SERIES No. 21, American Chemical Society : Washington, D.C., 1976; p.266.
2. White, J.L. ibid. p.282.
3. Fitzer, E.; Mueller, K.; Schaeffer, W. In "Chemistry and Physics of Carbon"; Walker, P.L., Jr., Ed.; Dekker: New York, 1971; Vol. 7, p.237.
4. Hüttinger, K.J. Proc. 3rd Conf. on Ind. Carbons and Graphite, 1971, p.136.
5. Hüttinger, K.J. Chemie IngenieurTechnik 1970, 42, 812.
6. Collett, G.W.; Rand, B. Thermochim. Acta. 1980, 41, 153.
7. Hayward, J.S.; Ellis, B.; Rand, B. Carbon '80, Ext. Abstr. 3rd Int. Carbon Conf., Baden-Baden, 1980, p.338.
8. Hayward, J.S. M.Sc. Thesis, University of Sheffield, 1981.
9. Shepherd, P.M. Ph.D. Thesis, University of Sheffield, 1980.
10. Shepherd, P.M.; Rand, B., ref. 7, p.342.
11. Berneis, K.; Wood, L.J. Chemistry and Industry 1955, 1186.
12. Giavarini, C.; Pochetti, F. J. Thermal Analysis 1973, 5, 83.
13. Stadelhofer, J.W. Carbon 1979, 17, 301.
14. Barr, J.B.; Lewis, I.C. Thermochim. Acta 1982, 52, 297.
15. Collett, G.W.; Shepherd, P.M.; Rand, B. Proc. 5th London Internat. Carbon and Graphite Conf. Vol.1., 1978, p. 280.
16. Traxler, R.N. In "Bituminous Materials"; Hoiberg, A.J., Ed.; Interscience: New York, 1964; Vol. 1, p. 143.
17. Sakai, M.; Inagaki, M. Carbon 1981, 19, 37.
18. Collett, G.W.; Rand, B. Fuel 1978, 57, 162.
19. Balduhn, R.; Fitzer, E. Carbon 1980, 18, 155.
20. Jongepier, R.; Kuilman, B. Rheol. Acta 1970, 9, 102.
21. Wada, Y.; Hirose, H. J. Phys. Soc. Japan 1960, 15, 1885.
22. Vinogradov, G.V.; Isayev, A.I.; Zolotarev, V.A.; Verebskaya, E.A. Rheol. Acta 1977, 16, 266.
23. Busch, R.; Fitzer, E. Ext Abstr of 16th Conf. on Carbon, San Diego, 1983, p. 48.
24. Sorensen, I.W.; Diefendorf, R.J., Ibid., p.48.
25. Bhatia, G.; Fitzer, E.; Kompalik, D. Carbone '84, Ext. Abstr. Intern. Carbon Conf., Bordeaux, 1984, p. 330.
26. Taylor, G.I. Proc. Roy. Soc. 1932, A138, 41.
27. Sherman, P. In "Emulsion Science"; Sherman, P., Ed.; Academic: New York, 1968, p. 217.
28. Goodwin, J.W. Colloid Science, Vol.2, Specialist Periodical Report, Chem. Soc. (London), 1975, p. 246.
29. Frisch, H.L.; Simha, R. In "Rheology, Theory and Applications"; Eirich, F., Ed.; Academic: New York, 1956, Vol.I, p. 525.
30. Nazem, F.F. Fuel 1980, 59, 851.
31. Nazem, F.F., Carbon 1982, 20, 345.
32. Whitehouse, S.; Rand, B. Carbon '82, Ext. Abstr. 6th London Int. Carbon and Graphite Conf., SCI., London, 1982, p. 183.
33. Rand, B.; Whitehouse, S. ref. 23, p. 30.
34. Riggs, D.M.; Diefendorf, R.J. ref. 7, p. 326.
35. Rand, B.; Whitehouse, S. ref. 23, p. 102.
36. Benn, M.; Rand, B., unpublished results.

RECEIVED June 28, 1985

4

Microstructure Formation in Mesophase Carbon Fibers and Other Graphitic Materials

J. L. White and M. Buechler

Materials Sciences Laboratory, The Aerospace Corporation, P.O. Box 92957, Los Angeles, CA 90009

> Carbon fibers spun from mesophase pitch illustrate how mesophase mechanisms—deformation and oxidation stabilization—can be used to produce filaments with moduli approaching the theoretical limit for the graphitic layer plane. The work reviewed here has sought insights into the ways in which microstructure forms in deformed mesophase products, such as needle coke and carbon fiber, using hot-stage observations and deformation and carbonization experiments. Three critical mechanisms are the creation of strong preferred orientation by deformation, interactions of disclinations brought into proximity by the deformation, and trapping of the oriented and disclinated microstructures by quenching and chemical reaction. Highly oriented fibrous morphologies are not thermally stable, and chemical stabilization is essential for preventing relaxation to less oriented and less disclinated structures of lower inherent modulus.

Cokes and manufactured graphites are unique among structural materials, because the liquid crystalline (mesophase) state governs the formation of their microstructures. The lamelliform morphologies of the discotic nematic liquid crystal (1) are locked into place as the carbonaceous mesophase hardens, and the microconstituents thus produced differ fundamentally from those of conventional metals or ceramics; e.g., mesophase-based materials do not have conventional polycrystalline grain boundaries, and disclinations are prominent structural features. Most mesophase products of practical importance form while undergoing deformation, and the microstructures trapped by hardening are often in deformed nonequilibrium states that would relax if hardening had not intervened. Structurally, therefore, most graphitic materials formed by liquid pyrolysis may be regarded as mesophase fossils with nonequilibrium microstructures.

In the previous symposium, we reviewed mesophase mechanisms involved in the formation of petroleum coke (2). Since 1975, two significant developments have been the use of hot-stage microscopy to observe the dynamic behavior of the carbonaceous mesophase in its fluid state (3-6), and the emergence of carbon fibers spun from mesophase pitch (7-9) as effective competitors in applications in which high elastic modulus or good graphiticity is important. This paper focuses on mesophase carbon fibers as an example of how the plastic mesophase can be manipulated to produce fibers with intense preferred orientations and elastic moduli that approach the theoretical limit for the graphite crystal in the a-direction.

Figure 1 shows where mesophase carbon fibers fit in the rapidly developing field of structural carbon fibers. At present, fibers produced from polyacrylonitrile (PAN) constitute the bulk of the carbon fiber produced; in 1981 all commercial PAN-based fibers fell below and to the left of the "1981 limit" for tensile strength and modulus. Since 1981, there have been major advances in the development of PAN-based fibers to achieve high strengths, particularly at the lower modulus levels (near 40 Mpsi, 276 GPa).

Carbon fibers spun from mesophase pitch were commercialized in 1976 and by 1980 had been brought to strength levels competitive with PAN-based fibers. The major advances in mesophase carbon fibers have been toward high modulus levels. Such fibers are suitable for special applications, e.g., spacecraft structures, in which the negative thermal expansivity, which is a concomitant of high modulus, can be exploited to design composites with zero thermal expansion. It may be noted that carbon fibers can also be prepared by spinning isotropic (nonmesophase) pitch, and these pitch-based fibers are now being commercialized for concrete reinforcement or asbestos substitutes (9,11). However, from Figure 1, the mechanical properties of those fibers are clearly inadequate as structural reinforcements for high-performance composites.

Carbon fiber development is highly competitive, and proprietary considerations limit information on the technical bases for the improvements being achieved. Figure 1 includes some data on experimental fibers, including mesophase carbon fibers produced from solvent-extracted pitch (10). If the strength levels in Figure 1 for the latter fibers can be attained in production fiber and if higher moduli can be attained by more severe heat treatment, the solvent-refined mesophase carbon fibers may prove to be a second-generation mesophase fiber with significant property improvements.

This paper commences with evidence for lamelliform morphologies in mesophase carbon fiber, summarizes relevant information on disclination structures in the carbonaceous mesophase, and then reviews what we learn of disclination behavior from hot-stage observations and from deformation and carbonization experiments. The results indicate that disclination interactions that occur before the mesophase is fully hardened play an important role in determining the microstructures of mesophase carbon fibers, as well as those of cokes and graphites that form through the carbonaceous mesophase.

Microstructure of Mesophase Carbon Fiber

Early grades of commercial mesophase carbon fibers often displayed the radial structure depicted in Figure 2. On heat treatment after spinning, the wedge opens for the same reason that shrinkage cracks form in the calcining of petroleum coke: Shrinkage is greater perpendicular than parallel to the mesophase layers.

The most common morphology observed in current mesophase carbon fibers of moderate modulus (55 to 75 Mpsi, 379 to 517 GPa) is a cylindrical filament with a random-structured core and a radial rim (12). Given the fracture section of Figure 3, with its scroll-like features, the core appears to be an array of $+2\pi$ and $-\pi$ disclinations. The radial rim of heavily wrinkled layers usually constitutes half or more of the cross section.

In single-filament tensile testing of high-modulus fibers (13), three types of fracture surface are common (Figure 4). The high-modulus filaments (E ⩾ 100 Mpsi, 689 GPa) fail with extensive shear on the well-oriented layers, resulting in strongly serrated fracture surfaces with good structural definition. The structural types include the open-wedge radial and round random-core filaments, as well as an oval-shaped filament with an oriented core; these are summarized schematically in Figure 5. The oriented-core microstructure may be described as a radial rim with oriented core layers lying between $+\pi$ wedge disclinations.

Easy shear on strongly oriented layers also accounts for the filament splitting and segmented fracture frequently observed in single-filament tensile tests on the mesophase carbon fibers of high modulus. This mechanism also appears to be involved in the yield phenomenon found in Sinclair loop tests on high-modulus fibers (13) and no doubt contributes to the poor flexural and compressive strengths exhibited by epoxy-matrix composites with mesophase carbon fiber reinforcements (14). Figure 6 illustrates tensile fracture in an oriented-core P120 filament (E = 120 Mpsi = 827 GPa); easy shear on the central layers of the oriented core has produced a long sliver that resembles a single crystal of graphite.

The Carbonaceous Mesophase and Its Disclination Structures

Identification of the molecular structures most conducive to the formation of needle coke or the spinning of mesophase fiber is a task on which Japanese workers have been particularly active (15). The model of carbonaceous mesophase sketched in Figure 7 is based on the molecular architecture proposed by Mochida et al. (16). In contrast to conventional nematic liquid crystals, the molecular units are disk shaped and range widely in size, even when the mesophase is produced by the pyrolysis of pure organic compounds (17). Many of the molecules are volatile or reactive in the temperature range over which the mesophase is fluid, and the evolution of gaseous species usually causes the mesophase to be extensively deformed by bubble percolation before it congeals to a solid semicoke.

The disk-shaped molecules are not rigorously oriented to parallel arrays; the mesophase state constitutes only a preferred orientation, but with virtually all molecular layers lying within ±25° of the director representing the average orientation (18).

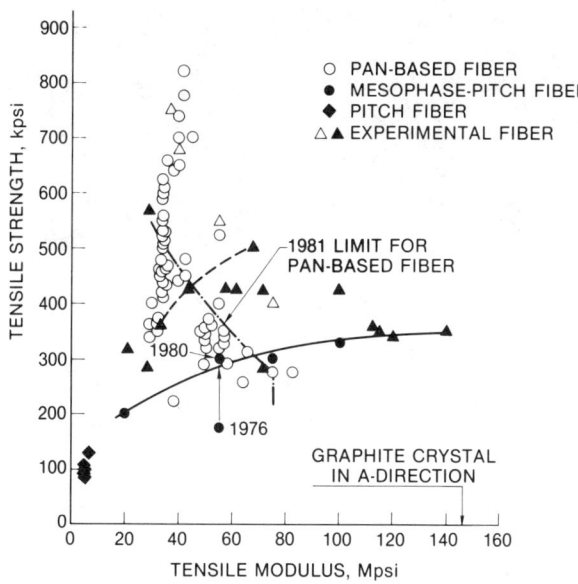

Figure 1. The status of carbon fiber development in 1984 based primarily on product data supplied by manufacturers. The solid line refers to commercial grades of fiber spun from mesophase pitch (7); the dashed line refers to an experimental mesophase fiber spun from solvent-extracted pitch (10). Experimental (noncommercial) fibers are indicated by triangular symbols.

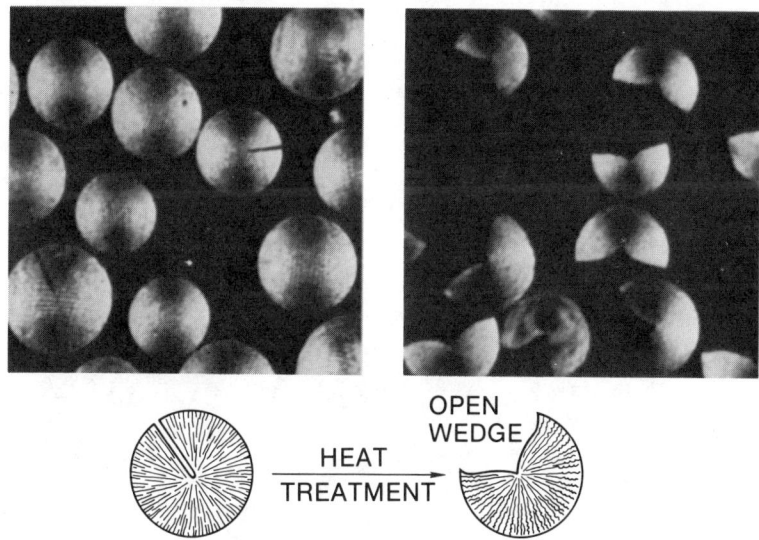

Figure 2. Mesophase carbon fiber with radial structure and open-wedge shape. Polarized light.

FRACTURE SECTION OF RANDOM FILAMENT

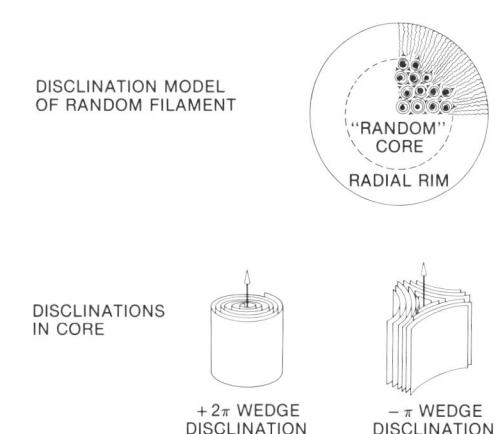

Figure 3. Mesophase carbon fiber with random-core structure and round shape.

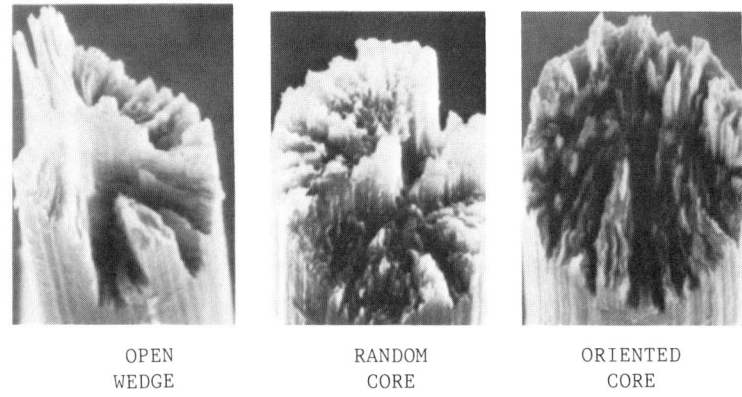

Figure 4. Tensile fracture surfaces for three structural types of high-modulus mesophase carbon fibers (E = 100 Mpsi = 690 GPa).

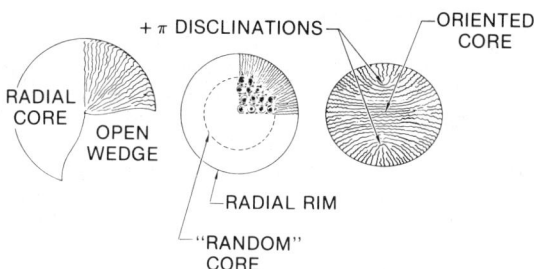

Figure 5. Structural models for the morphology of open-wedge, round, and oval filaments spun from mesophase pitch.

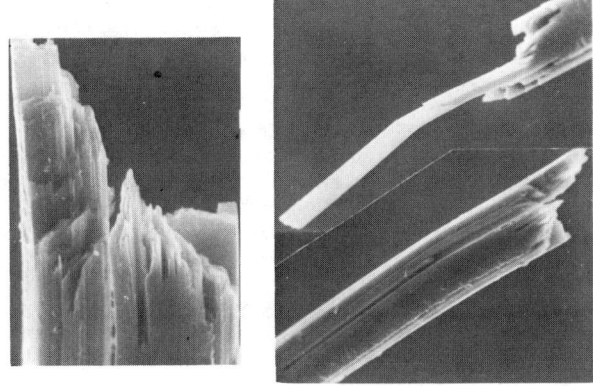

Figure 6. Tensile fracture surfaces for a mesophase carbon filament with tensile modulus of 120 Mpsi (827 GPa).

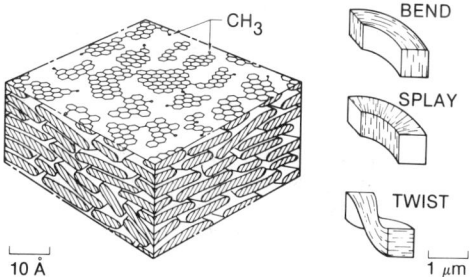

Figure 7. Schematic models for the carbonaceous mesophase [after Mochida et al. (16)] and its curvature strains of bend, splay, and twist. The molecular and strain models differ in scale by ~1000×.

Thus the mesophase tolerates easily the bend, twist, and splay essential to disclination structures, and both wedge and twist disclinations are readily identified in petroleum coke by polarized-light micrography (19). Figure 8 illustrates why disclinations do not occur in ordinary crystalline materials. The distortion energies are so large as to prohibit such disclinations from forming except by entrapment mechanisms such as the hardening of a liquid crystal. Figure 8 also illustrates how a Nabarro circuit (analogous to a Burgers circuit for a crystal dislocation) can be followed to define the rotational strength of a disclination.

Models for the wedge and twist disclinations commonly present in the carbonaceous mesophase, given in Figure 9, indicate the difference between them to inhere in the orientation of the rotation vector, defined by the Nabarro circuit, relative to the disclination line: In pure wedge disclinations, the vectors are parallel (or antiparallel); in pure twist disclinations, the vectors are perpendicular. Evidence for these disclination structures has been summarized in a recent review (1). Disclinations of higher order, e.g., -3π, have been found in the mesophase matrix of carbon-fiber-reinforced composites (20); however, these disclinations are forced by the "sheath effect" (the tendency for mesophase layers to align parallel to many substrates) acting between particular filament groupings (21), and disclinations of greater than second order have not been reported in bulk mesophase.

Fundamental to mesophase-fiber spinning and needle-coke formation is the mesophase flow behavior. Elementary studies have shown that uniaxial deformation produces fibrous morphologies with nearly pure wedge disclinations (2). Biaxial deformation, as experienced by the mesophase in the wall of an expanding bubble, produces lamellar morphologies with folded regions; the folds are bounded by π disclinations that vary in character from pure twist to pure wedge (19), as sketched in Figure 10. Although the π disclinations formed in bulk undeformed mesophase may be mixed disclinations, their character is readily altered by deformation.

Measurements by concentric-cylinder viscometers (22,23) on mesophase pitch pyrolyzed at constant heating rates have indicated that the viscosity rises sharply when volatilization and aromatic polymerization reactions become rapid. The isothermal observations of Figure 11, made in an industrial rheometer to high torque (24), reveal that the mesophase hardens by a continuous increase in viscosity (after passing an unstable region during the mesophase transformation). The polarized-light micrographs show that mechanical deformation in the rheometer effects a rapid refinement of microstructure, as indicated by the spacing of extinction contours, and can readily carry the microstructure beyond the limit of optical resolution.

Hot-Stage Observations

Since Hoover et al. (3) demonstrated that polarized light reflected from the free surface of a pyrolyzing liquid could be used to directly observe mesophase behavior, hot-stage microscopy has become a useful technique for studying the pyrolysis of carbon precursors. Provided that excessive volatilization does not

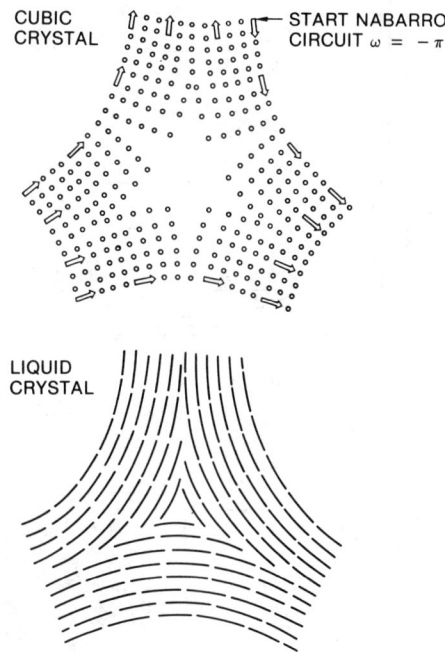

Figure 8. Wedge disclinations, with rotational strength $-\pi$, in a cubic crystal and in a liquid crystal. Note that the molecular scale of the liquid crystal is much larger than the atomic scale of the cubic crystal.

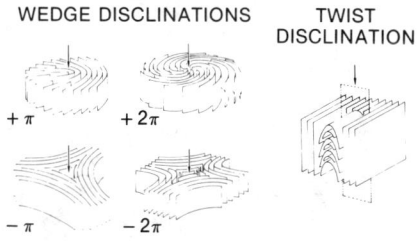

Figure 9. Schematic models for the wedge and twist disclinations of the carbonaceous mesophase.

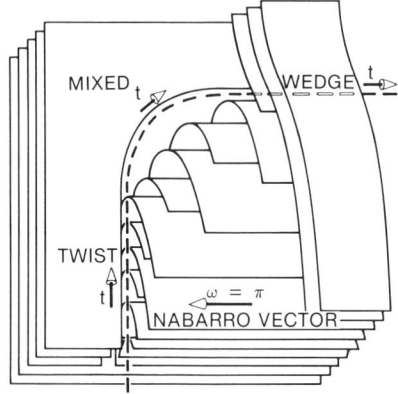

Figure 10. Schematic model for a disclination loop bounding a mesophase fold (1).

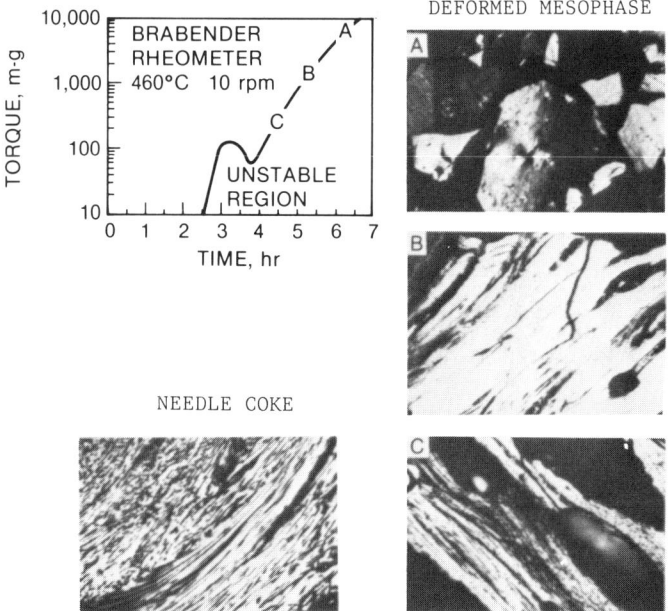

Figure 11. The increase in viscosity and refinement in microstructure as a mesophase pitch (from A240 petroleum pitch) is pyrolyzed within a rheometer (24). Crossed polarizers.

interfere by fogging the observation window or by forming a viscous mesophase skin (5), such methods enable qualitative evaluation of the dynamic behavior of the mesophase.

We constructed a hot-stage with quenching capability (25), see Figure 12, to relate free-surface observations to the three-dimensional morphology of bulk mesophase. A specimen can be pyrolyzed to a point of interest observed on the free surface, then quenched to a solidified body that can be sectioned and polished for detailed micrographic study. A simple probe was included in the hot-stage design so that the mesophase could be deformed at various stages of pyrolysis. A 32× objective with 6-mm working distance was adequate to resolve orientational fluctuations similar to those observed in nematic liquid crystals (26). The optical system remained stable during quenching; fine cracks could be seen to develop after the temperature dropped below the softening point of the mesophase.

The phenomena of coalescence, disclination reactions, and deformation response illustrated by film frame sequences in Figures 13-15 and 17 were observed on a petroleum pitch (Ashland A240) that had been thermally treated to reduce the evolution of volatiles. At observation temperatures near 440°C, the phenomena occurred at rates convenient for filming and for identifying disclination signs by rotating the plane of polarization.

Figure 13 depicts a coalescence event that took place in 7 sec. Like other coalescence events in which new disclinations are produced, a 2π disclination appeared at the position of the former boundary between the mesophase spherules. Frenkel (27) has pointed out that the time constant τ for such coalescence phenomena is given approximately by

$$\tau = \frac{\eta d}{\sigma} \qquad (1)$$

where η is the viscosity, d is the diameter of the coalescing bodies, and σ is the interfacial energy—the driving force for coalescence. Smith et al. (28) applied Equation (1) to several filmed coalescence events and reported that the pitch-mesophase interfacial energy is small, probably less than 0.1 dyne/cm.

Another mechanism by which new disclinations appear at the free surface is illustrated by the micrographic sequence in Figure 14. Two 2π disclinations of opposite sign are spontaneously generated from a region in which the polarized-light extinction contours pinch down below optical resolution; the reaction may be written as an annihilation reaction in reverse,

$$0 \longrightarrow (+2\pi) + (-2\pi) \qquad (2)$$

Reactions of this type are also observed to generate π disclinations of opposite sign. Reactions between disclinations of different order also occur; in the micrographic sequence of Figure 15, the reaction is

$$(+\pi) + (-2\pi) \longrightarrow (-\pi) \qquad (3)$$

In his classic work on liquid crystals, Friedel (26) summarized his observations of disclination reactions in nematic liquid crystals

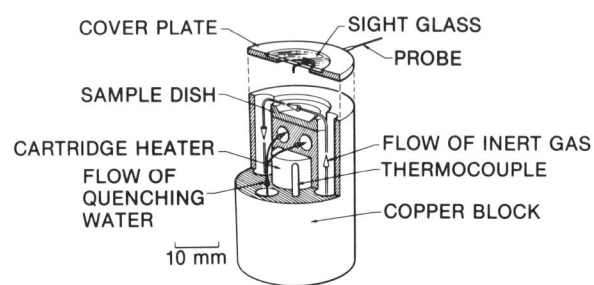

Figure 12. A microscope hot stage designed for quenching and for deformation of a specimen by a horizontal probe (25). Reproduced with permission from reference 25.

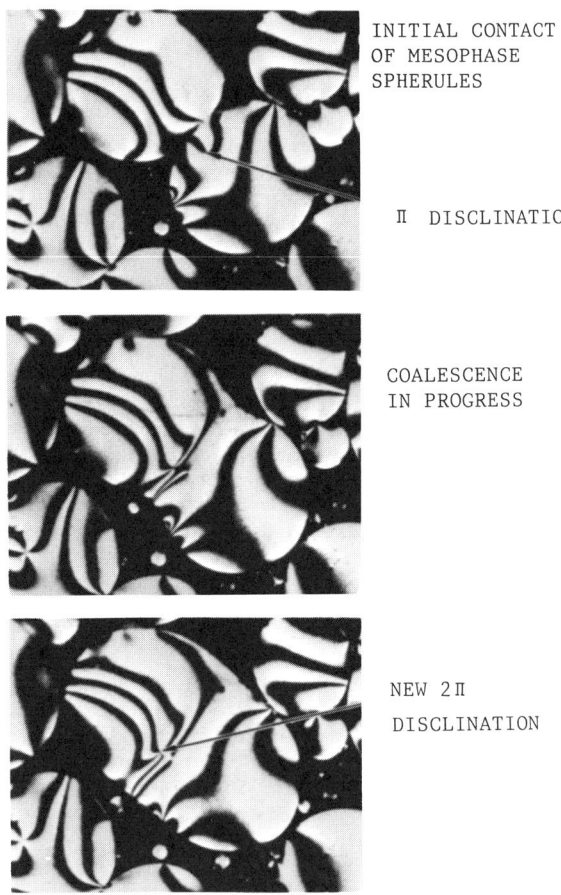

INITIAL CONTACT
OF MESOPHASE
SPHERULES

π DISCLINATION

COALESCENCE
IN PROGRESS

NEW 2π
DISCLINATION

Figure 13. Coalescence of mesophase spherules to produce a new 2π disclination. Hot-stage microscopy, crossed polarizers.

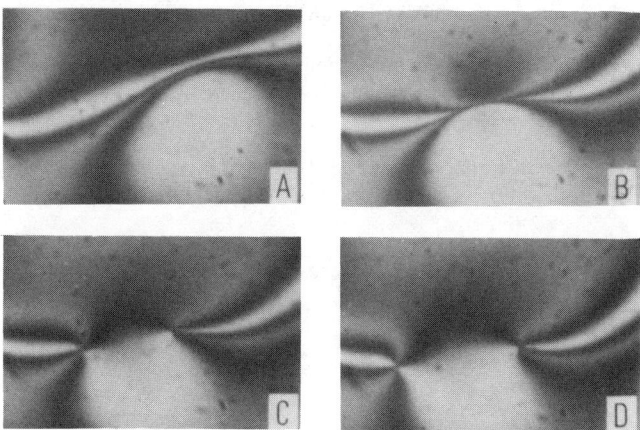

Figure 14. Pinch-off reaction to form 2π disclinations of opposite sign. Hot-stage microscopy, crossed polarizers. Reproduced with permission from reference 14.

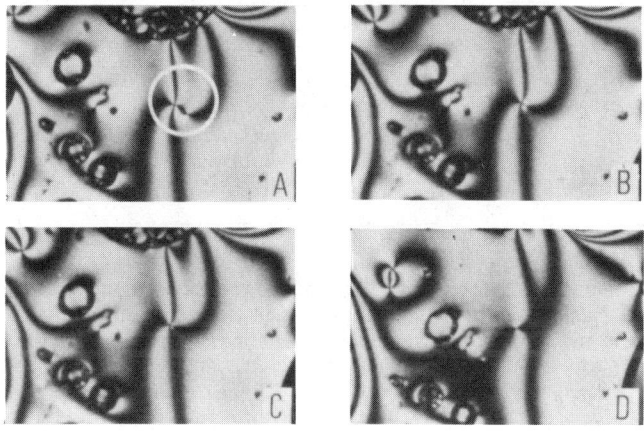

Figure 15. Reaction between disclinations of opposite sign. Hot-stage microscopy, crossed polarizers.

by a table similar to Table I. Our surface observations indicate that, as long as the viscosity remains low, the dynamic behavior of the carbonaceous mesophase is quite similar to that of nematic liquid crystals (6). However, as pyrolysis of the mesophase proceeds, the viscosity rises, and the disclination reactions slow well before the mesophase loses its deformability.

Table I. Disclination Interactions in Nematic Liquid Crystals

$(+2\pi) + (-2\pi) \rightleftarrows 0$	
	Annihilation and formation reactions
$(+\pi) + (-\pi) \rightleftarrows 0$	
$(+2\pi) + (-\pi) \rightleftarrows (+\pi)$	
	Reactions between disclinations of different order
$(-2\pi) + (+\pi) \rightleftarrows (-\pi)$	
$(+\pi) + (+\pi) \rightleftarrows (+2\pi)$	
	Combination and dissociation reactions
$(-\pi) + (-\pi) \rightleftarrows (-2\pi)$	

The spatial geometry of disclination reactions in bulk mesophase has recently been presented by Zimmer and Weitz (29). Working with coarse-structured mesophase prepared by lengthy pyrolysis of A240 petroleum pitch at 400°C, they defined the disclination arrays on a succession of polished sections spaced at about 7 μm. In this way a +π disclination was traced through a branching point (i.e., a reaction point) to form a -π and a +2π disclination. Thus a reaction

$$(+\pi) \rightleftarrows (+2\pi) + (-\pi) \tag{4}$$

may be visualized spatially as sketched in Figure 16, its direction determined by the motion of the branching point on the +π/+2π disclination line.

Fine deformed microstructures with strong preferred orientations could be produced by a single stroke of the wire probe (25), as illustrated by Figure 17. A vertical section made on a specimen quenched immediately after deformation confirmed that the underlying structure was fibrous with ±π and ±2π wedge disclinations (30). The relaxation or coarsening after deformation indicated that the mesophase was sufficiently fluid for disclination motion

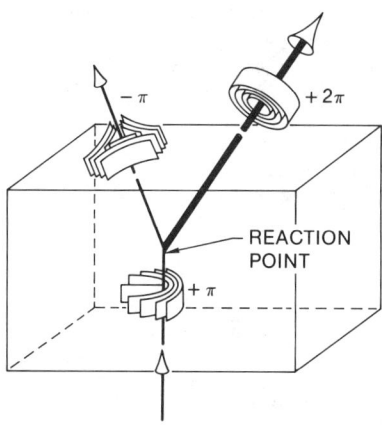

Figure 16. Spatial sketch of a disclination reaction in bulk mesophase.

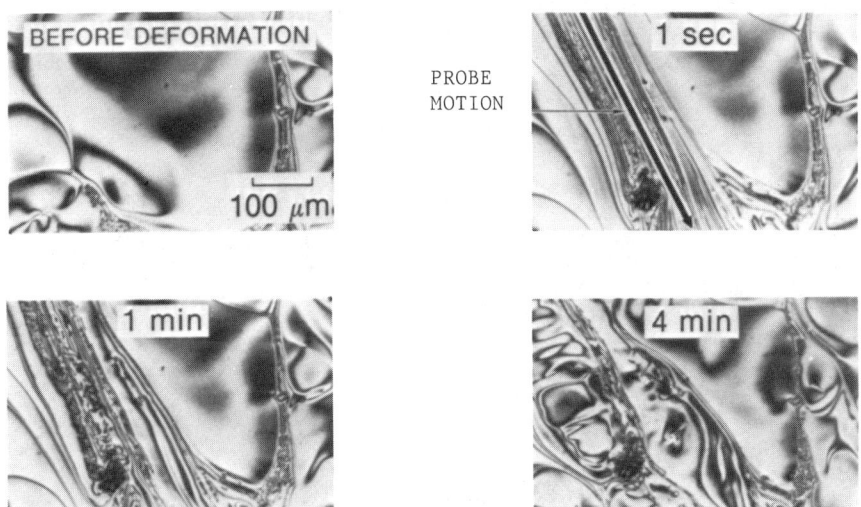

Figure 17. Formation and relaxation of deformed microstructure, after a single stroke by the wire probe on the hot stage (25). Crossed polarizers.

and reaction to occur, at least over the short distances between disclinations in the fine fibrous microstructure. At less advanced pyrolysis levels, relaxation was more rapid, and the mesophase quickly regained its original coarse microstructure. At more advanced pyrolysis levels, the extent of recovery was severely limited by the viscosity of the mesophase, and large residual densities of disclinations were left in the hardening mesophase. The capability of the mesophase to be deformed at pyrolysis levels well beyond those at which disclinations can interact accounts for the fine microstructures found in most products fabricated from the carbonaceous mesophase.

Deformation and Extrusion Experiments

Mesophase rods were extruded and drawn with the device shown in Figure 18, which is similar to equipment for spinning mesophase monofilaments (7,31). Jenkins and Jenkins (32) employed a similar device to produce extruded mesophase; in our studies, spinnerette diameters (0.2 to 2 mm) were smaller. The mesophase pitch was prepared from A240 petroleum pitch by the Chwastiak procedure (33), with extensive stirring and sparging with nitrogen. Full transformation to mesophase was attained by a 20-hr treatment at 400°C. A penetrometer test indicated the temperature range for extrusion and draw; the mesophase pitch softened at 309°C, and bubbling by pyrolysis gases was evident above 340°C.

Drawing by hand from a 2-mm spinnerette at about 340°C demonstrated the effectiveness of draw in orienting even a heavily bubbled rod (see Figure 19). From these observations, more controlled drawing experiments were devised, and a weight was attached to the extruding rod as depicted in Figure 18.

Figure 20 presents mesophase rods produced by extrusion alone and by a light draw after extrusion. As Jenkins and Jenkins observed (34), the strong preferred orientation induced by extrusion was easily disturbed by pyrolysis bubbles or even by small flow irregularities. However, modest draws (e.g., draw ratio = 2) after extrusion produced fibrous morphologies with good uniformity. At these draw levels, the nodes and crosses characteristic of wedge disclinations could be resolved on transverse sections.

Selected rods of drawn mesophase were subjected to carbonization runs to 1000°C so that the annealing behavior of the fibrous mesophase could be analyzed. The rods were first given an oxidation-stabilization treatment similar to that used in the manufacture of mesophase carbon fiber (7-9) to ensure that the rods would not collapse when the mesophase softened in carbonization and to learn the depth to which oxidation is effective in locking the microstructure into place. The mesophase rods displayed in Figure 21 were oxidized in air at 300°C for 8 hr, a treatment that induced some deep shrinkage cracks into which oxygen penetrated. From the high-magnification micrographs, a small degree of structural coarsening during the oxidation step can be inferred. The carbonization treatment provided good definition of the stabilized region. Mesophase within 10 μm of a surface or crack that had access to air retained the fibrous microstructure present after stabilization. At greater depths, disclination reactions proceeded

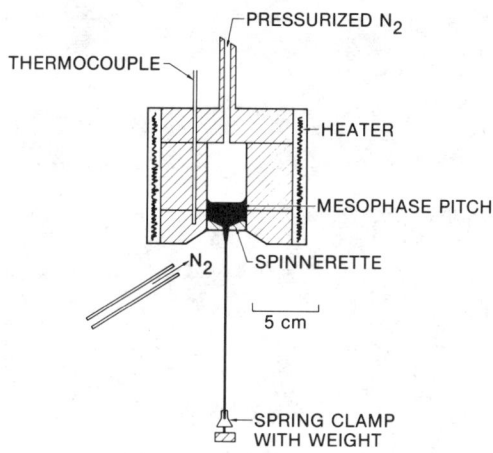

Figure 18. Apparatus for extrusion and drawing of mesophase.

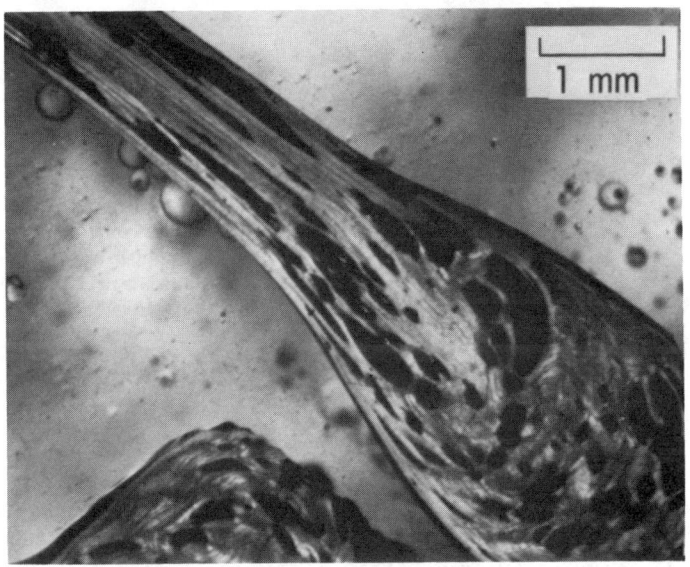

Figure 19. Strong preferred orientation produced by light draw of a bubbled mesophase rod. Crossed polarizers.

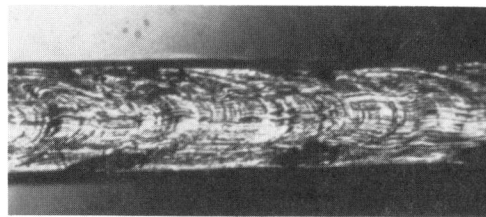

EXTRUDED
MESOPHASE
RODS

200 μm ORIFICE

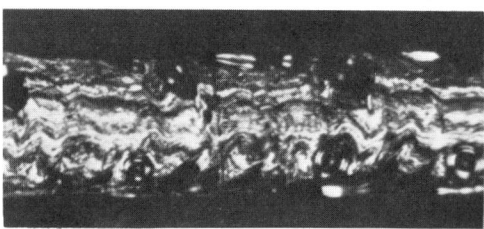

LONGITUDINAL
SECTIONS

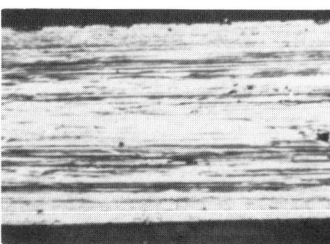

EXTRUDED AND DRAWN
MESOPHASE ROD

900 μm ORIFICE

LONGITUDINAL

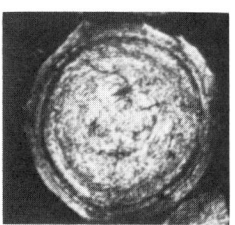

TRANSVERSE

Figure 20. The microstructures of mesophase rods, as extruded (above) and as drawn after extrusion (below). Crossed polarizers.

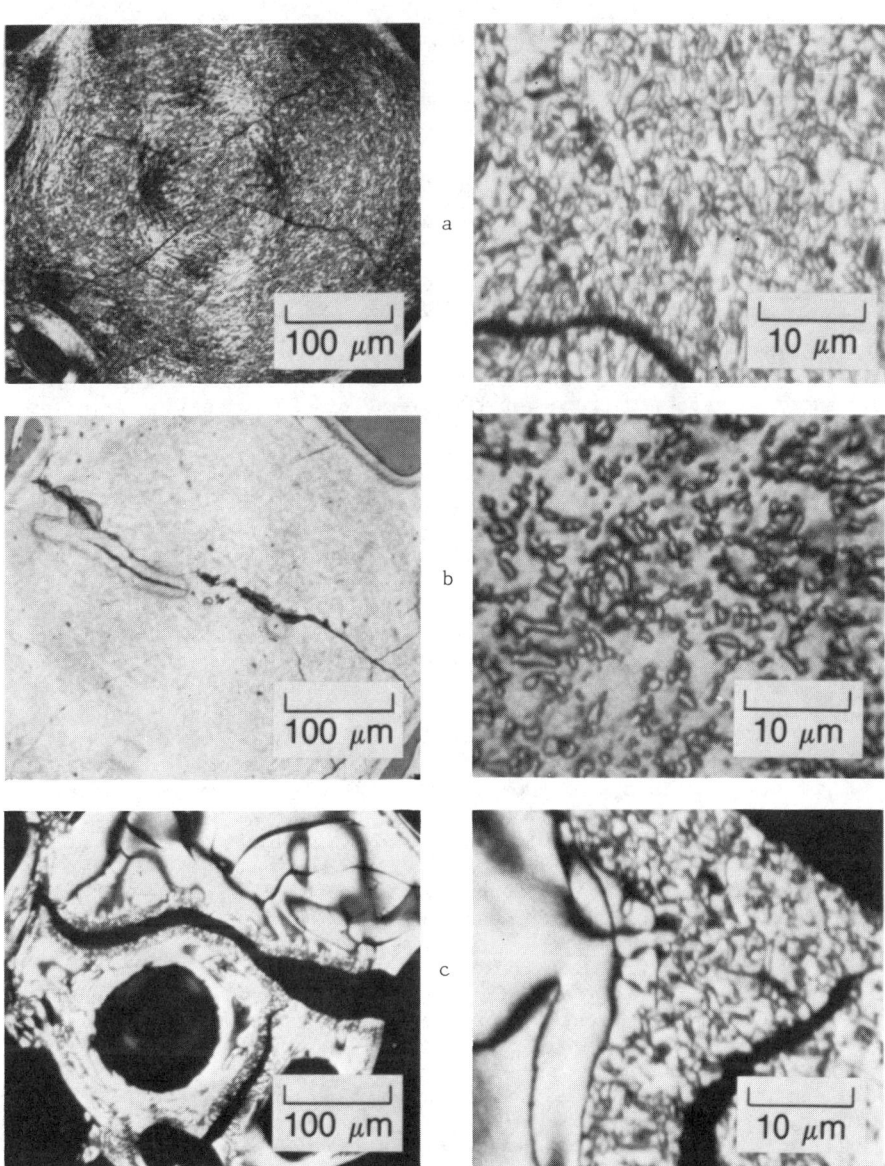

Figure 21. Transverse microstructures of drawn mesophase rods: (a) as extruded and drawn to a draw ratio of 3.6, (b) after oxidation at 300°C, and (c) after carbonization under inert atmosphere. Crossed polarizers.

to nearly complete annihilation, and in many rods the unstabilized mesophase had softened sufficiently to be ejected from the interior by the pressure of pyrolysis gases.

The carbonization experiments emphasize the need to fix or stabilize an oriented mesophase microstructure that has been imposed by mechanical deformation. The array of wedge disclinations imposed by drawing during cooldown is not thermally stable, and disclination reactions will proceed to coarsen the microstructure and reduce the preferred orientation if the mesophase softens sufficiently during carbonization.

Formation of Microstructure in Mesophase Products

Although the microstructural studies described here have been limited to structures that can be resolved by polarized light, the results suggest a pattern for the formation of the various filament morphologies shown in Figure 5. According to the Singer patent (7), the geometry of fiber spinning is essentially that shown in Figure 18. But the draw ratios and wind-up speeds are quite high, e.g., draw ratios near 1000 and speeds of the order of 3 m/sec. These values represent severe conditions of deformation and quench (of the order of 100°C/sec or higher), and the scale of mesophase structure must lie well below resolution by optical microscopy. Assuming that mesophase behavior at this scale is not substantially different from that observed at the scale of the hot-stage and deformation experiments, extensive disclination reactions may be expected to accompany the deformation of mesophase in filament spinning.

The manner in which a disclination loop can be converted by uniaxial deformation in the draw-down region to a pair of parallel wedge disclinations of opposite sign spaced within easy reaction distance (35) is diagrammed in Figure 22. The $\pm\pi$ and $\pm 2\pi$ disclinations composing the total disclination array of a fibrous microstructure may be expected to interact to varying degrees of completion as a function of mesophase viscosity and rate of filament cooling. The radial filament, which develops the open wedge upon heat treatment, may be expected to form under conditions of low viscosity and slow cooling that permit full annihilation of disclinations to leave a single $+2\pi$ disclination at the center. The random-core filaments are expected to form under conditions of higher viscosity and more rapid cooling, so that the disclination reactions are quenched to entrap an array of wedge disclinations before annihilation reactions can reach completion. The oval fiber, then, represents an intermediate state of disclination reaction in which just two $+\pi$ wedge disclinations remain, separated by the oriented core; the oval shape results from the anisotropic shrinkage of the mesophase upon heat treatment. The radial array of layers in the rim of the three filament structures corresponds to the equilibrium orientation of mesophase layers at a free surface. [The good contrast of polarized-light extinction contours observed on the free surface of mesophase in the hot stage results from the tendency for the mesophase layers to align perpendicular to this surface (4).] However, it appears doubtful that this alignment would prevail under the dynamic flow conditions existing as the mesophase congeals, and the mechanism of radial orientation near the filament surface might better be sought by investigating the flow dynamics of a discotic liquid crystal.

The potential for reactions between wedge disclinations composing a fibrous microstructure implies that caution must be used in the oxidation-stabilization step of fiber manufacture to avoid reactions that may tilt mesophase layers out of load-bearing alignment. For example, ±2π wedge disclinations formed under quiescent conditions are known to have continuous cores in which the mesophase layers tilt to form cup-shaped or saddle-shaped central regions (4). Thus the temperature employed for filament oxidation should be kept low to avoid relaxation reactions of the type sketched in Figure 23. The layers tilted out of load-bearing alignment will detract from the tensile modulus and may constitute flaws for the initiation of fracture.

On the basis of this discussion, the mechanisms of mesophase carbon fiber formation are closely related to those of needle coke, the principal differences being the extent to which the deformation and relaxation mechanisms are able to act. Because delayed coking involves relatively gentle but random deformation processes by bubble percolation and the long dwell times in the coke drum afford opportunity for extensive disclination annihilation and microstructural relaxation, the structure of needle coke can be well defined by polarized-light microscopy (2,36).

By comparison with delayed coking, fiber spinning imposes severe deformation, and cooling of the 10-μm filaments constitutes a rapid quench. Although the uniaxial stretch may produce well-defined arrays of disclinations, their spacing lies below optical resolution, and we must surmise from hot-stage quenching experiments (6) that the disclination structures in the partially reacted arrays may deviate appreciably from the equilibrium structures of Figure 9. Microscopic methods of higher than optical resolution are required to define such disclination structures and the extent to which they react or relax in stabilization and carbonization; such methods may also be useful in finding mesophase precursors or filament treatments that produce more structural cross-linking and layer-wrinkling in the finished filament, and thus greater resistance to shear and buckling failure when the high-modulus filaments undergo compressive or flexural loading in a composite.

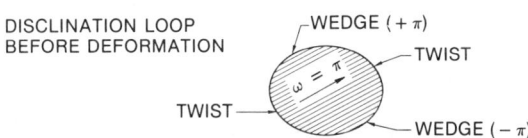

Figure 22. Uniaxial deformation of a disclination loop bounding a mesophase fold. Reproduced with permission from reference 35.

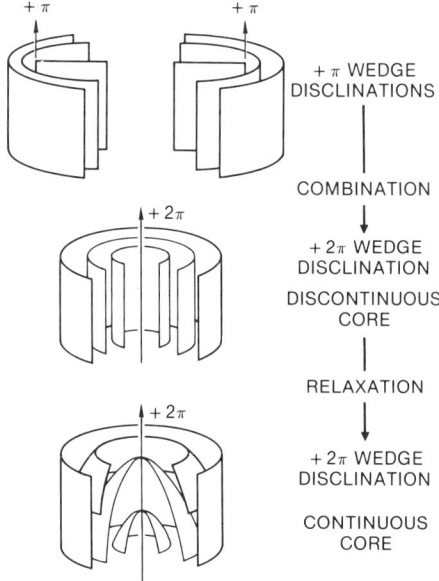

Figure 23. Combination of two +π disclinations may tilt mesophase layers out of fibrous alignment by formation of a continuous core in the +2π disclination.

Acknowledgments

We thank the Office of Naval Research for supporting the preparation of this review and Dr. L. H. Peebles, Jr., for encouragement and criticism. This paper also reviews selected aspects of work supported by the Space Division of the U.S. Air Force and the Naval Surface Weapons Center. We also wish to thank several coworkers who contributed in significant ways to the work reviewed here: J. E. Zimmer, C. B. Ng, G. W. Henderson, and P. M. Sheaffer.

Literature Cited

1. Zimmer, J. E.; White, J. L. Adv. Liq. Cryst. 1982, 5, 157.
2. White, J. L. In "Petroleum Derived Carbons"; Deviney, M. L.; O'Grady, T. M., Eds.; ACS SYMPOSIUM SERIES No. 21, American Chemical Society: Washington, D. C., 1976; p. 282.
3. Hoover, D. S.; Davis, A.; Perrotta, A. J.; Spackman, W. Ext. Abstr., 14th Conf. Carbon, 1979, p. 393.
4. White, J. L.; Buechler, M.; Ng, C. B. Carbon 1982, 20, 536.
5. Uemura, S.; Hirose, T.; Takashima, J.; Kato, O.; Harakawa, M. Ext. Abstr., 16th Conf. Carbon, 1983, p. 78.
6. Buechler, M.; Ng, C. B.; White, J. L. Carbon 1983, 21, 603.
7. Singer, L. S. U.S. Patent 4 005 183, 1977.
8. Diefendorf, R. J.; Riggs, D. M. U.S. Patent 4 208 267, 1980.
9. Otani, S.; Oya, A. This volume.
10. Riggs, D. M. This volume.
11. Tatsuhana, M. In "Progress in Science and Engineering of Composites"; Hayashi, T.; Kawata, K.; Umekawa, S., Eds.; ICCM-IV: Tokyo, 1982; Vol. 1, p. 79.
12. White, J. L.; Ng, C. B.; Buechler, M.; Watts, E. J. Ext. Abstr., 15th Conf. Carbon, 1981, p. 310.
13. Ng, C. B.; Henderson, G. W.; Buechler, M.; White, J. L. Ext. Abstr., 16th Conf. Carbon, 1983, p. 515.
14. Townsend, H. N. Proc. 3rd Int. Conf. on Composite Materials, 1980, Vol. 1, p. 453.
15. White, J. L. Sci. Bulletin, ONR Far East 1984, 9, 32.
16. Mochida, I.; Maeda, K.; Takeshita, K. Carbon 1978, 16, 459.
17. Lewis, I. C. Carbon 1982, 20, 519.
18. Auguie, D.; Oberlin, M.; Oberlin, A.; Hyvernat, P. Carbon 1980, 18, 337.
19. White, J. L.; Zimmer, J. E. Carbon 1978, 16, 469.
20. Zimmer, J. E.; Weitz, R. L. Ext. Abstr., 16th Conf. Carbon, 1983, p. 92.
21. Zimmer, J. E.; White, J. L. Carbon 1983, 21, 323.
22. Collet, G. W.; Rand, B. Fuel 1978, 57, 162.
23. Balduhn, R.; Fitzer, E. Carbon 1980, 18, 155.
24. Buechler, M.; Ng, C. B.; White, J. L. Ext. Abstr., 14th Conf. Carbon, 1979, p. 433.
25. Buechler, M.; Ng, C. B.; White, J. L. Ext. Abstr., 15th Conf. Carbon, 1981, p. 182.
26. Friedel, G. Ann. Phys. 1922, 18, 273.
27. Frenkel, J. J. Phys. (Moscow) 1945, 9, 385.
28. Smith, G. W.; White, J. L.; Buechler, M. Carbon 1985, 23, 117.

29. Zimmer, J. E.; Weitz, R. L. Ext. Abstr., Carbone 84 (Int. Carbon Conf., Bordeaux, France), 1984, p. 386.
30. White, J. L.; Buechler, M. Ext. Abstr., Carbone 84 (Int. Carbon Conf., Bordeaux, France), 1984, p. 344.
31. Singer, L. S. Fuel 1981, 60, 839.
32. Jenkins, J. C.; Jenkins, G. M. Fuel 1981, 60, 883.
33. Chwastiak, S. U.S. Patent 4 209 500, 1980.
34. Jenkins, J. C.; Jenkins, G. M. Carbon 1983, 21, 473.
35. Buechler, M.; Ng, C. B.; White, J. L. Ext. Abstr., 16th Conf. Carbon, 1983, p. 88.
36. Marsh, H.; Latham, C. S. This volume.

RECEIVED October 3, 1985

Electron Microscopic Observations on Carbonization and Graphitization

A. Oberlin, S. Bonnamy, X. Bourrat, M. Monthioux, and J. N. Rouzaud

Laboratoire Marcel Mathieu, Equipe de Recherche du CNRS No. 131, U.E.R. Sciences, Université d'Orleans, 45046 Orleans Cedex, France

> The physico-chemical properties of carbon materials depend both on their crystalline structure, which can be determined by electron diffraction, and on their microtexture, observable by transmission electron microscopy. Examples are chosen among heavy petroleum products and their derived carbons to describe the thermal behavior and the evolution of the carbon crystalline arrangement during carbonization and graphitization processes. At low pyrolysis temperatures, these materials contain basic structural units (BSU) comprised of aromatic ring structures, generally less than 10 Å in extent, and occurring singly or in stacks of two or three. The BSU's are randomly oriented almost up to the stage of semi-coke. Just prior to this stage, they orient parallel. The ability of carbonaceous matter to graphitize depends on the extent of the local molecular orientation (LMO) and on the chemical composition (functional groups) acquired before this stage.

This paper presents a general review of studies by our laboratory on the carbonization and graphitization of carbonaceous materials (1), with particular reference to petroleum-derived carbons.

All heavy petroleum products and carbonaceous materials are partially aromatic and thus contain planar aromatic ring structures linked to each other by functional groups to form macromolecules. Thermal treatment under inert atmosphere causes progressive release of the functional groups, carbonization of the material, and, dependent on the carbonized structure, transformation into graphite (graphitization). The work of this laboratory (2-7) has shown that both the graphitizability and the physicochemical properties depend entirely on the microtexture acquired during the first stage of carbonization, i.e., on the initial spatial arrangement of the planar aromatic ring structures. Such a microtexture can be studied at nearly the atomic scale by various modes of conventional transmission electron microscopy (TEM).

Techniques of Transmission Electron Microscopy

The principle of lattice fringe imaging consists in forming an image with two beams. For example, the unscattered beam and the (002) beam will produce (002) lattice fringes. Dark-field imaging, on the other hand, consists in forming an image with a single (hkl) scattered beam. Rather than use a goniometer stage in the microscope, which limits TEM resolution, the relaxation of the Bragg conditions appropriate to poorly crystallized products is employed (2,8).

The techniques that we have employed can be performed with any modern electron microscope having an optimum point-to-point resolution better than 5 Å and a line resolution better than 2 Å. Among other factors due to apparatus, the line resolution (9) is dependent on the size of the diffracting elementary domain (the equivalent of the crystallite for a two dimensional crystal). It decreases to less than optimum for poorly crystallized products. The incident beam tilt used for dark-field imaging has to occur in polar (Figure 1) rather than Cartesian coordinates. To have optimum resolution in bright field, an objective aperture 0.6 Å^{-1} in diameter must be used. In dark-field, to select only one beam at a time, an aperture 0.2 Å^{-1} in diameter is necessary.

The image formed by a convergent lens (Figure 2) is the double Fourier transform of the object O. The electron diffraction pattern is formed in the image focal plane A and is the first Fourier transform. In the Gauss plane G, the second Fourier transform is produced (image of the focal plane) which restores the object space in the image. With X-ray diffraction patterns as well as with electron diffraction patterns, this second Fourier transform is determined by simple calculations. Thus an $\underline{average}$ is obtained of all diffracting elementary units, i.e., the basic structural units, or BSU. An enormous number of BSU's have their shape and dimensions averaged because the minimum diffracting volume is about 10^{18} Å^3 for X-rays and 10^{10} Å^3 for electrons. Thus, neither X-ray nor electron diffraction patterns can reveal the microtexture (the arrangement in space of the BSU's). The advantage of electron diffraction is to obtain patterns in the micron range instead of a few hundredths of a millimeter as for X-rays. For example, a lamella folded parallel to an axis, or a pack of parallel ribbons, will produce the same X-ray and electron diffraction patterns. In a TEM, on the other hand, each elementary unit in the image can be $\underline{localized}$, thus individualizing it relative to its neighbor. To do that, the size of the BSU has only to be greater than the resolving power of the microscope. This particular property of the TEM allows, at a scale near that of the unit cell:
- measurement of the size, and determination of the shape and orientation in space, of each BSU;
- determination of the mutual orientation in space of the BSU, i.e., the microtexture (microstructure).

In this rests the advantage of TEM, which is the only technique yielding new structural information beyond that obtained by X-ray and electron diffraction.

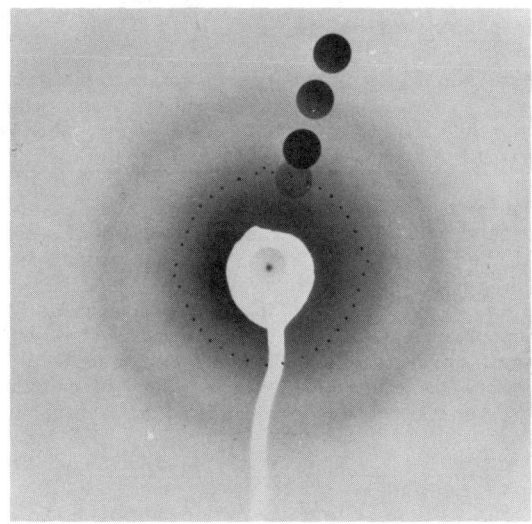

Figure 1. Test micrograph showing the displacement of the unscattered beam (small dots) in the selected area diffraction (SAD) pattern when it occurs in polar coordinates (Philips EM 300). The tilt has been fixed at the 002 Bragg angle for carbon ($\sim 0.3°$) and the azimuth changed by small increments. The 000 spot displaces along a practically perfect circle which corresponds to the 002 Debye Scherrer ring. Such a device allows exploration of any position in the SAD pattern, even when neither sharp nor intense hkl reflections are visible. The SAD pattern of an asphaltene heat-treated at 500°C has been superimposed to the test micrograph. Various positions of a 0.13 $Å^{-1}$ aperture are shown.

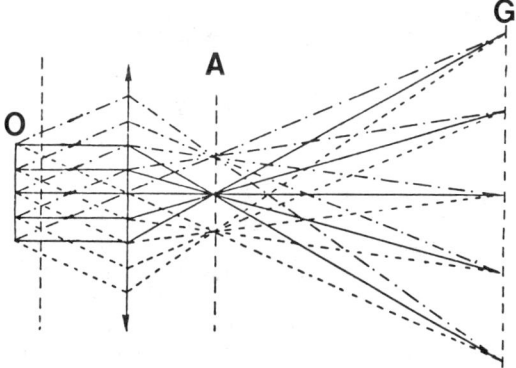

Figure 2. Ray path in a converging lens. In the image focal plane A, the electron diffraction pattern is formed (first Fourier transform of the real space where the object is localized). In the Gauss plane G, the image is formed (Fourier transform of the diffraction pattern, i.e., second Fourier transform) which restores the real space O magnified by the lens.

First Stage of Carbonization

All carbonaceous materials are composed mainly of carbon, hydrogen, and oxygen and in some cases contain sulfur and nitrogen. When they are pyrolyzed under an inert atmosphere, they release oxygenated functional groups and aliphatic CH groups, as tars and gases (10).

The carbonization behavior of a graphitizing material similar to light petroleum products (such as an atmospheric residue or even a pitch) will be described first. These materials are characterized by the formation of mesophase spheres. Single mesophase spheres nucleate, then grow, coalesce, and finally form material called bulk mesophase (11-13). This stage immediately precedes the semi-coke stage, which results from the sudden transition from a viscoelastic solid to a brittle solid. Then comes the further gas release leading to coke characterized by a high carbon content.

The Concept of a Basic Structural Unit (BSU). The elementary scattering units, or basic structural units (1,2,14), of pitches are similar to those of other carbonaceous materials. They are made of very small planar aromatic ring structures. An aromatic structure, even if it is very small, is sufficiently periodic to scatter electrons. When two of them stack together, at least two intense 002 beams and six weak 10 and 11 beams are obtained; the latter are due to the 10 and 11 reciprocal lines (hexagonal symmetry of the polyaromatic structure). The BSU's can thus be imaged either in 002 dark-field (BSU edge-on) or in 10 or 11 darkfields (8). In all cases, bright domains less than 10 Å in extent are obtained. Thus the stacks do not exceed two or three layers in thickness, and the structures are less than 12 fused benzene rings in area. Figure 3 shows some of the possible aromatic molecules (14). The BSU's in parallel array correspond to turbostratic stacking.

The Concept of Local Molecular Orientation (LMO). When observed by the naked eye a piece of semi-coke has a macroporous texture. When broken into fragments the pore walls appear as lamellae. After polishing, the pore walls appear as isochromatic areas when observed by optical microscopy.

The bulk mesophase of graphitizing materials observed by TEM appears as flat lamellae due to the grinding of the sample to particle sizes that are smaller than the optically isochromatic areas. Within a lamella, the aromatic BSU layers lie approximately parallel to the lamellar plane and thus constitute a region of local molecular orientation (LMO) (1-7). This molecular texture is nematic (15). It arises from disruption of the columnar structure of the initial mesophase spheres by the coalescence process (16,17). In lattice fringe images, the LMO appears as stage 1 in Figure 4 (1,18). For materials that graphitize well, the size of the lamella observed by TEM does not represent the real extent of the LMO since this is given by the entire isochromatic area. Thus the extent of the LMO in graphitizable materials can be measured only by means of the optical polarizing microscope (19-21).

Between 600°C and 1500°C the BSU's do not change much, either in diameter or in thickness, but they associate into distorted

columns which can trap single misoriented BSU's between them. The
formation of these columns (stage 2 in Figure 4) is the inevitable
consequence of the closer packing of the BSU's caused by the departure of heteroatoms and some of the defects. The model of Figure 5
(borrowed from 22,23) shows that such columns result naturally from
the improvement of the stacking of anisotropic BSU's as thermal
energy is added (like coins shaken in a box).

Above 1500°C the single misoriented BSU's disappear entirely. The distorted columns then quickly coalesce into stacks of
layers distorted in zigzag fashion (stage 3 in Figure 4). The
zigzag structures seen in the lattice fringe images have a period
of about 10 Å and are the memory of the single columns. The continuous layers retain their initial misorientation in the form of
the distortions left from coalescence of the columns. These distortions prevent growth in diameter, but the coalescence favors the
growth in thickness. Between 1500 and 2000°C (stage 3, pregraphitization), the number of layers per stack increases rapidly whereas
the diameter of the layers remains small (24).

The graphitization stage (HTT > 2000°C) corresponds to complete dewrinkling of the layers, which become stiff and nearly
perfect within each LMO area (stage 4 in Figure 4). Rapid development of three dimensional order then occurs because the defects
situated at the boundaries of the zigzag domains have been suddenly
removed.

Some materials contain more oxygen, nitrogen, sulfur, and less
hydrogen than those described above (e.g., heavy petroleum residues
and asphaltenes). When pyrolyzed, they follow the same microtexture stages as shown in Figure 4. Hence we will retain the term
bulk mesophase (1), in the broad sense, for the stage of local
molecular orientation (LMO) which is common to all carbonaceous
materials. It always occurs just before the passage to the brittle
solid (stage 1 in Figure 4). However, each product, having different chemical composition, is now characterized by a different
LMO extent, which may range from a hundred microns to less than a
hundred angstroms.

Small LMO regions correspond physically to deformed lamellae
which vary from flat to increasingly crumpled with decreasing radii
of curvature. As shown in Figure 6, a model of stacked paper
sheets can be used to simulate the progressive crumpling of the
lamellae. The radii of curvature of the paper folds decreases from
b to d. Entangled irregular pores are formed as the paper sheets
are increasingly crumpled. In coarse-structured materials, the
macropores are so large relative to the scale of TEM that the pore
walls, after grinding, appear as lamellae with radii of curvature
approaching infinity. However if the sample is naturally microporous, the pores can be much smaller than the ground fragment, and
the radii of curvature of the crumpled lamellae become measurable.

Each carbonized material is characterized by the size and
distribution of the pores. We have found it convenient to distinguish between large pores ($\geq$ 2000 Å) and smaller ones with sizes
down to 50 Å. Furthermore, the pore size depends on the chemical
composition of the material. The smallest pores are due to the
crosslinking effects of some heteroatoms (such as oxygen and
sulfur) or to the effect of defects such as tetrahedral bonds,
etc., which reduce BSU mobility during the soft stage. As the

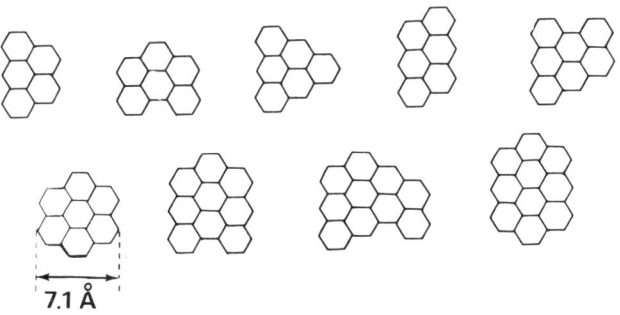

Figure 3. Some of the possible aromatic ring structures whose sizes fit those evaluated by TEM in 002 dark-field (14,34).

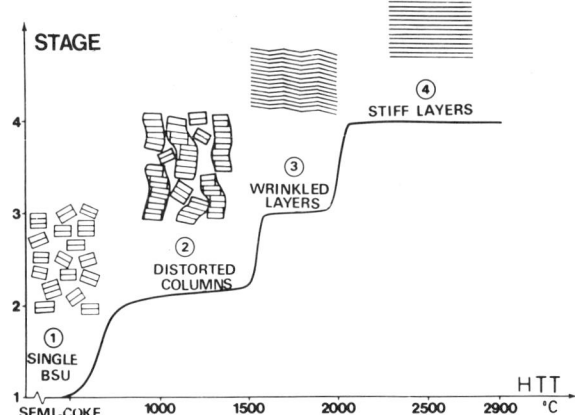

Figure 4. Structural stages of carbonization and graphitization, sketched from 002 lattice fringe images (1,18).
Reproduced with permission from reference 1.
Copyright 1984 Pergamon Press, Inc.

5. OBERLIN ET AL.   *Electron Microscopic Observations*   91

Figure 5. Model of distorted columns formed by BSU's (23).

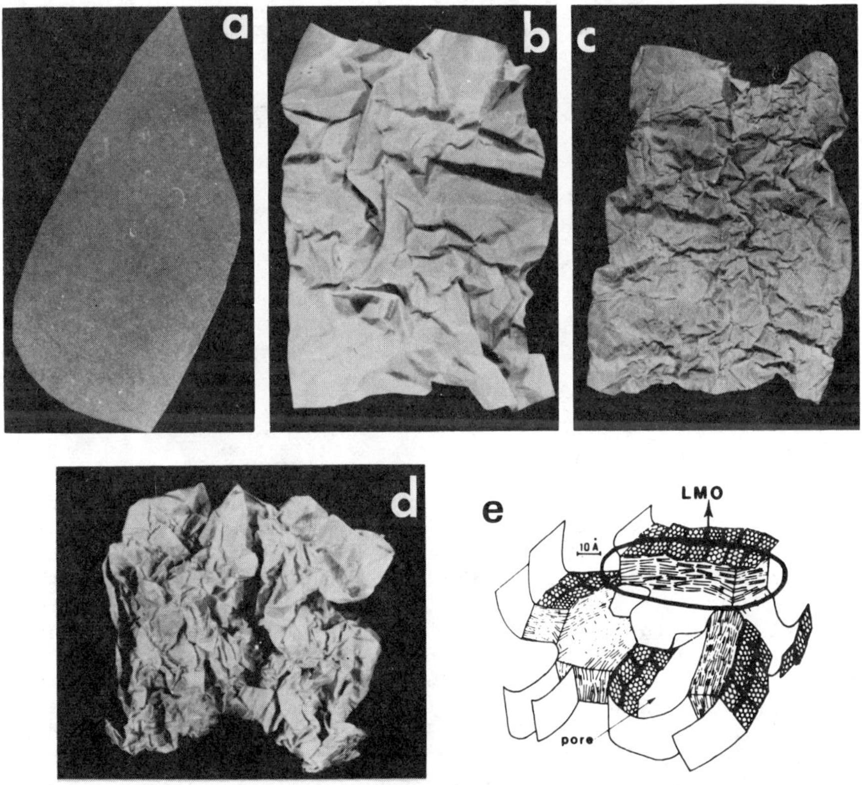

Figure 6. Progressive crumpling of flat lamella, illustrated with paper sheet models (35). The sketch **e** represents a porous fragment with intense crumpling. An area of homogeneous LMO is circled (4,5,36). Reproduced with permission from reference 1. Copyright 1984 Pergamon Press, Inc.

radii of curvature of the crumpled lamellae decrease, so do the pore sizes. In the end, the elemental domains of bulk mesophase can be less than 50 Å in extent. The structural sketch in Figure 6e shows the three dimensional association of a pore with the BSU's that constitute the local molecular orientation (LMO). The BSU's are represented by small packs of short parallel lines when seen edge-on, and by a hexagonal structure when their plane is seen. The size of the pore wall is equal to the extent of the local molecular orientation. Since the diameters of the pores cannot be measured by TEM, they can be classified by the LMO extents.

By TEM, the LMO extent is measured by the size of the corresponding cluster of bright domains in the 002 dark-field image (compare the circled area in Fig. 6e and the area LMO circled in the micrograph of Fig. 7a). Once this extent is evaluated from the micrographs, it can be classified into eight categories for pores and two supplementary ones for lamellae (1-5, 25-28). The flat lamellae represented in Fig. 8a are class 10, the crumpled lamellae are class 9 (Fig. 8b), and the pores (Fig. 8c) are class 8 to 1 according to decreasing size, from $\geq$ 2000 Å to < 50 Å, as summarized in Table I.

Table I. Size Classification for Regions of Local Molecular Orientation (LMO)

Histogram Class	LMO Extent (Å)
8	$\geq$ 2000
7	500-1000
6	350-500
5	250-350
4	150-250
3	100-150
2	50-100
1	< 50

Nearly all carbonaceous materials are structurally heterogeneous, composed of various fractions of the LMO classifications. For each carbonaceous material (26) there is a corresponding histogram that can be constructed for the frequencies of the different LMO sizes: flat lamella FL, crumpled lamellae CL, and pore classes 8 to 1. Thus in Figure 9, two examples are given of petroleum products after heat treatment at 1000°C. Figure 9a represents an asphaltene (fraction insoluble in heptane) from Athabasca tar sands, and Figure 9b represents an asphaltene obtained from non-biodegraded crude oil from the same basin (28). In this way a kind of fingerprint can be used to characterize the structural composition of each carbon precursor.

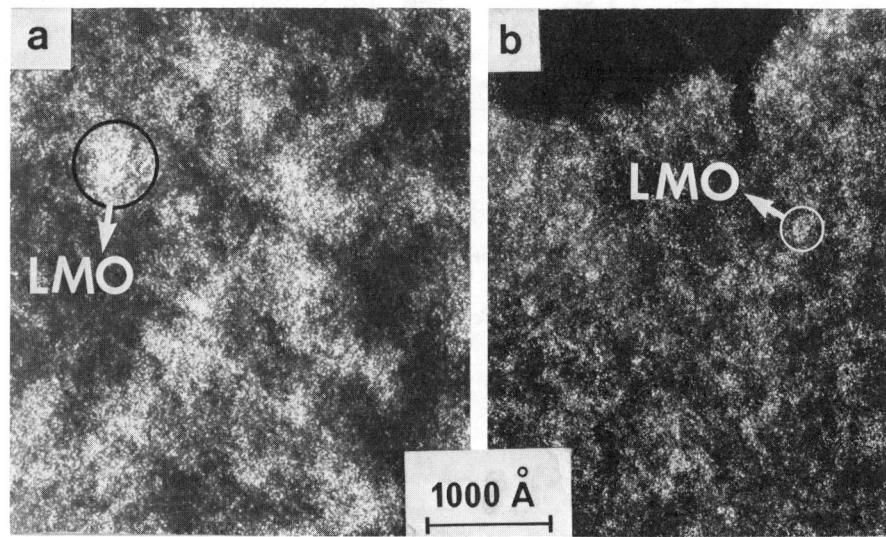

Figure 7. Micrographs showing how the LMO appear in 002 dark-field images.
(a) Class 7 (~ 600 Å). A homogeneous area is circled.
(b) Class 5 (~ 300 Å).

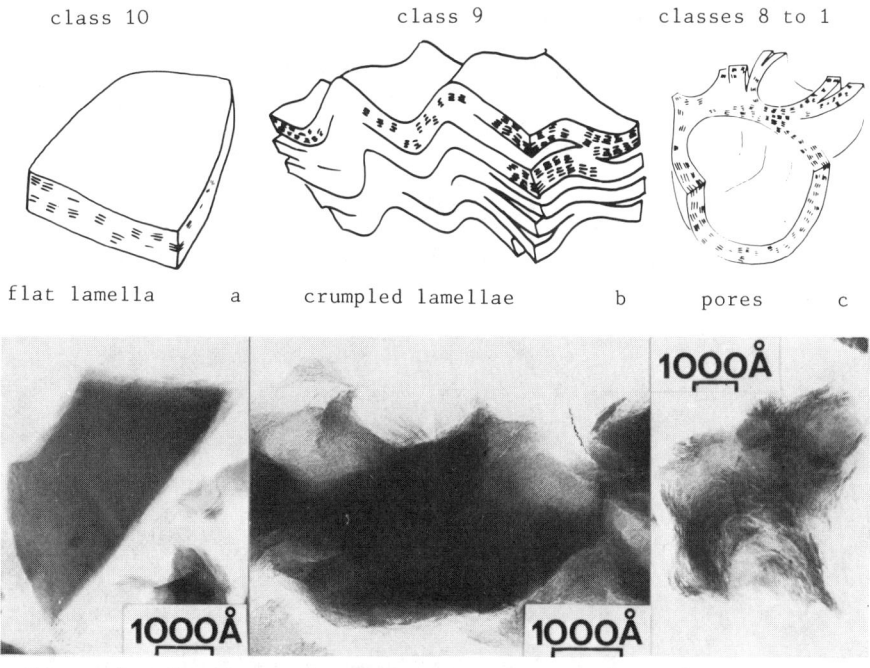

Figure 8. Sketch of the different LMO types (above) and micrographs showing their morphology in bright-field (below).
(a) Flat lamellae (class 10).
(b) Crumpled lamellae (class 9).
(c) Pores (class 8, ~ 2000 Å).
Reproduced with permission from reference 1.
Copyright 1984 Pergamon Press, Inc.

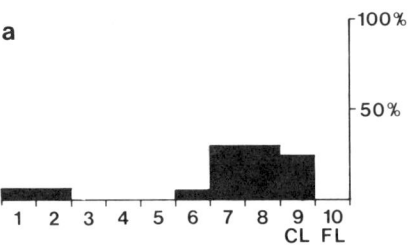

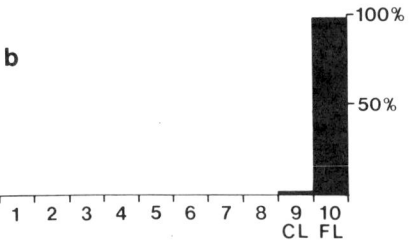

Figure 9. Histograms showing the structural distribution of LMO sizes for
(a) an asphaltene from Athabasca tar sands, and
(b) an asphaltene from a non-biodegraded oil from the Athabasca basin (28).

Graphitization

When carbonaceous materials are heat treated at higher temperatures, e.g., up to 2900°C, flat lamellae (class 10) and crumpled lamellae (class 9) give flat polycrystalline graphite (graphitizing carbons). The curved pore walls give polyhedral pores. Regions from LMO classes 4 to 8 become partially graphitized carbons, whereas the classes 1 to 3 give non-graphitizing carbons (1,6,7). The smaller the elemental domains of the initial bulk mesophase (pore walls), the smaller will be the polyhedral pores and the final degree of graphitization reached at 2900°C. The process is a progressive and statistically homogeneous transformation during which the probability P for finding a pair of layers with the graphite AB stacking sequence and the 3.35 Å interlayer spacing increases as the heat treatment temperature increases (29). For graphitizing carbons, P tends to 1, and $\bar{d}_{002}$ to 3.35 Å. However, these ideal values are never reached, even for the best graphitizing samples. For non-graphitizing carbons, P_{max} remains 0, and $\bar{d}_{002}$min. ~ 3.44 - 3.41 Å after treatment at about 2900°C. The P_{max} and $\bar{d}_{002}$min. of all carbonaceous materials studied up to now are distributed between the two extremes representing graphitizing and non-graphitizing carbons; hence all intermediates exist (1,6,7), particularly in petroleum products (asphaltenes) (25), as shown in Figure 10.

For materials containing only C, H, and O, oxygen and hydrogen have opposite effects. Therefore, an important parameter is the atomic ratio O/H (1,2,6,7). As oxygen content increases, the size of the elemental domains of bulk mesophase (local molecular orientation, LMO) decreases. As hydrogen content increases, LMO increases. It is necessary to measure oxygen and hydrogen at the point when LMO occurs. This takes into account both the possible dehydrogenation by oxygen into H_2O and the normal release of aliphatic CH into tars. The oxygen which remains in the material at the LMO stage corresponds to a strongly fixed oxygen which acts as a cross-linking agent. Correspondingly the residual hydrogen, and also the CH groups fixed on the aromatic structures, form a solvolytic suspensive medium for the BSU's (30).

When materials contain sulfur in addition to C, H, and O, the process is more complex because sulfur may form three kinds of bonds (27). One kind acts as oxygen does by forming H_2S before the semi-coke stage; another kind is unstable between 1300°C and 1700°C; and, for some products, there is a third kind of residual sulfur that remains firmly fixed to the carbon above 1700°C. In the latter case, the sulfur acts as a strong cross-linker. In contrast, unstable sulfur acts as a modifier between 1300°C and 1700°C, suddenly transforming carbon into graphite (31). Therefore, the material is not cross-linked. For these products, we calculated an atomic ratio factor F_{LMO}:

$$F_{LMO} = \frac{(O) + (S)}{(H)}$$

where the O/H atomic ratio applies to the point when the LMO is just formed, and the sulfur content is obtained after a 1700°C heat treatment to obtain a measure of the cross-linking sulfur. The linear relationship shown in Figure 11 fits the data well for a variety of petroleum products.

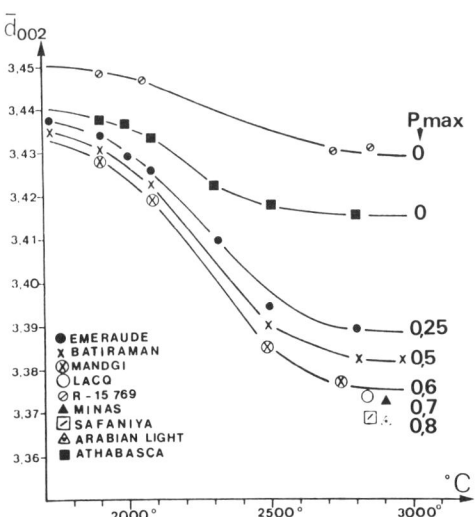

Figure 10. Examples of partially graphitizable carbonaceous materials (asphaltenes), ranging from $P_{max} \sim 0$ (non-graphitizing carbon) with $\bar{d}_{002}$min. ~ 3.44 Å, to $P_{max} \sim 0.8$ (highly graphitizing carbon) with $\bar{d}_{002}$min. ~ 3.36 Å. $\bar{d}_{002}$ is plotted as a function of heat treatment temperature (25,37).

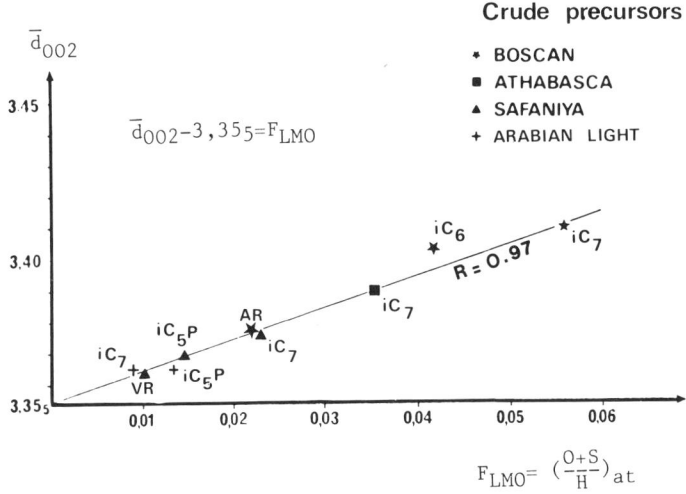

Figure 11. Relation between the graphitizability ($\bar{d}_{002}$ min.) and the atomic ratio F of cross-linking atoms (O and S) to hydrogen, measured just before the LMO appearance (27,38). Various fractions of a crude oil are reported: atmospheric residue AR, vacuum residue VR, and insolubles in n-alkanes (iC_x), e.g., iC_5P is an insoluble in pentane from a pilot plant.

Conclusion

Despite their heterogeneity, carbonaceous materials in general and heavy petroleum products in particular can be characterized for their graphitizability using the extent of local molecular orientation LMO (size of the elemental domains of bulk mesophase) as a criterion. For petroleum materials, the LMO is closely related to the elemental analysis. The factor F_{LMO} takes into account the influence of the various heteroatoms (hydrogen, oxygen, sulfur,...) and their evolution during carbonization. Thus the correlation between $\bar{d}_{002}$ (2900°C) and F_{LMO} is good enough to classify the graphitizability of most petroleum products.

Concerning the microtextures of carbons derived from petroleum products, we find that a range of intermediate textures exist between non-graphitizing and graphitizing carbons. The mechanism of graphitization is important to understand because the worth of carbonaceous products, among which are refinery residues, is a function of the ability to graphitize.

Finally, the determination of both microtexture and graphitization degree is useful to explain, and even to predict, the physical properties of carbons (32).

By combining all of the possible TEM modes, we are able to explain the behavior of carbonaceous materials primarily in terms of the local molecular orientations established in the final stages of liquid-phase pyrolysis. The models established from these observations are supported by the results of other techniques, such as infrared analyses (33), optical microscopy (27), X-ray diffraction (24), and Raman spectroscopy (22).

Literature Cited

1. Oberlin, A. Carbon 1984, 22, 521.
2. Oberlin, A.; Boulmier, J. L.; Villey, M. in "Kerogen"; Durand, B., Ed.; Technipress: Paris, 1979; p. 191.
3. Boulmier, J. L.; Oberlin, A. Proc. 7th Intern. Mtg. Org. Geochem., 1975, p. 781.
4. Villey, M.; Oberlin, A.; Combaz, A. Carbon 1979, 17, 77.
5. Oberlin, A.; Villey, M.; Combaz, A. Carbon 1983, 21, 565.
6. Joseph, D.; Oberlin, A. Carbon 1983, 21, 559.
7. Joseph, D.; Oberlin, A. Carbon 1983, 21, 565.
8. Oberlin, A. Carbon 1979, 17, 7.
9. Hirsch, P. B.; Howie, A.; Nicholson, R. B.; Pashley, D. W.; Whelan, M. J. "Electron Microscopy of Thin Crystals"; Butterworths: London, 1965; 549 pp.
10. Van Krevelen, D. W. "Coal"; Elsevier: Amsterdam, 1961, 514 pp.
11. Ihnatowicz, M.; Chiche, P.; Deduit, J.; Pregermain, S.; Tournant, R. Carbon 1966, 4, 41.
12. Brooks, J. D.; Taylor, G. Chem. and Phys. of Carbon 1968, 4, 243.
13. White, J. L. Prog. in Solid State Chem. 1974, 9, 59.
14. Oberlin, A.; Boulmier, J. L.; Durand, B. Geochim Cosmoch. Acta 1974, 38, 647.
15. Gasparoux, H. J. Chim. Phys. 1984, 81, 759.
16. Billard, J. Rev. Palais de la découverte 1978, 7, 19.

17. Auguie, D.; Oberlin, M; Oberlin, A.; Hyvernat, P. Carbon 1980, 18, 337.
18. Rouzaud, J. N.; Oberlin, A. Submitted to Carbon.
19. Alpern, B. Rev. Ind. Min. 1954, 593, 359.
20. Ragan, S.; Marsh, H. J. Phys. D: Appl. Phys. 1980, 13, 983.
21. Sugimura, H.; Kumagai, M.; Kimura, H. J. Fuel Society Japan 1970, 49, 744.
22. Rouzaud, J. N.; Oberlin, A.; Beny-Bassez, C. Thin Solid Films 1983, 105, 75.
23. Blayden, H. E.; Gibson, J.; Riley, H. L. Proc. Conf. Ultrafine Struct. Coals and Cokes, BCURA (London) 1944, p. 176.
24. Franklin, R. E. Proc. Royal Soc., 1952, 209, 196.
25. Monthioux, M.; Oberlin, M.; Oberlin, A.; Bourrat, X.; Boulet, R. Carbon 1982, 20, 167.
26. Bensaid, F. Doc. Sci. Thesis, Univ. Orléans, 1983.
27. Bourrat, X.; Oberlin, A.; Escalier, J. C. C. R. Acad. Sci. Paris 1984, 298(II), 695.
28. Bonnamy, S. To be submitted to Org. Geochem.
29. Mering, J.; Maire, J. in "Les Carbones"; Pacault, A., Ed.; Masson: Paris, 1965; vol. I. pp. 129-192.
30. Chermin, H. A. G.; Van Krevelen, D. W. Fuel 1957, 36, 85.
31. Christou, N.; Fitzer, E.; Kalka, J.; Schafer, W. J. Chim. Phys. 1969, edition spéciale, 50.
32. Guigon, M.; Oberlin, A.; Desarmot, G. Fibre Sci. and Technol. 1984, 20. 177.
33. Rouxhet, P. G.; Villey, M.; Oberlin, A. Geochim. Cosmochim. Acta 1979, 43, 1705.
34. Boulmier, J. L. Doc. Sci. Thesis, Univ. Orléans, 1976.
35. Goma, J. Doc. Sci. Thesis, Univ. Orléans, 1983.
36. Bonijoly,M.; Oberlin, M.; Oberlin, A. Int. J. Coal Geol. 1982, 1, 283.
37. Monthioux, M. Doc. Sci. Thesis, Univ. Orléans, 1980.
38. Bourrat, X. Submitted to Fuel.

RECEIVED January 21, 1986

# Residual Oil Processing
Predicting Slurry Oil and Coke Yields

W. P. Hettinger, Jr., D. P. Wesley, and R. H. Wombles

Ashland Petroleum Research and Development Department, Ashland Oil, Inc., Ashland, KY 41114

Decant or slurry oil is a highly aromatic material, which is a bottoms product obtained from catalytic cracking. In the past, catalytic crackers have been fed gas oils which are distillate materials boiling in the neighborhood of 650-1000°F. With the use of resid processing units such as Ashland's RCC process units, catalytic crackers are now fed a mixture of distillate feed and the vacuum bottoms portion of the crude. With the inclusion of the high molecular weight, high metals, and high heteroatom content vacuum bottoms in the catalytic cracker feed, the composition of the decant oil obtained from the resid processing units will be different from the decant oil obtained from an ordinary FCC unit. In this paper we will discuss the differences in composition of FCC and RCC decant oils as they relate to their use as specialty carbon precursors.

With world crude oil supplies shifting in quality from lighter and sweeter crudes to heavier and higher sulfur crudes, and with that trend expected to continue, most refiners are looking to residual processing to increase the amount of liquid transportation fuel which is obtained from a barrel of crude.(1)

During initial planning stages for a residual oil process, Ashland considered a number of residual processing schemes. Among the processes considered were visbreaking, coking, solvent deasphalting, and direct residuum desulfurization followed by either hydrocracking or catalytic cracking.(2) After a detailed examination of these processes, Ashland determined that none met its requirements which included capital and operating cost limitations.

With this decision made, Ashland set out to develop a new residual oil conversion process which could effectively produce a greater amount of transportation fuel from each barrel of crude processed. It was concluded that a new process would take the best features from the fluid catalytic cracking process and couple them with innovative improvements in related key areas such as unique process conceptions, hardware innovations, and advanced catalyst systems. The final goal of the process was the conversion of low valued, high boiling, low hydrogen/carbon ratio, high coke forming, atmospheric reduced crude oil feedstocks into lower boiling, high valued gasoline and middle distillate fuels and petrochemical feedstocks.

The residual processing scheme which Ashland decided upon was the Reduced Crude Conversion Process. The Reduced Crude Conversion Process (RCC) achieves selective yield patterns of transportation fuels and other light products through the direct conversion of atmospheric petroleum residuum in a fluidized catalytic system. During Ashland's development of the RCC process, extensive investigation of process variables and methods of handling heavy metals and carbon were conducted with a 200 B/D demonstration unit. The following variables were investigated on the 200 B/D unit:

- feedstock quality
- reaction zone residence time
- catalyst activity
- catalyst deactivation by heavy metals accumulated on the circulating catalyst
- reaction zone temperature
- pressure
- catalyst/oil ratio
- catalyst type
- catalyst/oil mixing

Among other activities, one assignment was to determine methods of evaluating feedstock quality. In the following pages a discussion of the basic differences between RCC and FCC operations, techniques of evaluating reduced crude feedstocks, some aspects of catalytic cracking chemistry, and finally some techniques which have been developed to predict the performance of a reduced crude in the RCC unit by knowing the molecular composition of the reduced crude will be given.

Discussion

The RCC process developed by Ashland is a catalytic conversion process. Catalytic conversion or catalytic cracking is a refinery process designed to reduce the molecular weight and boiling point of higher boiling petroleum fractions. In normal catalytic cracking operations gas oil is used as the feed to the cracking unit. Generally gas oil for catalytic cracking has a boiling point of about 630-1025°F. Gas oil is produced by vacuum distillation of atmospherically topped crude. Figure 1 gives a schematic diagram of

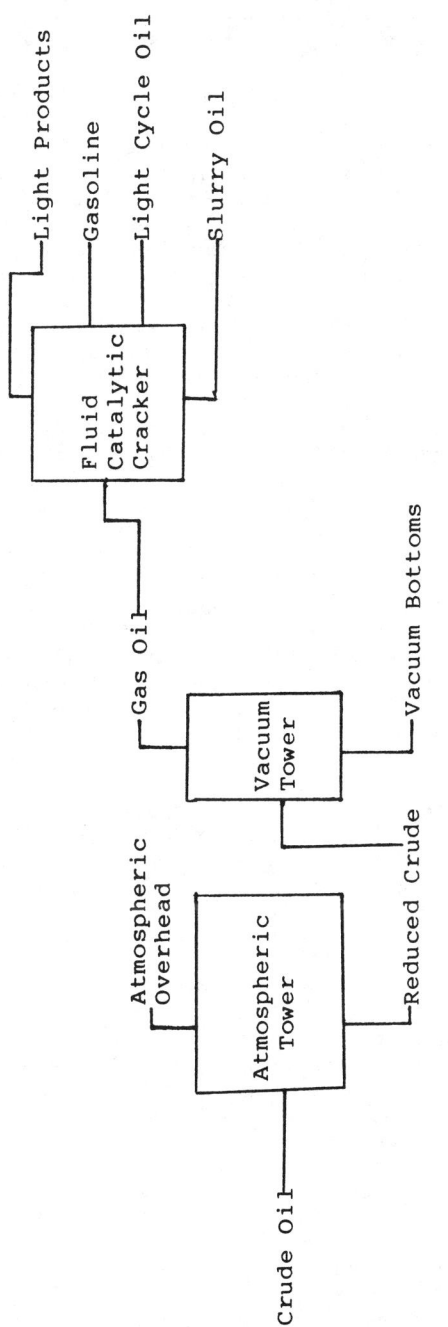

Figure 1. Schematic diagram of FCC operation.

a normal FCC operation. With Ashland's RCC process the feed to the cracking unit is reduced crude. Reduced crude is produced by the atmospheric distillation of crude oil. Reduced crude has a boiling point of about 630°F+. Figure 2 is a schematic diagram of the RCC operation.

As can be seen from both Figures 1 and 2, the major products from a cracking operation are light products, gasoline, light cycle oil, slurry or decant oil, and coke. The desire in a cracking operation is to maximize the production of high valued gasoline and light cycle oil while minimizing the production of light products, slurry oil and coke. In other words there is a desire to maximize the production of transportation fuel.

One of the most important factors which affects the product distribution of a cracking unit is feedstock quality. One of the process variables studied during the development stages of the RCC process was feedstock quality. In particular, the quality of the RCC feedstock plays a very important role in the amount of slurry oil and coke which the unit yields. K-factor, API gravity, boiling range, Conradson carbon, hydrogen, sulfur, nitrogen, metals and asphaltene contents are physical and chemical properties which are usually used to characterize cracker feedstocks.

For the characterization of RCC feedstocks, it was determined that a more detailed molecular description of the feedstock was necessary. The more detailed molecular description of RCC feedstocks involves dividing the feedstock into six molecular types 1) saturates 2) monoaromatics 3) diaromatics 4) greater than diaromatics 5) polar aromatics and 6) asphaltenes. This separation of the RCC feedstock is accomplished by using high performance liquid chromatography.

In the program of feedstock quality analysis, the goal was to predict the product distribution that can be expected from the reduced crude conversion of each feedstock. In this paper the results of slurry oil and coke yield predictions will be discussed. The next section gives a brief overview of the chemistry of catalytic cracking and some of the basic concepts which were used to guide the development of the slurry oil and coke yield predictions.

Overview of Cracking Chemistry

Catalytic cracking is a refinery process designed to reduce the molecular weight and boiling point of higher boiling petroleum fractions. In this section the possible fate of the six molecular types of which a reduced crude is considered to be composed will be discussed. These molecular types are 1) saturates 2) monoaromatics 3) diaromatics 4),diaromatics 5) polar aromatics and 6) asphaltenes. Models of each molecular type are suggested in Figure 3 for the purpose of discussion. (These models are only for the purpose of discussion and may not represent actual molecules found in reduced crude.)

Figures 1 and 2 show that there are five major products obtained from a catalytic cracking operation. These products are 1) light products 2) gasoline 3) light cycle oil 4) slurry or decant oil and 5) coke. The light products are those which have a boiling point below 170°F, gasoline has a boiling point of 170°F-430°F, light cycle oil has by definition here a boiling point of 430°F-630°F, slurry or

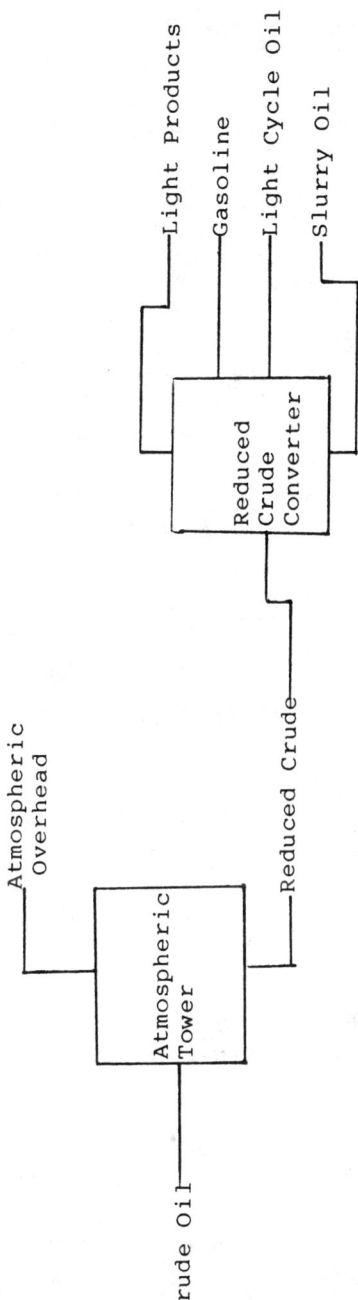

Figure 2. Schematic diagram of RCC operation.

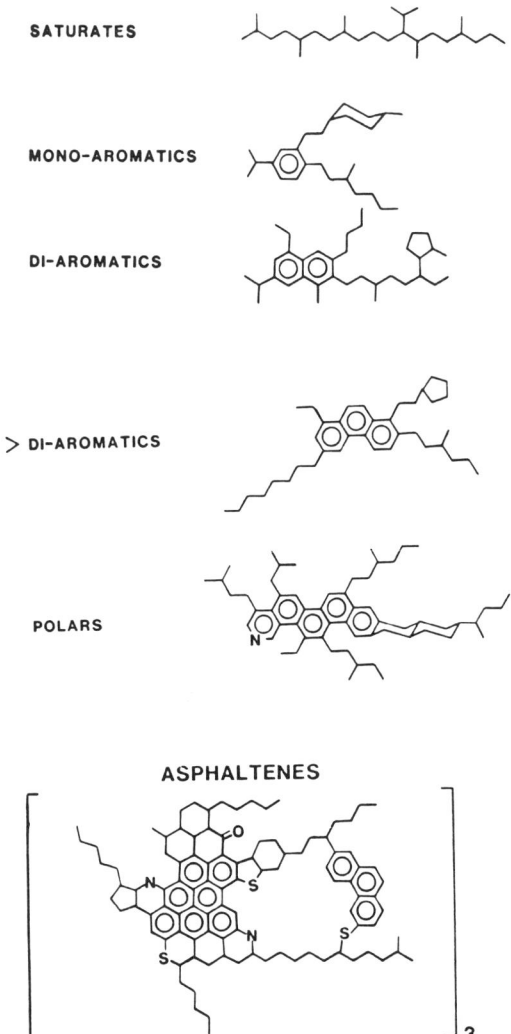

Figure 3. Models of molecular types.

decant oil has a boiling point greater than 630°F, and coke is the product which does not volatize and remains on the catalyst. It is the goal of any catalytic cracking operation to maximize the production of transportation fuels, that is gasoline and light cycle oil, while minimizing the production of the lower valued heavy products such as slurry oil and coke.

With the stated desire to minimize the production of slurry oil and coke, the generalized chemical structure of each of the six molecular types will be examined to investigate the probability of obtaining slurry oil and coke from each molecular type. A good "rule of thumb" is that carbon-carbon alkyl bonds are subject to catalytic cracking while carbon-carbon aromatic bonds are resistant to catalytic cracking. During catalytic cracking one can only expect the cracking of alkyl substituents whether they are present as saturates or as substituents on aromatic molecules.

Refer to Figure 3 for the following discussion of the possible fate of each of the six molecular types present in reduced crude during catalytic cracking.

The saturates are composed of entirely carbon-carbon alkyl bonds. Since all the bonds in a saturate molecule are susceptible to catalytic cracking, it is theoretically possible that saturates can be completely cracked to products other than slurry oil and coke.(3)

The monoaromatics consist of benzene rings substituted with a number of alkyl groups. The major reaction taking place in the catalytic cracker will be removal of alkyl groups. Since benzene has a boiling point of 176°F, it is also anticipated that the monoaromatics will be cracked to products other than slurry oil and coke.(4)

The diaromatics are composed of a naphthalene ring system substituted by alkyl groups. Again the major reaction in the catalytic cracker will be the removal of alkyl groups. Since naphthalene has a boiling point of 424°F, it is quite likely that these diaromatic containing molecules can be cracked to products other than slurry oil and coke.

The $>$diaromatics are composed of aromatic ring systems of three or more fused rings substituted by alkyl groups. The major catalytic cracking reaction occurring is again the cracking of the alkyl groups. Since phenanthrene has a boiling point of 644°F, the aromatic ring systems in the $>$diaromatics along with alkyl substituents remaining after catalytic cracking go to slurry oil and coke. Only the products obtained from the cracking of the alkyl substituents can go to the desired transportation fuels.

The polar aromatics obtained by this chromatographic separation technique have been shown to be composed of large heteromolecule containing aromatic ring systems substituted by alkyl groups. A considerable amount of nitrogen and sulfur is found in aromatic rings of the polar aromatics. The polar aromatics are similar to the $>$diaromatics in that the aromatic ring system in the polar aromatics along with alkyl substituents remaining after catalytic cracking go to slurry oil and coke. Only the products obtained from the cracking of the alkyl substituents will go to the desired products.

The chemical structure of asphaltenes is still not well understood. Ashland Research is currently studying the catalytic cracking of asphaltenes. At the present time these studies have shown that the asphaltenes yield mostly slurry oil and coke when catalytically cracked.

From the discussion above it is possible to draw three conclusions which will be used as a foundation to develop a slurry oil plus coke yield prediction equation. These conclusions are 1) the saturates, monoaromatics, and diaromatics are conceivably completely convertible to products lighter than slurry oil and coke during catalytic cracking 2) only the aliphatic portion of the $\rangle$diaromatics and polar aromatics are convertible to products lighter than slurry oil and coke, while the aromatic ring systems along with alkyl substituents remaining after catalytic cracking go to slurry oil and coke, and 3) for prediction purposes it is assumed that asphaltenes go completely to slurry oil and coke.

Of course, the reduced crude conversion process is not 100% efficient. By this it is meant that to date no catalyst and operating conditions have been developed which completely remove saturates, monoaromatics, diaromatics, and alkyl substituents of polynuclear aromatics from the slurry oil. Therefore, to predict slurry oil plus coke yield one must determine what proportion of each molecular type present in the reduced crude feedstock remains in the slurry oil and coke.

Reduced Crude and Slurry Oil Analysis

In order to develop the slurry oil plus coke yield prediction, samples of reduced crude feedstock and 630°F+ slurry oil from tests run on Ashland's 200 B/D RCC demonstration unit were obtained. The first step of the analysis of the two streams involved a chromatographic separation. The separation was accomplished with a Waters Prep 500A HPLC. A flow scheme for the separation is given in Figure 4. For the reduced crudes, the first step of the analysis was the precipitation of asphaltenes using n-pentane as the precipitating solvent. In the case of slurry oils, asphaltenes were not precipitated prior to chromatographic analysis. Asphaltenes were not precipitated from the slurry oils because asphaltenes constituted such a small percentage of the slurry oils. The analysis was done with a diamine bonded silica and a silica column in series. The saturates, monoaromatics, and diaromatics were eluted from both columns using hexane. After the diaromatics were eluted, the silica column was bypassed and a MTBE/hexane solvent was used to elute the $\rangle$diaromatics from the diamine bonded silica column. After the elution of the $\rangle$diaromatics, the diamine bonded silica column was backflushed with methylene chloride to obtain the polar aromatics. Typical results of the chromatographic analysis of a reduced crude and slurry oils produced from that reduced crude are given in Table I.

After the chromatographic separation, each chromatographic fraction was analyzed for molecular weight by vapor pressure osmometry using chloroform as the solvent at 45°C. The instrument used was a Knauer vapor pressure osmometer. Carbon, hydrogen, and nitrogen analyses of each chromatographic fraction were obtained via a Carle-Erbe elemental analyzer. Each chromatographic fraction was also analyzed by nuclear magnetic resonance spectroscopy using a Varian XL-200 spectrometer. The NMR spectra were integrated to

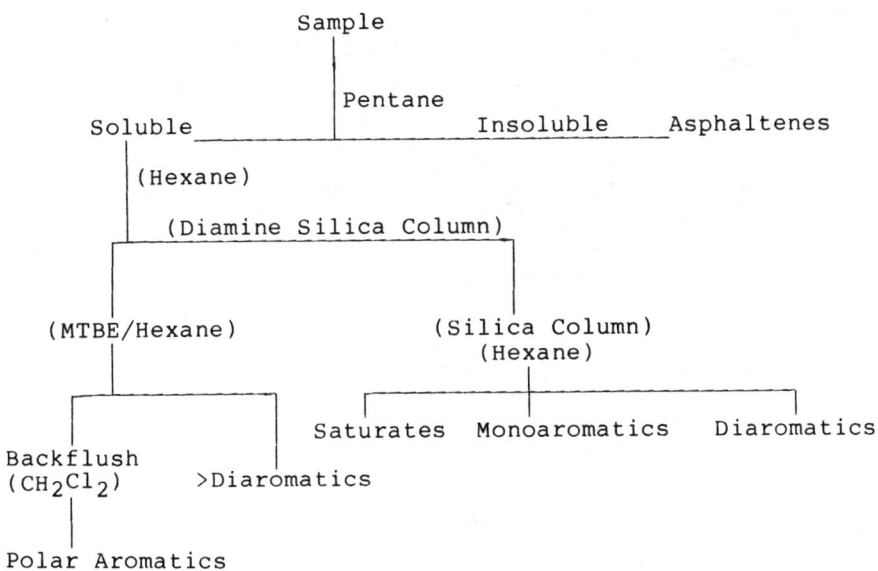

Figure 4. Flow scheme for HPLC analysis.

Table I. Typical HPLC Analysis of Reduce Crude Feedstocks and Slurry Oils Produced from that Feedstock

	Feed	Slurry Oil		
Test Number	All	294	308	315
Saturates, wt. %	40.3	7.7	17.5	12.0
Monoaromatics, wt. %	12.6	3.3	5.0	1.6
Diaromatics, wt. %	9.9	33.4	33.4	36.3
>Diaromatics, wt. %	24.6	49.6	41.8	47.9
Polar Aromatics, wt. %	7.3	6.0	2.4	2.2
Asphaltenes, wt. %	5.2	-	-	-
API Gravity	18.7	-4.8	1.3	-2.2
Slurry Yield (wt. %)	-	8.5	9.6	9.2

determine the relative percentages of aromatic α, β, λ hydrogens present in each chromatographic fraction.

The carbon and hydrogen analyses, VPO molecular weights, and NMR results were used to calculate average molecular parameters for the ₓdiaromatics and polar aromatics obtained from both the reduced crude and slurry oils. The average molecular parameters were calculated by the method of Williams.(5)(6) Typical results of the average molecular parameter calculations for ₓdiaromatic and polar aromatic chromatographic fractions obtained from both the reduced crude and slurry oils are given in Tables II and III.

Development of Yield Predictions

The stated goal was to develop RCC product yield predictions from feedstock composition and properties. One of the first relationships investigated was a coke yield relationship. Table IV shows the relationship first observed between coke production and feedstock composition. As can be seen, for this particular feedstock the average coke make over a total of 24 tests was 13.5 wt. % while the amount of polar aromatics plus asphaltenes in the feedstock was 12.5 wt. %. This is a fairly close approximation of coke yield.

The polar aromatics plus asphaltenes relationship versus coke yield was investigated for a number of reduced crudes. Table V gives the results of findings for eight RCC feedstock reduced crudes. As can be seen in a number of instances the predictions are consistent.

The coke yield from a given feedstock is affected by many variables such as reactor temperature, metals on catalyst, and catalyst type. It was also found that coke yield predictions did not agree when the RCC feedstocks were high API gravity reduced crudes. By high API gravity reduced crudes it is meant reduced crudes with an API gravity of 20 or greater. Some of the variables in coke production were taking effect as can be seen in Table V, where predicted coke yield falls far short of the actual coke yield. Because of these variations in coke yield, it was decided that a slurry oil plus coke yield prediction would be a better, more consistent prediction equation. The following pages will give the details of the development of the slurry oil plus coke prediction equation.

In order to predict slurry oil plus coke yield one must determine what proportion of each molecular type present in the reduced crude feedstock remains in the slurry oil and coke.

The six molecular types present in reduced crude can be listed in two categories 1) those types whose removal from slurry oil and coke depend on the efficiency of unit operation namely; saturates, monoaromatics, and diaromatics and 2) those types which must produce some slurry oil and coke namely; ₓdiaromatics, polar aromatics, and asphaltenes.

A measure of unit operating efficiency is the API gravity of the slurry oil produced. As a "rule of thumb" the higher the API gravity of the slurry oil the less efficient the operation. With this fact in mind an attempt was made to link the saturates, monoaromatics, and diaromatics slurry oil plus coke producing factor to slurry oil API gravity. In addition to the above, two assumptions were made in order to develop the slurry oil plus coke yield prediction.

Table II. Average Molecular Parameters of
Diaromatics from Reduced Crude and Slurry Oil

	Feed		Slurry Oil	
Test No.	All	294	308	315
Aromaticity	0.41	0.83	0.79	0.81
No. Aromatic Carbons	16.9	18.9	22.7	20.6
No. Aromatic Rings	3.9	4.5	5.7	4.8
No. Alkyl Substituents	5.5	3.0	3.7	3.4
No. Carbons Per Alkyl Substituent	4.5	1.3	1.6	1.4
No. Alkyl Carbons	24.8	3.9	5.9	4.8

Table III. Average Molecular Parameters
of Polar Aromatics from Reduced Crude and Slurry Oil

	Feed		Slurry Oil	
Test No.	All	294	308	315
Aromaticity	0.35	0.77	0.74	0.76
No. Aromatic Carbons	22.8	16.7	13.6	19.7
No. Aromatic Rings	4.4	4.3	3.9	5.0
No. Alkyl Substituents	8.3	3.0	2.3	3.3
No. Carbons Per Alkyl Substituent	5.2	1.7	2.1	1.9
No. Alkyl Carbons	43.2	5.1	4.8	6.3

Table IV. Initial Coke Yield Relationship for Reduced Crude

Feedstock	Number of Tests	Coke Yield, wt.%	Polar Aromatics + Asphaltenes, wt.%
#1	24	13.5	12.5

These assumptions were 1) a very small amount of the coke produced is derived from the saturates, monoaromatics, and diaromatics and 2) the building blocks of the coke produced are the asphaltene, ⟩diaromatic and polar aromatic molecules in the slurry oil. The following will discuss the methods used to determine what proportion of each molecular type present in the reduced crude feedstock remains in the slurry oil and coke.

Table I shows that there is a direct relationship between the saturates content of a slurry oil and the API gravity of a slurry oil. As the saturates content of a slurry oil increases, the API gravity of the slurry oil increases. The relationship found is shown graphically in Figure 5. This relationship was developed by the analysis of a number of slurry oils. During this same program it was found that there is a linear relationship between the saturates content of a slurry oil and the weight percent yield of a slurry oil. This relationship is shown in Figure 5. Using these two relationships a saturates slurry oil plus coke factor for a slurry oil of a given API gravity was developed. This saturates factor changes as the API gravity of the reduced crude feedstock changes. The saturates slurry oil plus coke factors for 10-20 and 20-30 API gravity reduced crudes are given in Table VI.

Analysis of many feedstocks and slurry oils indicated that the monoaromatics followed a pattern similar to the saturates. Therefore, a monoaromatics slurry oil plus coke factor was developed. The monoaromatics slurry oil plus coke factors for 10-20 and 20-30 API gravity reduced crudes are given in Table VII.

The diaromatics were found to follow a different pattern. No relationship was found between the diaromatics content of a slurry oil and its API gravity. It was determined, however, that approximately one third of the diaromatics in the reduced crude feedstock remained in the slurry oil.

Table VIII shows the method used to calculate the ⟩diaromatics factor. The data for these calculations are derived from Table II, which shows that the ⟩diaromatic fractions from the three slurry oils have an aromaticity of approximately 0.8. The calculation shows that approximately 50.2% of the ⟩diaromatic molecules remain in the slurry oil. Similar calculations with a number of reduced crudes and slurry oils derived from those reduced crudes has indicated that approximately 51.5% of the ⟩diaromatic molecules remain in the slurry oil.

Table IX shows a similar method of calculating the polar aromatics factor. The data for these calculations are derived from Table III. Table III indicates that the polar aromatic fractions from the three slurry oils have an aromaticity of approximately 0.75. The calculation shows that approximately 45.7% of the polar aromatic molecules remain in the slurry oil. Similar calculations with a number of reduced crudes and slurry oils derived from those reduced crudes has indicated that approximately 46% of the polar aromatic molecules remain in the slurry oil.

At the present time the assumption is being made that asphaltenes go quantitatively to slurry oil and coke.

From the above discussion one can propose an equation for the prediction of slurry oil and coke yields in the RCC unit from a particular reduced crude feedstock. The equation is the following:

Table V. Results of Coke Yield Predictions for Reduced Crudes

	Asphaltenes + Polar Aromatics	Number of Runs	Average Coke
Feedstock No. 1	12.9	16	13.1
Feedstock No. 2	12.3	39	14.4
Feedstock No. 3	16.1	21	15.3
Feedstock No. 4	14.5	11.	14.7
Feedstock No. 5	11.6	14	15.8
Feedstock No. 6	12.5	24	13.5
Feedstock No. 7	14.3	25	14.3
Feedstock No. 8	13.9	16	14.8

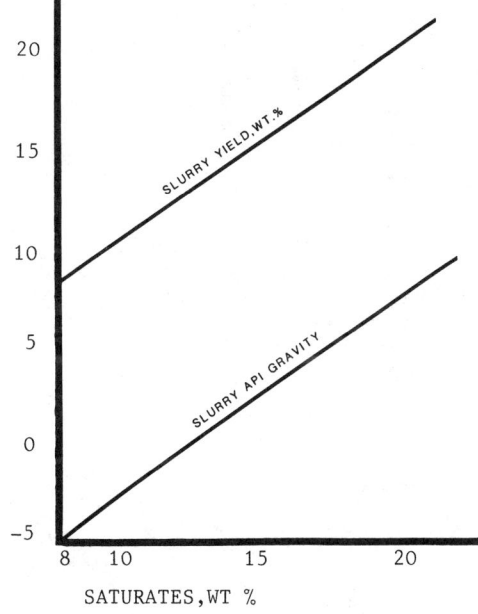

Figure 5. Slurry oil saturates content versus slurry oil API gravity and yield.

Table VI. Saturates Slurry Oil Plus Coke Factor

API Gravity of Slurry Oil	Saturates Slurry Oil Plus Coke Factor 10-20 API Gravity Reduced Crudes	Saturates Slurry Oil Plus Coke Factor 20-30 API Gravity Reduced Crudes
-5	0.7	-
-4	0.8	-
-3	1.0	-
-2	1.3	-
-1	1.5	-
0	1.6	-
1	1.9	-
2	2.2	-
3	2.5	1.3
4	2.8	1.5
5	3.2	2.1
6	3.5	2.4
7	3.9	2.8
8	4.3	3.2
9	4.7	3.6
10	5.2	4.0
11	-	4.8
12	-	5.3
13	-	5.8

Table VII. Monoaromatics Slurry Oil Plus Coke Factor

API Gravity of Slurry Oil	Monoaromatics Slurry Oil Plus Coke Factor 10-20 API Gravity Reduced Crudes	Monoaromatics Slurry Oil Plus Coke Factor 20-30 API Gravity Reduced Crudes
-5	0.4	-
-4	0.4	-
-3	0.5	-
-2	0.6	-
-1	0.6	-
0	0.6	-
1	0.7	-
2	0.8	-
3	0.8	0.3
4	0.8	0.4
5	0.9	0.4
6	1.0	0.5
7	1.0	0.6
8	1.0	0.6
9	1.1	0.6
10	1.2	0.7
11	-	0.8
12	-	0.8
13	-	0.9

Table VIII. Calculation of Diaromatics Slurry Oil Plus Coke Factor

	Reduced Crude	Slurry Oil
Aromaticity	0.41	0.80
Number of Aromatic Carbons	16.9	16.9
Number of Alkyl Carbons	24.8	4.2
Molecular Weight	579	

20.6 Alkyl Carbons Removed = 247.2 MW

41.2 Alkyl Hydrogens Removed = 41.2 MW

Total Molecular Weight Removed = 288.4

49.7% Removed

50.2% Remaining in Slurry Oil

slurry oil + coke, wt.% = saturates factor + monoaromatics factor + 0.33 (diaromatics) + 0.515 (,diaromatics) + 0.46 (polar aromatics) + asphaltenes

Table X shows the results of calculations for two RCC tests made on the same feedstock. The predictions are consistent with actual yields. Tables XI and XII are also results of slurry oil plus coke yield predictions from various reduced crudes. All prediction made in Tables X-XII were made with reduced crudes having an API gravity of 10-20. The saturates and monoaromatics factors change as the API gravity of the reduced crude feedstock changes. Tables XIII and XIV give the results of slurry oil plus coke predictions for West Texas Intermediate and Illinois Basin reduced crudes. These two reduced crudes have API gravities of 20-30. Again consistent predictions were made.

Conclusions

A method has been developed to predict the amount of slurry oil plus coke that will be obtained from a given reduced crude in Ashland's RCC process under various operational efficiencies by knowing the molecular type composition of the reduced crude. This method is currently being used in Ashland's crude oil evaluation program. At the present time Ashland's commercial RCC unit is operational and adjustments are being made to the prediction equation in order to predict yields from the commercial unit. Work on developing equations to predict the yield of other RCC products continues.

It is our feeling that the next step in evaluating reduced crudes as feedstocks for the RCC unit will be in the area of further clarification of the structure of the alkyl carbons; that is, what percent of the alkyl carbons are paraffinic and what percent are naphthenic. In the RCC unit, which contains a catalyst with a high metals content there is a race between cracking of the naphthenic rings to lighter products and dehydrogenation of these same naphthenic rings to aromatic rings resistant to cracking as illustrated below:

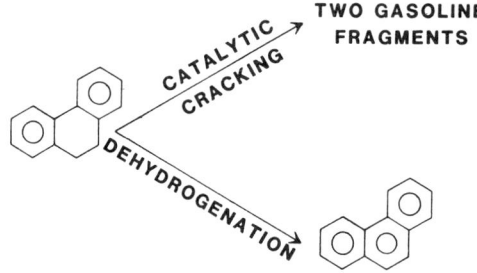

The method which has been illustrated in this paper will point out significant differences in the quality of RCC feedstocks, but only a more detailed understanding of the nature of the alkyl carbons will detect very subtle difference between the quality of two feedstocks.

Table IX. Calculation of Polar Aromatics Slurry Oil Plus Coke Factor

	Reduced Crude	Slurry Oil
Aromaticity	0.35	0.75
Number of Aromatic Carbons	22.8	22.8
Number of Alkyl Carbons	43.2	7.6
Molecular Weight	918	

35.6 Alkyl Carbons Removed = 427.2 MW

71.2 Alkyl Hydrogens Removed = 71.2 MW

Total Molecular Weight Removed = 498.4

54.3% Removed

45.7% Remaining in Slurry Oil

Table X. Slurry Oil Plus Coke Predictions for RCC Feedstock

	Run 368 Slurry 2.1	Run 373 Slurry −4.3
Saturates	2.2	0.8
Monoromatics	0.8	0.4
Diaromatics	3.0	3.0
>Diaromatics	10.9	10.9
Polar Aromatics	3.7	3.7
Asphaltenes	5.8	5.8
Total	26.4	24.6
Actual Yield	26.7	24.7

Table XI. Slurry Oil Plus Coke Predictions for RCC Feedstock No. 2

Run Number	Slurry API Gravity	Predicted Slurry + Coke	Actual Yield
351	−4.4	25.2	25.8
363	0.4	26.4	26.7
364	4.0	27.6	30.7

Table XII. Slurry Oil Plus Coke Predictions for RCC Feedstock No. 3

Run Number	Slurry API Gravity	Predicted Slurry + Coke	Actual Yield
294	-4.8	25.6	23.1
324	0.4	26.8	25.1
312	3.9	28.1	28.8

Table XIII. Slurry Oil Plus Coke Prediction for West Texas Intermediate Reduced Crude

Run Number	Slurry API Gravity	Predicted Slurry + Coke	Actual Yield
304	7.7	23.0	22.7
306	6.3	22.3	21.9
307	3.1	20.9	20.0

Table XIV. Slurry Oil Plus Coke Prediction for Illinois Basin Reduced Crude

Run Number	Slurry API Gravity	Predicted Slurry + Coke	Actual Yield
299	9.8	23.6	24.2
300	8.6	23.0	22.9
322	4.8	21.4	20.7
336	3.1	20.6	20.3

Acknowledgments

The authors gratefully acknowledge the assistance of Dr. Stan Smith and Mr. John Layton of the University of Kentucky Nuclear Magnetic Resonance Spectroscopy Center for their assistance in obtaining NMR spectra.

Literature Cited

1. National Petroleum Council, "Survey on Refinery Flexibility, an Interim Report," U. S. Department of Energy, January, 1980.
2. Busch, L. E., Hettinger, W. P., Zandona, O. J., "Reduced Crude Conversion: An Inexpensive Route to High Octane Gasoline," National Petroleum Refiners Association 80th Annual Meeting, San Antonio, Texas, March 21-23, 1982.
3. Greensfelder, B. S. Voge, H. H., Goad, G. M., "Catalytic and Thermal Cracking of Pure Hydrocarbons," Industrial and Engineering Chemistry, 41 (11), 2573-2584 (November 1949).
4. Greensfelder, B. S., Voge, H. H., Goad, G. M., "Catalytic Cracking of Pure Hydrocarbons," Industrial and Engineering Chemistry, 37 (12), 1168-1176 (December, 1945).
5. Williams, R. B., "Symposium on Composites of Petroleum Oils, Determination and Evaluation," ASTM Spec. Tech. Publ., 224, 169-94 (1958).
6. Clutter, D. R., et. al., "Nuclear Magnetic Resonance Spectrometry of Petroleum Fractions," Analytical Chem., 44, 1395-1405 (1972).

RECEIVED January 22, 1986

7

Synthetic Aromatic Pitch
Aromatic Pitches from the Asphaltene-Free Distillate Fraction of Catalytic Cracker Bottoms

G. Dickakian

Specialties Technology Division, Exxon Chemical Company, Houston, TX 77029

>   Catalytic cracking bottoms (CCB) is a widely used
>   aromatic feedstock for carbon production, such as
>   aromatic pitch, carbon black and carbon fibers.
>   We fractionated CCB by a high vacuum distillation
>   into several distillate fractions and undistillable
>   residue. The distillate fractions were subjected to a
>   two-stage high temperature thermal process to convert
>   them to aromatic pitches. The composition of the
>   pitches produced was determined by a solvent analysis
>   to define their suitability for synthetic carbon
>   production.

This chapter describes the preparations and characteristics of highly aromatic and highly anisotropic pitches from the distillate fraction of catalytic cracker bottoms (CCB). CCB is the aromatic residue from a catalytic cracking process.
    CCB is fractionated by a high vacuum distillation into a distillable fraction (around 50% yield) and a non-distillable residue. These CCB fractions, distillate and residue, vary significantly in their physical characteristics, chemical structure, asphaltene content, molecular weight and aromatic ring distributions. Highly aromatic and highly anisotropic pitches were prepared by a high temperature two-stage thermal treatment of CCB distillate fraction. A number of key reaction parameters effecting pitch production, yield, characteristics, and the formation of the highly anisotropic toluene insolubles were investigated. The toluene insolubles fraction was used for the production of carbon fibers.
    A middle CCB-distillate fraction (420-520°C/760 mm Hg) was produced by a high vacuum distillation (0.5-1.0 mm Hg 15/5 column). The CCB feed was selected carefully, only CCB feed with high aromaticity (aromatic carbon=65-73 atom %), which was produced from a high severity catalytic cracking operation, was used.
    The CCB-distillate fraction contains no ash or asphaltenes (n-Heptane insolubles at reflux). It has a low and narrow molecular weight distribution (Mn=290), and contains a very narrow aromatic ring distribution (3, 4, 5, and 6 rings). Table I gives the

characteristics of the CCB middle fraction which was used in this investigation for our pitch production.

CCB distillate was transformed into an aromatic pitch by a two-stage process. Initially, the CCB-distillate was thermally-treated at high temperatures, 400–450°C, under atmospheric (nitrogen) pressure and finally the unreacted fractions were vacuum stripped at 0.5-1-0 mm/760 mm Hg, leaving the aromatic pitch in 30–40% yield in which there is up to 74% of a highly anisotropic toluene insolubles fraction.

A number of important process parameters were investigated to find out their effect on pitch characteristics and yield of the toluene and quinoline insolubles. The pitches produced were characterized by solvent analysis, NMR, thermal, and elemental analysis. Insolubles in toluene, pyridine and quinoline were used because these fractions represent the fusable and infusable anisotropic liquid crystal fraction formed in the pitch.

Aromatic pitches were produced by treating CCB-distillate at 400°C, 410°C, 420°C, 430°C, and 440°C. We found that process temperature is a very important parameter in determining the rate of toluene and pyridine insoluble formation. It was also found that a relatively high temperature (around 430°C) is required to produce a pitch with a high liquid crystal content. Table II gives the composition of pitches produced at 400–440°C. Figure 1 illustrates graphically, the effect of process temperature on the rate of toluene, pyridine, and quinoline insolubles formation.

The effect of reaction time was investigated by thermally treating CCB distillate at 420°C and 430°C for 1, 2, 3, and 4 hours. We found that at both temperatures investigated increasing reaction time resulted in increasing the rate of toluene, pyridine, and quinoline insolubles. Table III gives the composition of pitches produced at 420°C and 430°C for varying time (1 to 4 hours), and Figure 2 illustrates, graphically, the effect of reaction time on the formation of the toluene, pyridine and quinoline insolubles.

The chemical structure of CCB-distillate pitches produced at varying temperatures (410-430°C) were determined by solid-state NMR, proton NMR, and carbon/hydrogen atomic ratio. We found that increasing the thermal treatment temperature resulted in increasing the aromatic carbon content in the pitch, increasing the aromatic protons, decreasing the benzylic and total aliphatic protons, and increasing the carbon/hydrogen atomic ratio. These changes in the chemical structure of the pitches are produced as a result of the varying degree of dealkylation of the alkylaliphatic side chains, polymerization, and condensation of the aromatic rings. Table IV gives the NMR data and carbon/hydrogen atomic ratio of CCB-distillate pitches produced at 410°C-430°C.

The thermal characteristics of the aromatic pitch is one of the key characteristics of a pitch which is used for high temperature applications such as carbon or graphite anode or carbon fiber production. We determined the TGA thermograms in nitrogen (10°C/min). TGA data indicates that as the thermal treatment temperature used for producing the pitch is increased, the pitch becomes more thermally stable as indicated by the decreased volatile content and increasing coke yield. Table V gives TGA data and coke yield at 550°C for CCB-distillate pitches produced at 400°C-440°C. Figure 3 gives a typical DTG thermogram (nitrogen, 10°C/min) of a CCB-distillate pitch prepared at 430°C.

CCB-distillate pitches produced using CCB-distillate and our high temperature two-stage process are highly anisotropic in structure as determined by polarized light microscopy. The anisotropic content of the pitch is very much dependent on the thermal treatment temperature. We found that anisotropic structure begins developing at a low rate when thermally treating the distillate at 400° - 410°C. We also found that a temperature between 430° - 440°C is required to produce a distillate pitch with 90-100% of optical anisotropy. Figures 4, 5, and 6 present the micrographs of CCB-distillate pitches produced at 410°C, 420°C and 430°C, respectively.

Rheological properties of pitches are key in defining their usefulness for high temperature applications or for spinning into carbon fibers. We determined the glass transition temperature (Tg) of CCB-distillate pitches prepared at 400°C, 410°C, 420°C, 430°C, and 440°C using a differential scanning calorimeter (DSC). We found that all the pitches prepared have a low Tg (217 to 229°C). It was also found that increasing the process temperature led to increasing the Tg of the pitch. DSC data are presented in Table VI.

The aromatic pitches produced from CCB-distillate are being developed for pitch carbon fiber production.

Table I. Characteristics of Middle Cut CCB-Distillate Fraction

Boiling Range (°C/760mm)	420-520
Ash (Wt. %)	0.0005
Asphaltenes (n-Heptane Insolubles) (Wt.%)	nil
Coking Value (Wt. % at 550°C)	1.0
Number Average Mol. Weight	290
Carbon/Hydrogen Atomic Ratio	0.90

NMR-Data

Aromatic carbon (%)	72.0
Benzylic Protons (%)	34.3
Total Aliphatic Protons (%)	37.2

Aromatic Ring Distillation

1	Ring	1.0
2	Rings	15.0
3	Rings	30.0
4	Rings	45.0
5	Rings	7.0
6+	Rings	1.0
3+4	Rings	75.0
4+5	Rings	52.0
3+4+5	Rings	82.0

Table II. Effect of Process Temperature on CCB-Distillate Pitch Composition

Pitch Process Conditions

	Feed	400	410	420	430	440
Temperature (°C)		400	410	420	430	440
Time (Hours)		1.0	1.0	1.0	1.0	1.0
n-Heptane Insolubles (%)	0	78.0	95.3	75.9	97.2	99.0
Toluene Insolubles (%)	0	0.1	1.9	4.9	32.2	60.0
Pyridine Insolubles (%)	0	0	0	0.4	18.7	38.1
Quinoline Insolubles (%)	0	0	0	0.1	3.3	13.4

Table III. Effect of Reaction Time on CCB-Distillate Pitch Composition

Pitch Process Conditions

	\<------420------\>				\<------430------\>			
Temperature (°C)								
Time (Hours)	1	2	3	4	1	2	3	4
n-Heptane Insolubles (%)	95.2	98.2	98.1	98.1	97.2	99.3	98.4	98.8
Toluene Insolubles (%)	13.2	24.6	31.0	38.0	42.2	53.5	62.4	73.5
Pyridine Insolubles (%)	1.7	4.1	10.1	17.3	18.7	25.0	35.9	54.0
Quinoline Insolubles (%)	0.1	0.5	1.3	2.7	3.3	6.0	11.2	22.2

Table IV. Effect of Process Temperature on CCB-Distillate Pitch Chemical Structure

	Feed	410	430
Temperature		410	430
Time (Hours)		1.0	1.0
Carbon Distribution			
Aromatic Carbon (%)	72.0	83.0	90.7
Total Aliphatic Carbon (%)	28.0	16.2	9.3
Proton Distribution			
Aromatic Protons (%)	34.3	46.5	58.7
Benzylic Protons (%)	37.2	39.0	34.1
Total Aliphatic Protons (%)	28.5	14.5	7.2
Carbon/Hydrogen Atomic Ratio	1.03	1.23	1.62

Table V. Thermogravimetric Analysis of CCB-Distillate Pitches

Pitch Process Conditions

	Feed	400	410	420	430	440
Temperature (°C)						
Time (Hours)		1.0	1.0	1.0	1.0	1.0

Temperature (°C)	Cumulative Loss, Wt.%					
200	1.2	0.4	0.3	0	0	0
300	40.5	2.1	1.3	1.8	0.5	0.3
400	95.6	17.2	20.8	10.0	3.7	2.7
500	96.4	56.4	50.0	43.3	18.8	16.9
600	97.0	66.9	60.0	56.4	37.9	30.0

Coke Yield (Wt. %) @ 500°C	5.6	43.6	50.0	56.7	81.2	83.1

Table VI. DSC Data of CCB-Distillate Pitches Prepared at Various Temperatures

Process Conditions

Temperature (°C)	400	410	420	430	440
Time (Hours)	1.0	1.0	1.0	1.0	1.0

Initiation Temperature (°C)	187	189	191	194	198
Glass Transition Temperature (Tg) (°C)	217	218	219	223	229
Termination Temperature (°C)	242	247	251	254	258

7. DICKAKIAN *Aromatic Pitches from the Asphaltene-Free Distillate Fraction* 123

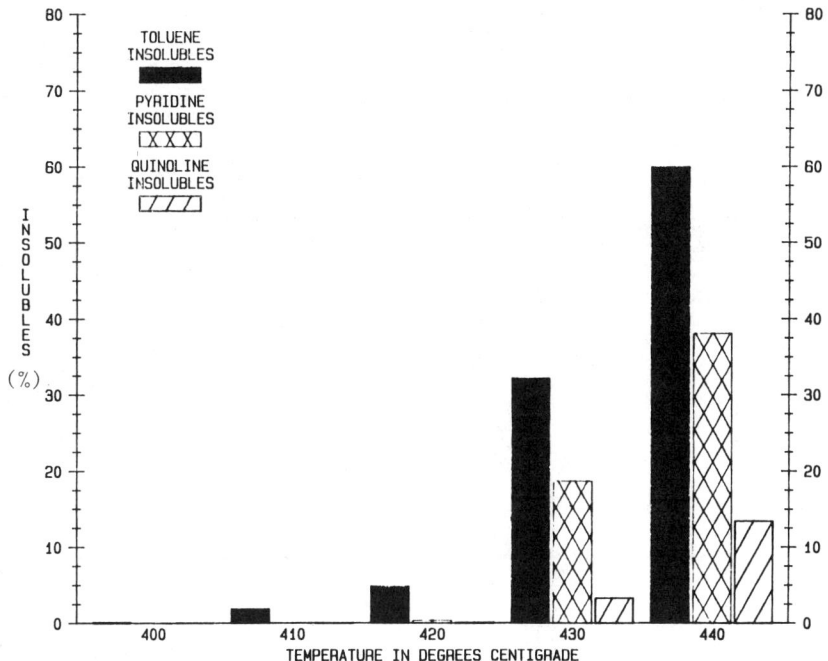

Figure 1. Effect of process temperature on the rate of toluene, pyridine, and quinoline insolubles formation.

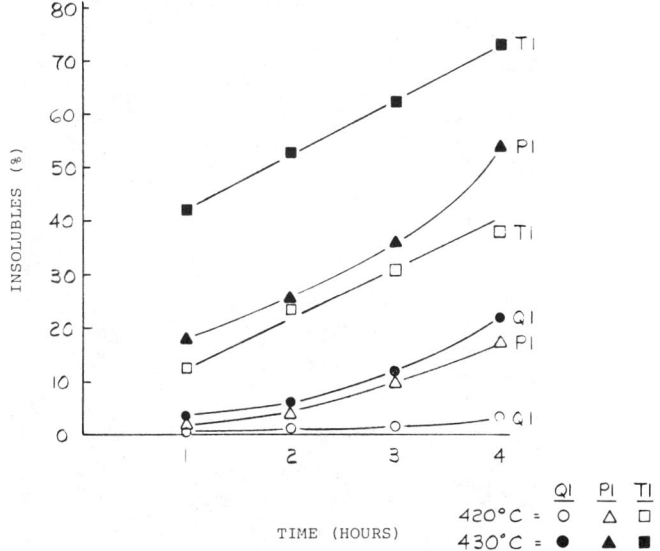

Figure 2. Effect of reaction time on toluene, pyridine, and quinoline insolubles formation.

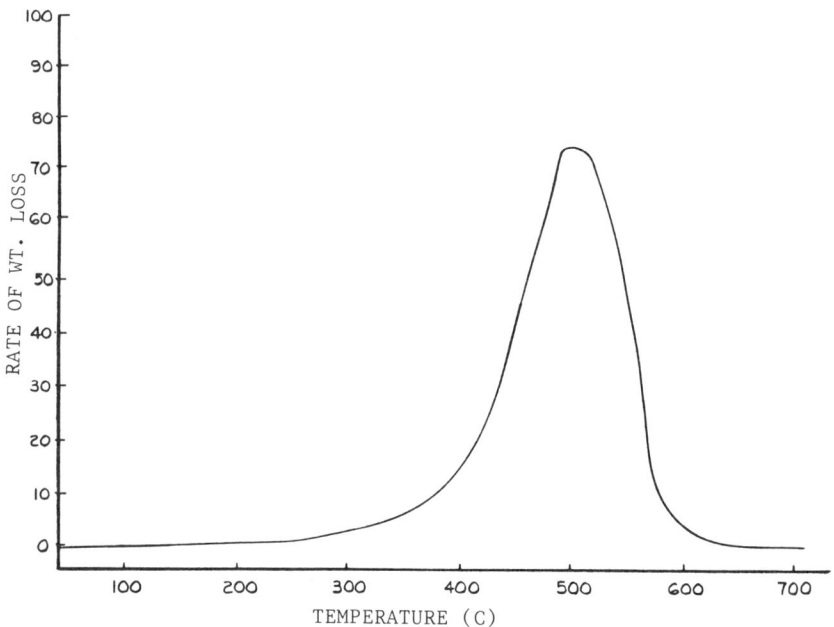

Figure 3. Typical DTG-thermogram of CCB-distillate pitch.

Figure 4. Micrograph of CCB-distillate pitch prepared at 410 C.

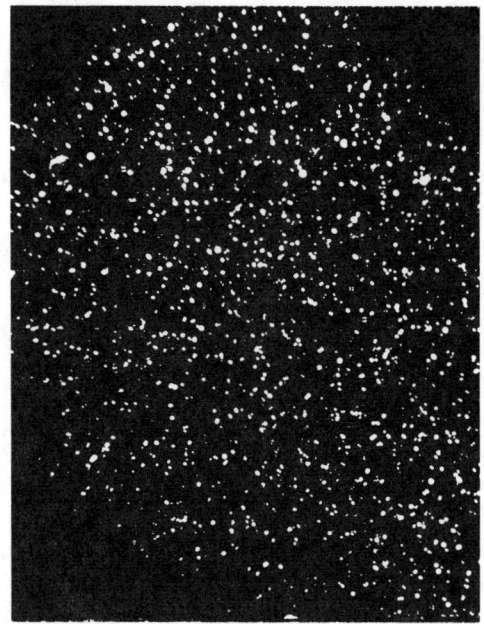

Figure 5. Micrograph of CCB-distillate pitch prepared at 420 C.

Figure 6. Micrograph of CCB-distillate pitch prepared at 430 C.

RECEIVED September 10, 1985

8

Synthetic Aromatic Pitch
Aromatic Pitches from the Distillate Fraction of Catalytic Cracker Bottoms and Residue Fractions

G. Dickakian

Specialties Technology Division, Exxon Chemical Company, Houston, TX 77029

> Catalytic cracking bottoms (CCB) is a widely used aromatic feedstock for carbon production, such as aromatic pitch, carbon black and carbon fibers.
> The distillate and residue obtained by high vacuum-distillation of catalytic cracking bottoms were converted into highly aromatic and anisotropic pitches. We used a two-stage high temperature process at 400°C. The effect of various process parameters on pitch composition and rheology was investigated.

Heavy aromatic feedstock such as the by-products from the petroleum and coal industries are used for the production of aromatic pitches. The characteristics of the pitches produced depend on the chemistry of the feedstock and the process type and conditions.

Catalytic Cracker Bottoms (CCB) which is the heavy residue from the catalytic cracking of petroleum distillate is a common aromatic feedstock used for synthetic carbons and pitch production. CCB, like other heavy aromatic feedstock, is composed of alkyl-substituted polycondensed aromatics with a very wide molecular weight distribution.

CCB was fractionated into six asphaltene-free distillate fractions of varying boiling ranges and an asphaltene-rich non-distillable residue. Characterization of the distillate and the non-distillable fractions indicate significant differences in the asphaltene, ash, aromaticity, molecular weight and aromatic ring distributions.

Both CCB fractions (distillate and residue) were transformed into aromatic pitches by a high temperature thermal process at atmospheric pressure followed by vacuum stripping. A number of reaction parameters effecting pitch yield and characteristics were investigated.

CCB was fractionated by high vacuum distillation (100-500 microns and 15/5 column) into six distillable fractions with boiling points ranging from 270°C to 520°C at 760 mm Hg and a non-distillable residue (510°C+). The boiling characteristics of the CCB distillate fractions are given in Table I. The physical and chemical characteristics of the distillate and residue fractions are given in Table II. The NMR data are given in Table III and the aromatic ring distributions (by Mass Spectroscopy) are given in Table IV.

The CCB fractions were thermally-treated at 420°-450°C at atmospheric pressure in a nitrogen atmosphere and then vacuum stripped (0.5-1.0mm Hg) to remove the unreacted fractions. The pitch chemical structure was determined by NMR, and the pitch composition was determined by solvent analysis with toluene (at reflux) and quinoline (at 75°C). Toluene insolubles were determined because it represents a fusable pitch fraction with a 100% optical anisotropy on melting.

Three distillable fractions (Numbers 4, 5, and 6, in Table I) were thermally-treated at the same conditions (430°C for 3 hours) to investigate the effect of the aromatic ring distribution of the fraction on the pitch yield and composition. We found that increasing the number of 4, 5, and 6 aromatic rings results in increasing the pitch yield and the rate of toluene insolubles formation. The yield and composition of pitches prepared from distillate fractions, Numbers 4, 5, and 6 are given in Table V.

Distillate fraction Number 4 (from Table I) was treated at 430°C for varying time (3, 4, and 5 hours). Increasing reaction time led to increasing pitch yield and the rate of toluene insolubles formation, but not quinoline insolubles. Table VI gives the yield and composition of pitches prepared from distillate fraction Number 4.

The temperature used for the thermal treatment is a very important factor in the dealkylation, polymerization, and condensation of the polycondensed aromatic molecules. The effect of temperature on the toluene and quinoline insolubles formation was investigated using two of the higher boiling distillate fractions (Numbers 5 and 6 from Table I). We found that increasing the thermal treatment temperature led to increasing the pitch yield and the rate of toluene and quinoline insolubles formation. Table VII presents details of the thermal treatment of distillate fractions Numbers 5 and 6, at 420°C, 430°C, 440°C and 450°C. Figures 1, 2 and 3 illustrate graphically the effect of temperature on pitch yield, and toluene and quinoline insolubles formation, respectively.

The undistillable CCB residue remaining after separating the distillable fractions from CCB has different characteristics in comparison to the distillate. The CCB residue has a higher boiling point (510°C), high ash content, higher asphaltene content (around 20 weight %), high coking characteristics (26-36% coke yield at 550°C), higher average molecular weight (339) and higher aromatic ring distribution (6+ aromatic rings). A comparison of the characteristics of a CCB-distillate fraction and CCB-residue is given in Table VIII.

CCB-residue was thermally-treated at 420°C for three hours at atmospheric pressure in a nitrogen atmosphere and then vacuum-stripped at 1.0 mm Hg to produce a pitch in a very high yield. We found that using CCB-residue as a feed for the thermal-treatment resulted in:
    (a) producing a pitch in a very high yield (63.2% vs.24.5% when using a distillate feed).
    (b) CCB-residue pitch has higher content of toluene insolubles (32.0% vs 18.0% when using a distillate feed).

(c) CCB-residue pitch has higher quinoline insolubles: toluene insolubles ratio than CCB-distillate pitch (about 10 times higher).

A comparison of the composition of pitches produced from CCB-distillate and residue (at the same conditions) is given in Table IX.

In conclusion, the ash-free and asphaltene-free CCB distillate fractions provide a potential aromatic feedstock for producing highly aromatic and highly anisotropic pitches with a high toluene insolubles and low content of the high molecular weight and infusible quinoline insolubles.

Table I. Distillation Characteristics of CCB-Fractions

CCB-Fraction	Boiling Point (°C/760mm Hg)	Wt.% of CCB Feed
Distillate No. 1	271 - 400	10.0
Distillate No. 2	400 - 427	23.8
Distillate No. 3	427 - 454	13.3
Distillate No. 4	454 - 471	11.7
Distillate No. 5	471 - 488	13.4
Distillate No. 6	488 - 510	10.0
Non-Distillable Residue	510+	17.5

Table II. Characteristics of CCB-Fractions

	Distillates			Non-Distillable Residue
	No. 4	No. 5	No. 6	
Asphaltenes (n-Heptane) insolubles) (Wt.%)	nil	nil	nil	22.0
Ash (Wt%)	0.0004	0.0005	0.0004	0.110
Number Average Mol. Weight	269	285	291	339
Carbon/Hydrogen Atomic Ratio	0.88	0.87	0.94	1.05
Coking Yield at 550°C (Wt.%)	nil	nil	nil	32.4
Coking Yield at 1000°C(Wt.%)	nil	nil	nil	17.0

Table III. Chemical Structure of CCB-Fractions

NMR – Data	Distillates			Non-Distillate Residue
	No. 4	No. 5	No. 6	
Aromatic Carbon (%)	62.5	62.8	65.8	68.8
Aromatic Protons (%)	26.4	25.7	26.7	35.4
Benzylic Protons (%)	27.6	28.5	28.8	31.6
Total Aliphatic Protons (%)	46.0	45.8	45.3	32.9

Table IV. Aromatic Ring Distributions of CCB-Fractions

Aromatic Rings	Distillates			Non-Distillable Residue
	No. 4	No. 5	No. 6	
2 rings	17.0	9.4	10.4	14.0
3 rings	35.7	25.1	28.6	15.2
4 rings	42.0	50.1	47.6	31.0
5 rings	3.1	12.2	8.7	15.7
6+ rings	0.6	1.1	2.2	9.0
3 + 4 rings	77.7	75.2	76.2	46.2
4 + 5 rings	45.1	62.3	56.3	46.7
3 + 4 + 5 rings	80.8	87.4	84.9	61.9

Table V. Aromatic Pitch Production from CCB-Fractions
Effect of Distillate Fraction Boiling Characteristics

Distillate Fraction No.	Boiling Range (°C/760mm Hg)	Thermal Treatment		Pitch Yield (%)	Pitch Composition	
		Temp. (°C)	Time (Hrs.)		Toluene Insolubles (%)	Quinoline Insolubles (%)
4	454 – 471	430	3	19.0	38.0	0.4
5	471 – 488	430	3	25.0	42.0	0.5
6	488 – 510	430	3	31.5	45.8	0.7

Table VI. Aromatic Pitch Production from CCB-Fractions

Effect of Reaction Time

Distillate Fraction No.	Thermal-Treatment Temp (°C)	Time (Hrs.)	Pitch Yield (%)	Pitch Composition Toluene Insolubles (%)	Quinoline Insolubles (%)
4	430	3	18.0	41.5	0.5
4	430	4	21.0	51.0	0.8
4	430	5	27.6	63.5	1.2

Table VII. Aromatic Pitch Production from CCB-Fractions

Effect of Reaction Temperature

Distillate Fraction No.	Thermal-Treatment Temp. (°C)	Time (Hrs.)	Pitch Yield (%)	Pitch Composition Toluene Insolubles (%)	Quinoline Insolubles (%)
5	420	3	22.5	27.0	0.1
5	430	3	25.0	40.0	0.3
5	440	3	32.0	61.5	1.3
5	450	3	36.6	73.0	4.5
6	420	3	24.4	37.0	0.2
6	430	3	31.4	55.0	0.4
6	440	3	37.1	73.0	10.5

Table VIII. Comparison of the Characteristics of CCB Distillate and Residue

	CCB-Distillate	CCB-Residue
Boiling Range (°C/760mm Hg)	471 - 488	510+
Asphaltenes (Wt%)		
n-Heptane Insolubles (Wt.%)	nil	18 - 22
Coking Yield @ 550°C (Wt%)	nil	26 - 32
Coking Yield @ 550°C (TGA)	nil	12 - 17
Aromatic Carbon Atom (%)	65	69
Carbon/Hydrogen Atomaic Ratio	0.87	1.05
Number Average Mol. Weight	285	339
% Molecular Weight (225 - 400)	94	77
% Aromatic Rings (3 + 4 + 5 rings)	87	10

Table IX. Aromatic Pitch Production from CCB Distillate and Residue

Feed	Thermal Treatment		Pitch Yield (%)	Pitch Composition	
	Temp. (°C)	Time (Hrs.)		Toluene Insolubles (%)	Quinoline Insolubles (%)
Distillate	420	3	24.5	18.0	0.3
Residue	420	3	63.2	32.0	7.4

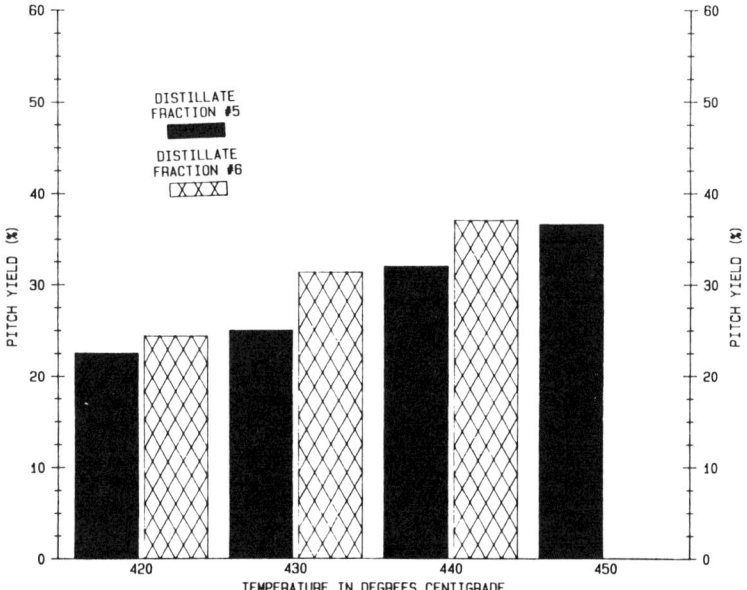

Figure 1. Effect of temperature on pitch yield.

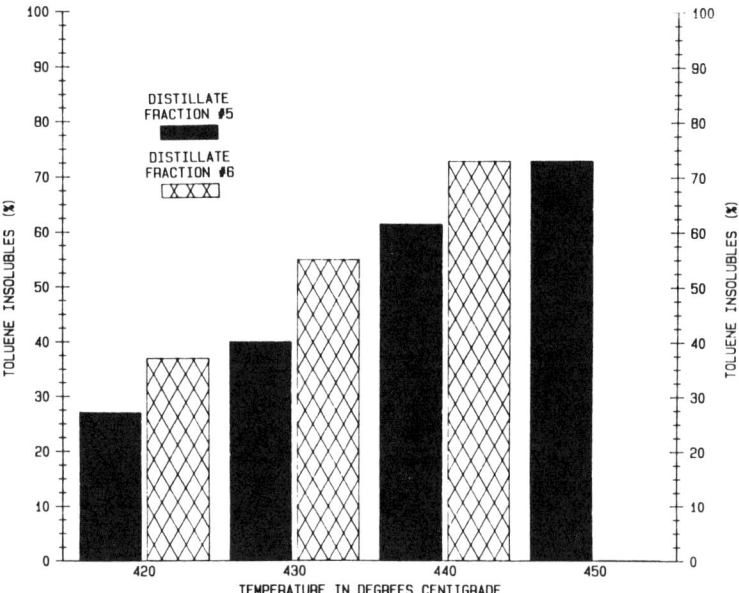

Figure 2. Effect of temperature on toluene insolubles formation.

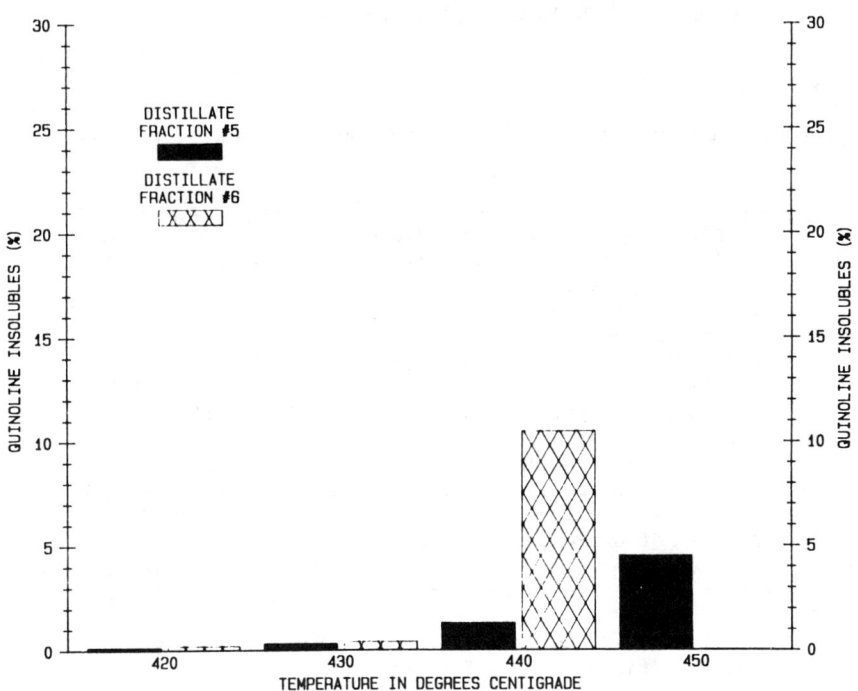

Figure 3. Effect of temperature on quinoline insolubles formation.

RECEIVED September 10, 1985

9

Synthetic Aromatic Pitch
Aromatic Pitch Production Using Steam-Cracker Tar

G. Dickakian

Specialties Technology Division, Exxon Chemical Company, Houston, TX 77029

Steam cracker tar (SCT) is a by-product from the steam cracking of naphtha or gas oils to produce ethylene. The characteristics and yield of SCT is dependent on the feed characteristics, the plant design and severity of cracking.

SCT, like other heavy aromatic materials, is composed of alkyl substituted low molecular weight polynuclear aromatic oils (Mn = 160) and high molecular fraction (asphaltenes), insoluble in paraffinic solvents (Mn = 700 - 1500). The characteristics of SCT, derived from naphtha, gas oil and desulfurized gas oil steam cracking, are given in Table I.

SCT can be converted into highly aromatic pitches by physical, thermal and chemical processes such as: vacuum or steam stripping, thermal or catalytic oxidative-polymerization at 229-260°C, or by a thermal process at 370-450°C at atmospheric nitrogen or hydrogen pressure. The physical or chemical characteristics of the pitches produced from SCT depend on the type of process and conditions used. Table II gives the characteristics of SCT pitches produced by distillation, catalytic air-oxidation and thermal process.

The most suitable process for transforming SCT into highly aromatic pitch is the thermal process at high temperature (380-430°C) at atmospheric, high or reduced pressure. The main chemical reactions taking place during the thermal process are dealkylation, aromatization, and the condensation of aromatic rings into high aromaticity pitch. The increase in the aromatic carbon atom (by carbon - NMR) during a typical thermal process of SCT at 380°C at atmospheric pressure is illustrated in Figure 1.

When using the thermal process for the production of SCT pitch, the temperature and time are important process parameters. The higher the temperature used, the higher is the aromaticity and condensation of the aromatic rings. The average carbon and proton distributions (determined by Nuclear Magnetic Resonance Spectroscopy) of SCT pitches prepared by thermal process at 390°C and 430°C are presented in Table III.

SCT pitches produced by a thermal process at appropriate conditions have high coking yields, high aromatic carbon, low viscosity, high carbon content and very low content of polar atoms. A comparison of the physical and chemical characteristics of two commercial petroleum pitches, two commercial coal tar pitches and a SCT pitch prepared by a thermal process is presented in Table IV.

Viscosity is an important characteristic of pitches used as binders for the production of carbon and graphite electrodes. We used a Haake balance to measure SCT, petroleum and coal tar pitch viscosity. SCT pitches have viscosity between 1000-4000 cps at 160°C. A comparison of the viscosity-temperature relationship of two SCT pitches prepared by thermal and catalytic processes, a commercial petroleum and a coal tar pitch used for the production of carbon anodes is given in Figure 2.

We used thermal analysis to determine the thermogravimetric analysis (TGA) and the differential thermogravimetric analysis (DTG) of SCT pitches to obtain information on volatility and coke yield at various temperatures up to 1000°C. DTG was found very useful in defining process modifications to reduce volatiles in the pitch and increase pitch coke yield. Figure 3 gives the DTG (in nitrogen) of several SCT pitches prepared by distillation, thermal and catalytic process, in comparison with petroleum and coal tar pitches.

SCT pitches like petroleum and coal tar pitches contain an asphaltene free polycondensed aromatic oil with 2-6 aromatic rings. The aromatic oil in the pitch can be quantitatively determined by using a high vacuum distillation with continuous agitation to avoid pitch cracking or coking. The oil in the pitch is important as it effects pitch volatility, viscosity and the development of anistropic structure when coking the pitch during the carbonization of the green carbon anodes. Figure 4 gives the vacuum distillation curves of SCT, petroleum and coal tar pitches.

The molecular weight distribution of SCT, petroleum and coal tar pitches were determined by Gel Permeation Chromatography at high temperature using 1,2,4-trichlorobenzene as the solvent and a UV-spectrophotometer at wave-length 320 mm as the detector. A comparison of the molecular weight distribution curves of SCT pitch and petroleum and coal tar pitches is presented in Figure 5.

In summary, high softening point, high coking value and high aromaticity pitches can be prepared from SCT. The physical, thermal, chemical and coking characteristics of SCT-Pitches is dependent on the type of process used, design of plant and the process conditions especially, temperature, time, presence of catalyst or oxygen and pressure.

Table I. Physical and Chemical Characteristics of Steam Cracker Tars from Naphtha and Gas Oil Cracking

	SCT from Naphtha Cracking	SCT from Gas Oil Cracking (1)	SCT from Gas Oil Cracking (2)	SCT from Desulfurized Gas Oil Cracking
1. Physical Characteristics				
Viscosity cst at 210°F	13.9	19.3	12.4	25
Coking Value at 50°F (%)	12	16	24	25
Toluene Insolubles (%)	0.200	0.200	0.250	0.100
n-Heptane Insolubles (%)	3.5	16	20	15
Pour Point (°C)	--	+5	-6	+6
Ash (%)	0.003	0.003	0.003	0.003
2. Chemical Structure				
Aromatic Carbon (atom %)	65	72	71	74
Aromatic Protons (%)	34	42	42	38
Benzylic Protons (%)	40	44	46	47
Paraffinic Protons (%)	25	14	12	15
Carbon/Hydrogen Atomic Ratio	0.942	1.001	1.079	1.44

Table II. Characteristics of SCT, SCT-Pitches Produced by Distillation, Polymerization and Thermal Processes

	SCT (Feed)	Processes		
		Vacuum Distillation	Catalytic (Air) Polymerization	Thermal
1. Physical Characteristics				
Sp. gr. (at 20°C)	1.110	1.160	1.191	1.265
Softening Point (R&B) °C	10	105	110	108
Coking Value at 550°C (wt%)	20	33	43	53
Benzene Insolubles (%)	0.05	0.20	18.0	28.0
Quinoline Insolubles (%)	0.05	0.10	0.20	2.5
Viscosity cst at 160°C	--	979	13140	3000
2. Chemical Structure				
Aromatic Carbon (%)	70	72	72	77
Aromatic Protons (%)	42	42	44	50
Benzylic Protons (%)	46	46	37	37
Paraffinic Protons (%)	12	12	18	13
Carbon/Hydrogen Atomic Ratio	1.079	1.27	1.34	1.37

Table III Effect of Thermal Process Temperature on the Carbon and Proton Distribution in SCT-Pitch

	SCT (Feed)	SCT-Pitch Prepared at 390°C	SCT-Pitch Prepared at 430°C
Aromatic Carbon (atom %)	71	76	78
Aromatic Protons (% of total protons)	42	50	53
Benzylic Protons (% of total protons)	46	37	36
Paraffinic Protons (% of total protons)	12	12	11

Table IV Characteristics of SCT, Petroleum and Coal Tar Pitches

	SCT Pitch	Petroleum Pitches		Coal Tar Pitches	
		(1)	(2)	(1)	(2)
1. Physical Characteristics					
Sp. gr.	1.265	1.223	1.260	1.270	1.259
Softening Point (R&B) °C	110	117	110	101	113
Coking Value at 550°C (wt %)	52.0	54.0	56.0	56.5	59.7
Ash Content (%)	0.100	0.150	0.21	0.200	0.300
Benzene Insolubles (%)	27.0	8.0	30.0	41.98	48.7
Quinoline Insolubles (%)	2.5	0.5	11.5	21.7	26.0
Viscosity (%) (cps) at 160°C	3000	1400	2050	1116	840
2. Chemical Characteristics					
Aromatic Carbon (atom %)	78	82	80	89	88
Aromatic Protons (%)	50	57	--	84	86
Benzylic Protons	37	34	--	13	11
Paraffinic Protons	12	9	--	3	3
Carbon/Hydrogen Atomic Ratio	1.37	1.44	1.57	1.77	1.76

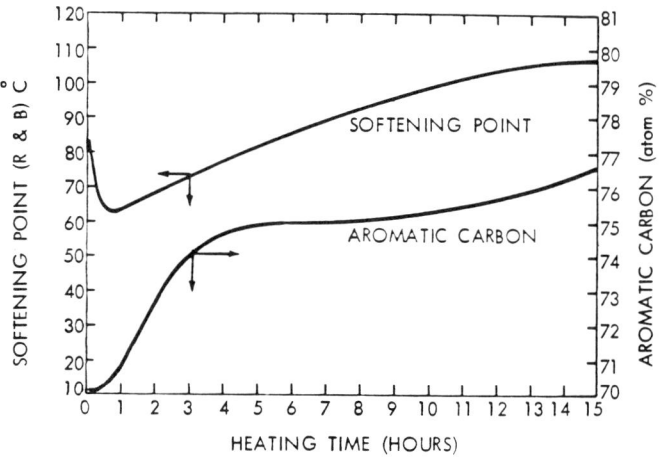

Figure 1. Carbon distribution during thermal treatment.

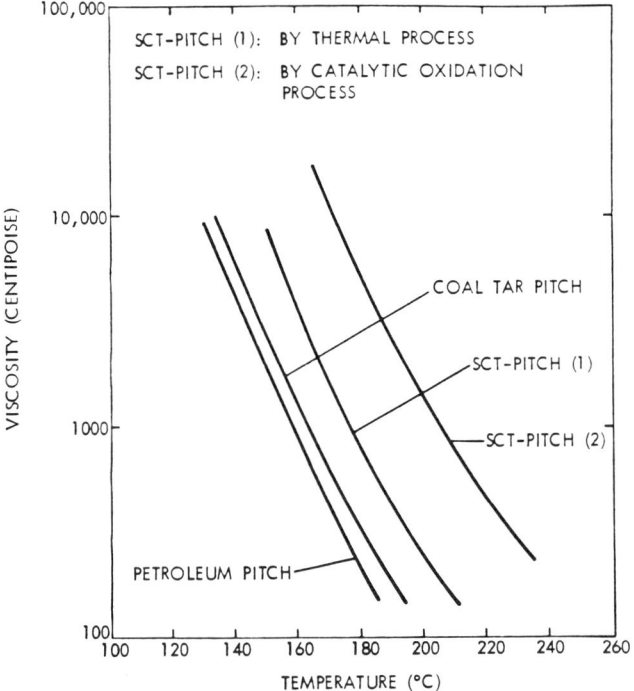

Figure 2. Viscosity-temperature curves for SCT-pitch, coal tar pitch, and a petroleum pitch.

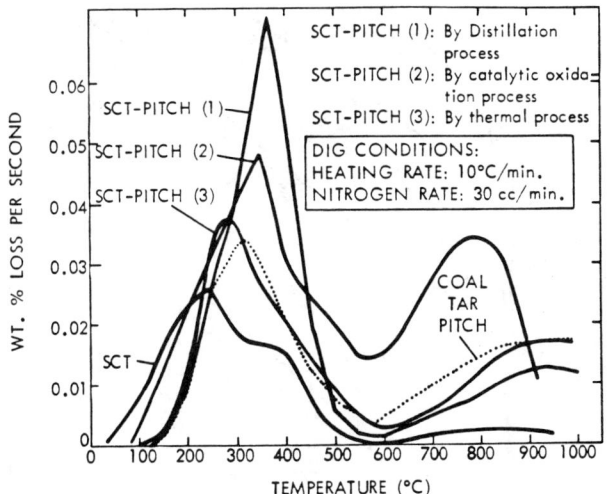

Figure 3. Differential thermogravimetric analysis (DTG) in nitrogen of SCT, petroleum, and coal tar pitches.

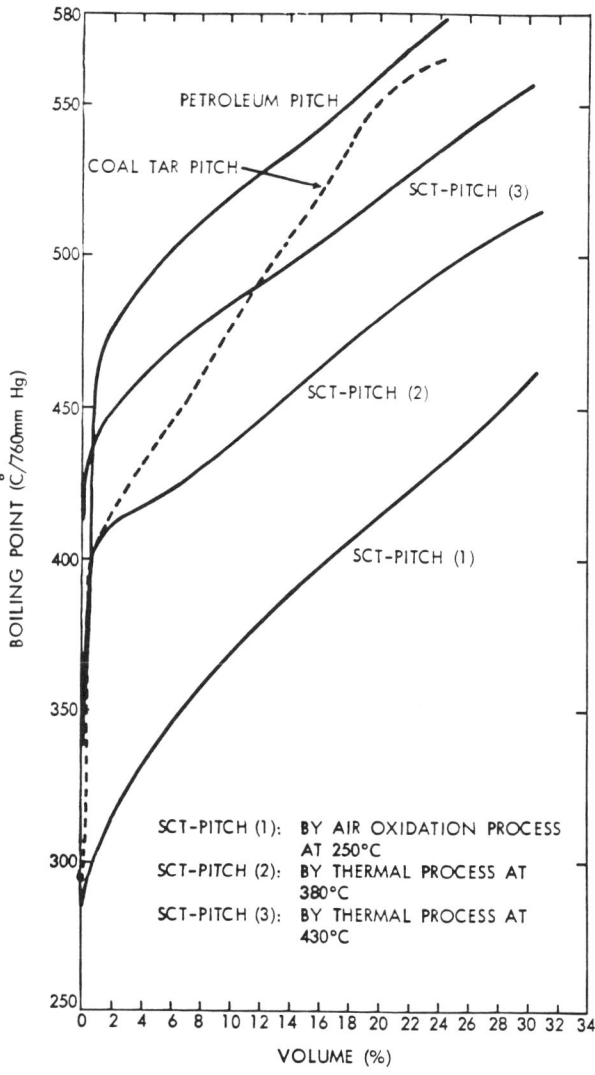

Figure 4. Vacuum-distillation of SCT, petroleum, and coal tar pitches.

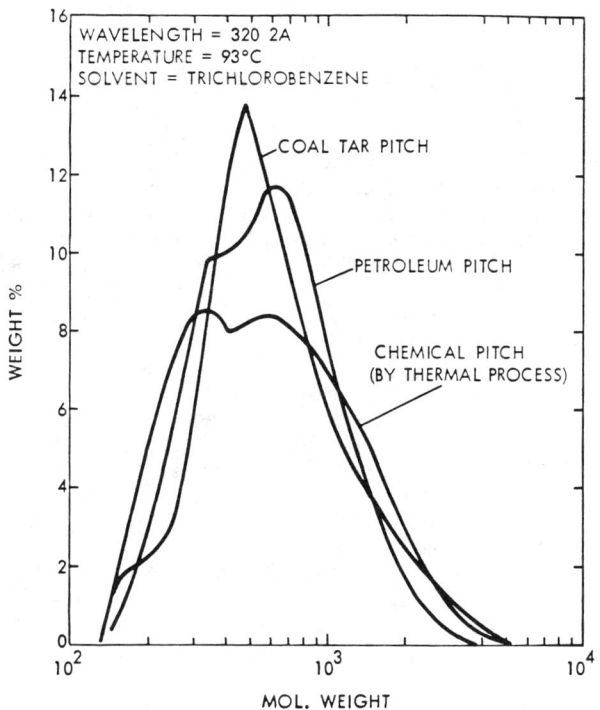

Figure 5. Molecular weight distribution of SCT, petroleum, and coal tar pitches.

RECEIVED September 10, 1985

10

Petroleum-Coke Overview

James H. Waller, Gary W. Grimes, and John A. Matson

The Pace Consultants, Inc., 5251 Westheimer, Houston, TX 77056

> Petroleum coke markets are complex due to coke's status as a refinery by-product and its use in a myriad of product applications, each influenced by unrelated economic forces. Heavier crude oils containing higher sulfur levels will increase the amount of 1,000+ Fahrenheit material available for coker feed, although coking units must compete with other bottom-of-the-barrel dispositions. However, the economic incentive to operate existing coking capacity will decrease and reduce coker operating rates at least through 1985.
>
> Petroleum coke markets are developed around a demand hierarchy that consists of several end-uses. Premium markets, including the use of calcined petroleum coke, will continue to be cyclical. As such, demand patterns for petroleum coke will remain volatile.
>
> Fuel grade petroleum coke will continue to experience incremental demand in Western European markets. As such, prices will closely follow steam coal sold in European markets. Petroleum coke will continue to be priced at a discount to coal.

Since the early 1950s, petroleum coke markets have changed radically. Once treated only as a refinery by-product by refiners, petroleum coke is now a permanent feature of several end use markets that are influenced by a variety of economic variables. Many refiners have begun to commit corporate resources to marketing their petroleum coke in hopes of realizing incremental revenues. Marketing was formerly left to a few specialized marketers who contributed to market viability by developing new customers and new uses of petroleum coke that are now considered as permanent market fixtures.

In 1983, over 17 million short tons of petroleum coke were produced in the United States, where approximately two-thirds of the world's coking capacity is located. Pace estimated the value of this production to be over $650 million before further processing.

0097-6156/86/0303-0144$06.00/0
© 1986 American Chemical Society

This presentation will focus on three basic market components: supply, demand (markets), and pricing. For petroleum coke, these components are not as straightforward as other petroleum product markets. Also, the following will discuss the use of petroleum coke by the utility power industry and new markets on the horizon.

Supply

Coking is the most economical method used to convert heavy residual fuel oil and heavy crudes to lighter, more valuable refined products. The buildup of coking capacity in the United States has resulted from economic forces that dictated a balance of light and heavy products. Coke production increases have generally followed increasing trends for light refined products such as gasoline and declining demand for residual fuel oil, as natural gas became a preferred fuel.

Similar to other refining and industrial processes, the decision to construct and operate a coking unit is dependent upon unique economic factors. The price differential between residual fuel oil and crude oil is a major variable, although the relative prices of light and heavy crude, the demand for refined products, as well as the amount and type of conversion unit capacity also affect coking economics. Notice the value of petroleum coke was not included. Cokers have been justified in most cases even though the coke product was assumed to have zero value.

Since petroleum coke is a by-product, normal supply/demand analysis of coke markets are insufficient to forecast production. The methodology used at Pace for forecasting both the quantity and quality of coke production (Figure 1) is:

1. Forecast the demand for refined products using relationships tied to our economic forecasts and expected efficiency factors (such as miles per gallon of gasoline).
2. Forecast the crude slate that would likely be used to meet the forecast demand. This is determined by an analysis of domestic reserves, historical production trends, and estimates of new production. The shortfall between domestic production and refinery crude runs is met by imported crudes which are selected on the basis of production/export capabilities, logistical factors, and historical trends.
3. Once the crude slate is established, the supply of 1,000+ material is determined from our database of crude assays. 1,000+ material is that part of the barrel of crude oil which has a boiling point of 1,000°F or greater. This is typically the feed material to a coker.
4. Forecast the quantity of 1,000+ material required to meet the demand for other products such as residual fuel oil and asphalt. The remaining 1,000+ material is assumed to be coker feed.
5. Use the coke yields typical for each forecast crude to determine the total production and quality of green coke.

Our forecast is developed on a regional basis and then consolidated to a total U.S. forecast. We have computerized this forecasting methodogy, building a model that allows us to examine the effect of varying economic growth, product demands, and new crude discoveries on coke production and quality.

The Pace forecast for refined products is shown graphically in Figure 2. The most notable features are increasing demand for middle distillates (diesel and jet A), continued weakness in gasoline demand through 1990 followed by a slight upturn, and continued decline in demand for residual fuel oil through the remainder of this century.

Figure 3 shows the crude runs necessary to meet our refined product demand forecast. The forecast includes two scenarios: one of normal trendline growth portrayed by several industry observers and Pace's own outlook for refinery crude runs based on a "cyclical" economic forecasting model. Our forecast shows runs to crude stills will remain below 12.5 million barrels per day for the remainder of the decade.

The expected origin of crude oils to be processed in the future is shown in Figure 4. The crude oil actually processed, as mentioned earlier, has a pronounced effect on both coke quantity and quality, since individual crudes vary in terms of viscosity, sulfur content and other contaminants such as trace metals.

As shown in Figure 5, average crude gravity is expected to drop another degree API by the end of the century. This is a less dramatic change than the almost two degree drop of the last six years. Sulfur content is expected to increase from about one percent to 1.2 percent by 2000. That portion of the average crude mix with a boiling point greater than 1000 degrees Fahrenheit is expected to increase approximately one percent by 2000.

These and other factors are incorporated into Pace's coke production model. Our original forecast indicated peak coke production in 1982 of approximately 17 million short tons, followed by a sharp decline to 1985. The predicted decline never occurred for several reasons. We believe the major reason was the production momentum that occurred following the completion of several coker projects in late 1983. Many of the projects, originally planned at a time of attractive coking economics but started up under less favorable conditions, were operated regardless of the economics. It is difficult to complete an expensive refinery upgrade and let it sit because the economics no longer justify its operation.

Figure 6 represents our current forecast of coke production and average sulfur content. Coke production through the first half of 1984 occurred at an annualized rate of 19.7 million short tons (about 86 percent of calendar day capacity basis). However, monthly production rates have recently dropped considerably. From April's rate of about 56,000 short tons per day, which was the highest ever incurred by the United States refining industry, production has slipped to 52,000 short tons per day.

Coke production should continue to follow a downward trend throughout 1984 and 1985 before increasing through the remainder of the century. The basis for the forecast in the near term is an expected narrow price differential between both heavy crude and residual fuel and light crudes. Although several factors may be cited for narrow price differentials over the next 18 months to two years, the most obvious is the refining industry's (both at home and abroad) increased ability to process heavy crude oil into lighter refined products. Between 1980 and 1983, approximately 1.5 million barrels per day of new capacity was installed throughout the free world to destroy heavy oil; currently, another 1.9 million barrels per day of capacity is either announced or under construction. The buildup in heavy oil processing operations has been detrimental to coker economics.

Figure 7 shows expected coke production levels in the United States by sulfur content category. Shown is Pace's current petroleum coke production forecast. Most of the additional coke production will be in the category of four percent sulfur or greater.

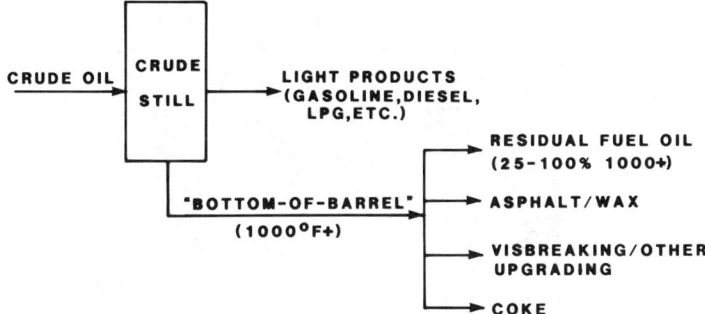

1. FORECAST DEMAND/PRODUCTION OF RESID, ASPHALT
2. FORECAST CRUDE SLATE/1000+ PRODUCTION
3. 1000+ AVAILABLE FOR COKER FEED CALCULATED BY DIFFERENCE AFTER OTHER DEMANDS ARE MET
4. COKE PRODUCTION FROM TYPICAL YIELDS

Figure 1. Methodology.

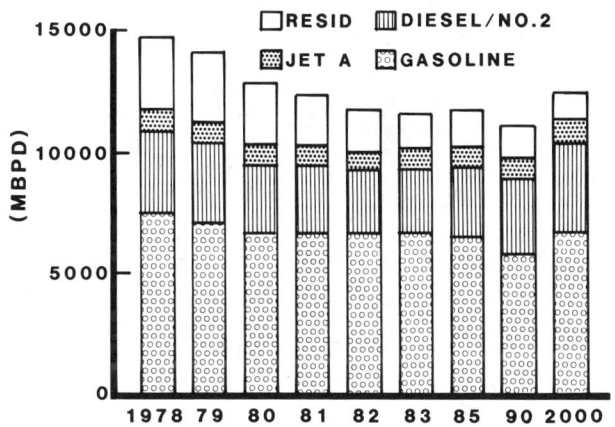

Figure 2. Major refined product demand.

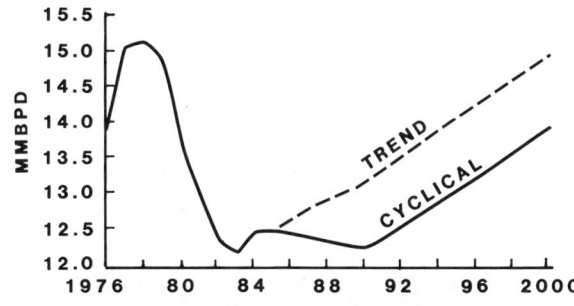

Figure 3. Runs to crude stills.

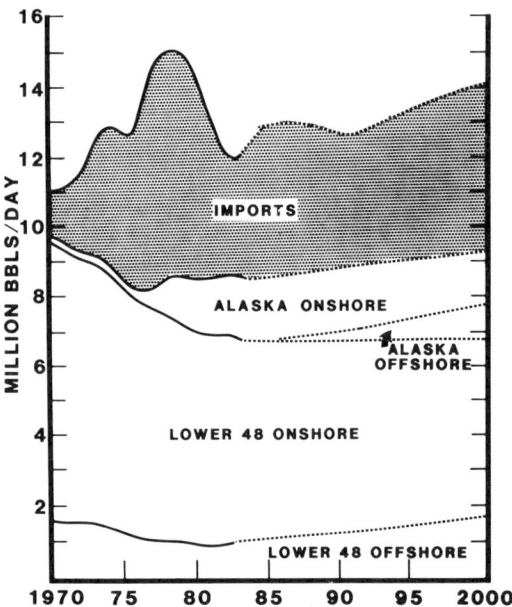

Figure 4. U.S. crude oil supply.

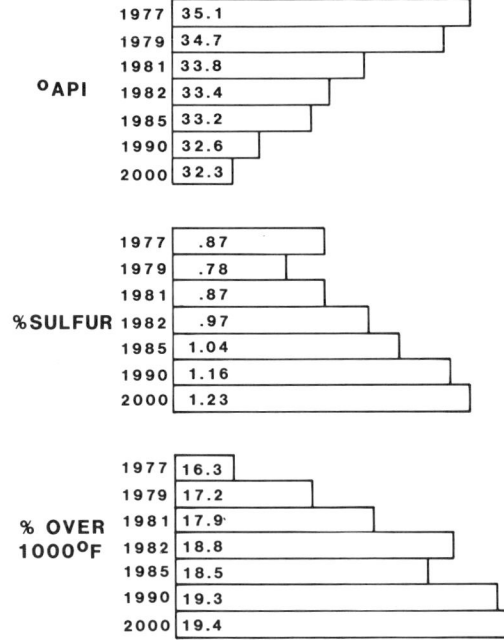

Figure 5. United States crude quality.

This figure also indicates the expanded coke capacity base and how under-utilized this equipment will be in the mid-1980s. No additional coking units are expected for 1986 through the end of this century.

Demand.

Several characteristics of petroleum coke markets should be recognized to fully understand market behavior. Succinctly, these include:

- Price inelastic supply
- Coke is a refinery by-product
- Marketing functions are performed by specialized marketers
- Several distinctive end uses exist
- Demand follows an identifiable price hierarchy.

The first two characteristics of today's coke markets were alluded to earlier. Coke prices rarely, if ever, influence the decision to produce coke or construct a coking unit. Therefore, the entire coke marketing scheme has been one of maximizing revenues from coke disposal requirements.

Figure 8 illustrates 1983 coke markets according to consuming industries worldwide. Most of the petroleum coke consumed by the steel industry occurs outside of the United States, since metallurgical coal is widely available here. Most calcined coke is consumed by the aluminum industry in the production of primary aluminum. Cement producers burn a coke/coal mix in their cement kilns. Finally, utility coke consumption is relatively a small portion of total demand.

Figure 9 describes petroleum coke markets in terms of consuming world regions. Western Europe consumes the largest share of United States petroleum coke. The European steel and cement industries are consistent customers. In fact, the European fuel market (cement, ceramics, glass, utilities) can be considered as the market sump. If the price of coke declined to a hypothetical level, the European fuel market alone could probably absorb all of the United States coke production.

Several combined factors lead to this conclusion. They include:

- Europe is a net importer of solid fuels.
- The European market is price sensitive and flexible enough to switch much of the consumption to the cheapest available source.
- The low ash, high BTU characteristics of petroleum coke complement the high ash, low BTU characteristics of local coal.
- Most installations have fuel blending equipment to take advantage of economical fuels.
- The volume of coke consumption is small relative to coal consumption.

Petroleum coke consumption by the utility power industry has been minor, even though another solid fuel, coal, has become the industry's major source of fuel. Domestically, only 630,000 short tons of petroleum coke was consumed by the United States utility industry in 1983, or approximately three percent of total domestic petroleum coke production.

Four United States utilities burned petroleum coke in 1983: Delmarva Power and Light, Pennsylvania Power and Light, Northern States Power and Wisconsin Power and Light. The common incentive for burning petroleum coke was the reduction of fuel costs. Petroleum coke's high BTU content can

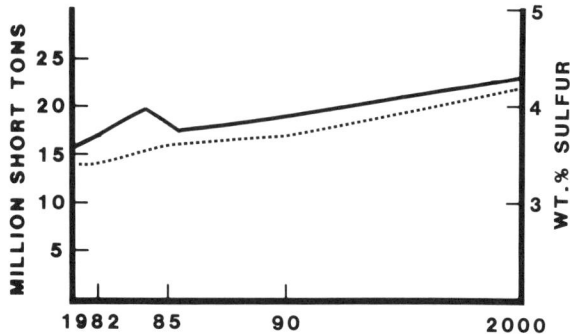

Figure 6. United States green coke supply and quality.

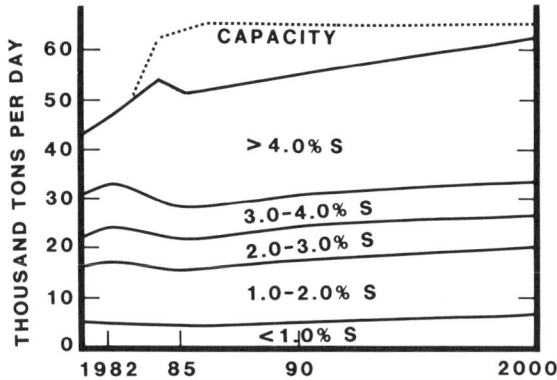

Figure 7. United States coke production.

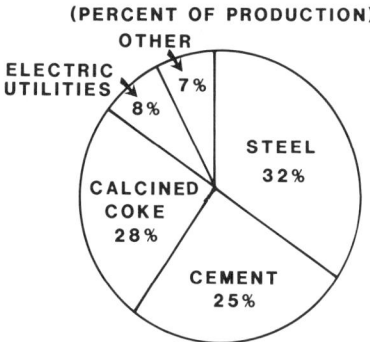

Figure 8. World markets for U.S. green petroleum coke - 1983.

contribute significantly to a lower fuel bill. Delmarva combines fuel oil with their coke, while the other utilities blend a coke-coal mixture for fuel.

Several reasons may be cited for the domestic utility power industry's overall lack of interest in petroleum coke. One is the well established use of coal, which is purchased under long-term contracts with relatively level prices. By comparison, both the availability and price of coke are typically volatile from year to year.

The high sulfur content of fuel grade petroleum coke has also restricted its use as utility power fuel. Additional capital costs would be incurred by a utility choosing to burn petroleum coke to reduce sulfur emissions.

Use of petroleum coke would also require greater logistical, handling, and purchasing flexibility by a utility. With current technology, petroleum coke is burned as a blend with coal, anthracite, fuel oil or natural gas. This requires additional mixing equipment and storage space which may not be compensated for by lower petroleum coke prices.

The final point for petroleum coke demand is new uses that are currently being developed. No single new application is expected to increase demand over the next few years. However, the combined effect could be significant. In fact, the existence of today's demand can be directly attributed to creative market development over the years for this refinery by-product.

Briefly, new uses may include improved coke-fuel oil mixtures, feedstock for petrochemical manufacture via gasification, lime kiln fuel for paper mills and as cogeneration fuel at refinery sites. The potential for coke usage with fuel oil is particularly high in Japan, where several industry participants estimate an additional 1.5 million short tons could be consumed annually.

Also, cogeneration project developers are seriously considering petroleum coke as a viable fuel source. The Delmarva Power plant in Delaware City, Delaware has been producing electricity and steam for Getty Oil's nearby refinery for several years. Atlantic Richfield is currently constructing a cogenerator at its Houston, Texas refinery with start up expected in 1986. The ARCO project will generate 135 MW and consume 500,000 tons of petroleum coke each year.

Pricing.
Petroleum coke prices are determined by several factors, including quality characteristics (e.g., sulfur, metals, and ash content), availability of competing carbon-based products, and requirements for further processing. Petroleum coke marketers and producers have continuously attempted to maximize the value of their available supplies by selling as much coke as possible to higher-valued end-use markets, or developing new markets where premium prices could be supported. Thus, a pricing hierarchy has evolved to the extent that each end-use market maintains unique pricing mechanisms for petroleum coke.

As mentioned previously, the European industrial fuels market is placed near the bottom of the petroleum coke market hierarchy. Home heating markets in Europe attract the highest prices for petroleum coke but consume a small fraction of annual production. The largest "premium" market by far is the domestic calcining industry. Market/pricing hierarchy is shown below.

Petroleum Coke Market Hierarchy

- European Space Heating
- United States Calcining
- Japanese Carbide
- Japanese Steel
- European Steel
- Japanese Industrial Fuel
- European Industrial Fuel
- United States Industrial Fuel

The petroleum coke market hierarchy arose because the value and distribution of petroleum coke has historically been limited by the quality demanded in each specific end use. As this market hierarchy became defined, a clearly stratified pricing scheme emerged for various coke qualities. However, the market hierarchy can crumble if demand for higher quality petroleum coke decreases, as it has several times in the past. When this occurs, petroleum coke that previously was consumed by upper tier markets, such as the domestic calcining industry or Japanese steel industry, is dumped into the lower tier markets.

Marketers who fail to anticipate cyclical downturns in their premium markets can suffer severe losses when forced to sell to fuel markets instead. To compound this fragile pricing structure, the supply keeps coming since petroleum coke production is inelastic to demand.

Figure 10 illustrates the consequences of the volatile petroleum coke pricing scheme. In 1981, the aluminum industry was absorbing large quantities of low sulfur petroleum coke. This resulted in a terrific price spread between low sulfur and high sulfur coke. Rather quickly, however, the price differential disappeared in 1982 as the aluminum industry suffered the effects of economic recession.

Since that time, several factors have combined to continue the price roller-coaster. Low sulfur petroleum coke prices rose considerably early in 1984 as aluminum production increased. By the second quarter, low sulfur prices reached $60 per short ton and higher at several Gulf Coast production points. Prices for fuel-grade coke were also strong, due to several factors, including:

- Strong demand in premium markets which prevented higher quality coke from being dumped into the fuels markets.
- Active export markets that consisted of buyers stocking up with solid fuel and suppliers over-extending supply commitments. The buildup in customer stockpiles was indirectly linked to the British coal strike and the threatened United States coal strike.
- A slight decrease in petroleum coke production that began in June.

Fuel grade prices reached $42 to $44 by late summer. At these levels, many offshore customers began to shun coke purchases in favor of coal. Exports should remain soft for the remainder of the year as a result.

The close pricing relationships between fuel grade petroleum coke and at least one coal sold in the world markets are shown in Figure 11. South African coal was chosen to illustrate this relationship because it is widely

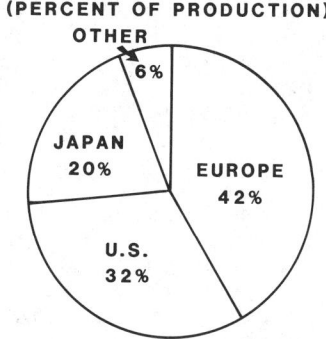

Figure 9. Markets for U.S. green petroleum coke - 1983.

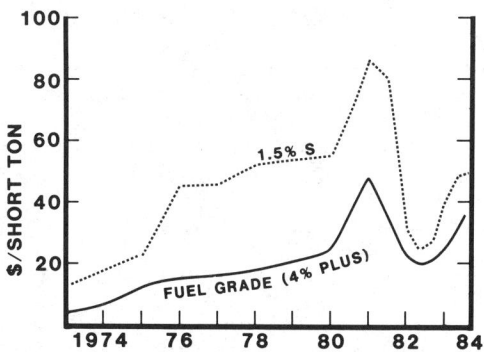

Figure 10. Gulf Coast green coke prices.

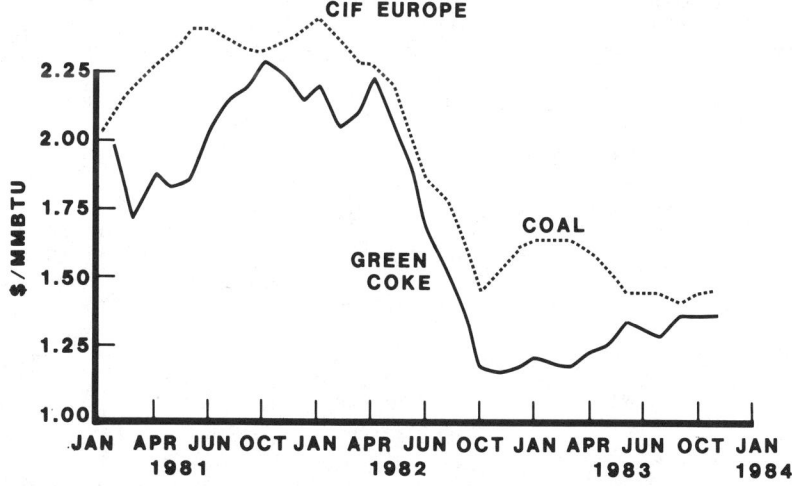

Figure 11. Price of U.S. green coke and South African coal.

available on a spot basis in several world markets. Petroleum coke is also very spot-oriented.

As shown, petroleum coke sells at a discount to South African coal (as well as coal from other regions that is sold primarily through spot transactions). There are several reasons for this discount:

- Petroleum coke has a higher sulfur content than South African coal.
- Some forms of petroleum coke (such as shot coke) can have additional handling costs.
- There is a "spot-on-spot" characteristic of coke—because of sudden declines in premium markets, more coke may be available for the spot fuel market—usually at an increased discount.

Petroleum coke's discount to coal can be at times very unstable. Coal prices are primarily cost-based while coke prices depend on several factors previously mentioned. Although we do not believe fuel grade petroleum coke prices will follow coal/coke pricing relationships to the letter, we feel that as markets continue to develop, the relationship will become more clearly defined to market participants and observers.

Summarizing petroleum coke pricing, there are as many pricing variables to consider as there are applications of petroleum coke. Mechanisms for fuel grade coke have been discussed briefly; however, this ignores several other pricing considerations for calcined petroleum coke, needle coke, metallurgical petroleum coke, and others. These markets have myriad pricing factors which must be considered in establishing current prices or forecasting future prices.

In spite of these complexities, two conclusions can be made. First, petroleum coke prices will remain volatile due to cyclical demand in premium markets, fragile market hierarchies, and the orientation toward significant spot purchasing arrangements. The second conclusion is that the absolute "floor" price of petroleum coke (regardless of quality) will be linked to steam coal in Western Europe.

These conclusions have notable implications for petroleum coke consumers. For example, a common response of solid fuel consumers when asked about their views on possible petroleum coke purchases is "it is not available when I need it and when it is the price is too high." This statement underscores petroleum coke's volatility and describes the inherent risks involved in the markets.

Fortunately for petroleum coke consumers, middle marketers exist who are willing to hold inventories and expose themselves to supply and demand risks. In return, marketers receive profit margins that could indeed be avoided if a consumer were to purchase petroleum coke directly from a producer (refiner). Middle marketers, therefore, absorb much of the risk involved in petroleum coke trade, stabilizing prices, and allowing more solid fuel consumers to consider and plan coke purchases.

RECEIVED February 8, 1985

Delayed-Coking Process Update

Robert DeBiase, John D. Elliott, and Thomas E. Hartnett

Foster Wheeler Energy Corporation, 110 South Orange Avenue, Livingston, NJ 07039

>   Important recent trends and new developments have
>   contributed to profitable, reliable, and safe operation
>   of delayed cokers. A typical delayed coker consists of
>   four sections: coking, fractionation, coker blowdown,
>   and coke dewatering and handling. The main types of
>   coke dewatering and handling systems are described as
>   pit, pad, railcar, and dewatering bin. General coke
>   types, feedstock considerations, pretreatment and
>   process variables are reviewed with emphasis on recent
>   trends towards minimizing production of fuel grade coke
>   from heavy feedstocks. Typical uses of petroleum coke
>   are discussed, including those for fuel grade coke.
>
>   Trends and developments on the design of modern delayed
>   cokers include improved heater design, larger coke
>   drums designed for longer life at short operating
>   cycles, extended range hydraulic decoking systems,
>   enclosed blowdown systems and improved energy
>   efficiency. Older delayed cokers can be revamped in a
>   number of ways to increase capacity and improve the
>   yield of desirable products.

Delayed coking is a processing technology that has been in use for over five decades. During this time, it has come into widespread use as an economic means for upgrading heavy crudes, residues, tars and decant oils to produce gas, gasoline, gas oil and coke. It is seen as an attractive residue upgrading process because of its moderate capital investment and its ability as a single unit, to process a wide variety of feedstocks. As more and more delayed cokers are built, new technology is being developed to create a more profitable, reliable and safe operation. This paper will briefly review the basic aspects of delayed coking and discuss recent trends and new developments.

In delayed coking, a residual feedstock is charged to a furnace where it is rapidly heated and thermally decomposed. The heater effluent then enters a coke drum where the reaction is

0097-6156/86/0303-0155$06.00/0
© 1986 American Chemical Society

completed and petroleum coke and overhead vapors are formed. The process mechanism for delayed coking is as follows (1):
  (1) Partial vaporization and mild cracking of the feed as it passes through the furnace.
  (2) Cracking of the vapor as it passes through the drum.
  (3) Successive cracking and polymerization of the heavy liquid trapped in the drum until it is converted to vapor and coke.

The coke produced is mostly elemental carbon and is used in applications described below. The gaseous and liquid products are valuable feedstocks for downstream processing or sometimes used as products.

Unit Description

A typical delayed coker unit consists of coking, fractionation and blowdown sections, along with coke handling facilities. Coker gas is either processed in a dedicated vapor recovery unit or may be sent for processing, together with other gases, to a centralized vapor recovery unit.

Coking Section. Figure 1 is a simplified process flow diagram of typical coking and fractionation sections. The major equipment included in the coking section are the coker heater, the coke drums and the hydraulic decoking equipment. The feedstock enters either hot from an upstream processing unit or cold from storage. It is often preheated within the delayed coker unit. The feed is charged to the bottom of the fractionator which is used for feed surge. In the bottom of the fractionator the feed combines with the condensed recycle. The resulting heater charge is pumped to the coker heater where it is rapidly heated to the desired coking temperature before flowing to the coke drum where the coking reaction is completed. The coke remains in the drum and the overhead vapors are directed to the fractionation section. A minimum of two coke drums are required, one drum is in coking service while the other drum is being decoked. After steaming and cooling of the coke, the upper and lower flanges of the coke drum are removed. Next, a pilot hole is bored through the coke using high pressure water and a hydraulic boring tool. The coke is then cut out with a hydraulic cutting tool. The coke falls from the drum to the dewatering facilities for separation of the coke from the water.

Fractionation Section. A typical fractionation section includes the coker fractionator and attendant heat exchange equipment, the light gas oil side stream stripper and the overhead system. The coke drum overhead vapors enter the fractionator under shed trays which are located below conventional wash trays. Hot induced gas oil reflux is pumped to the wash trays to condense recycle and to wash the product vapors. The light and heavy gas oil products are condensed as sidestream products. The light gas oil product is usually steam stripped in a sidestream stripper. The overhead vapors from the fractionator are partially condensed and the gas and gasoline products are directed to the vapor recovery unit.

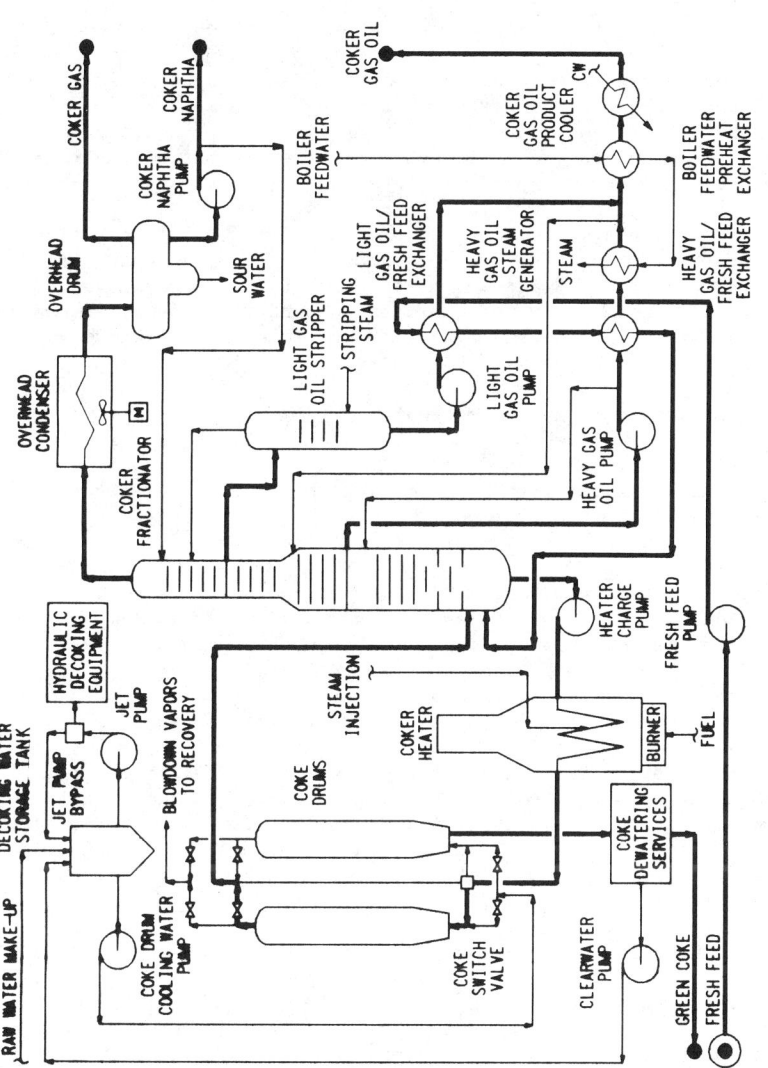

Figure 1. Typical Delayed Coker Process Flow Diagram.

A portion of the condensed gasoline is pumped back to the tower as reflux. The sour water collected in the overhead drum is sent to battery limits for treating.

Coker Blowdown System. Figure 2 shows a typical coker blowdown system. This system is utilized for both pollution control and for increased recovery of hydrocarbons. After a drum is switched from coking to decoking service, the coke is steamed out and then cooled by water injection. During this time, the hydrocarbons stripped from the coke are directed, together with the resultant steam, to the blowdown system. The steam and hydrocarbon from the steamout and cooling cycle operations flow to the coker blowdown drum where the heavy hydrocarbons are condensed by a circulating gas oil stream. These heavy hydrocarbons are pumped back to the coker fractionator. Steam leaves the top of the coker blowdown drum and is condensed in the blowdown condenser along with a small amount of oil. The oil and water are separated in the blowdown settling drum, with the water going to offsite treating facilities or to the decoking water tank, while the oil goes to slop for dewatering and recovery. The remaining vapors from the blowdown settling drum may be compressed and sent back to the coker fractionator overhead drum or treated and sent to the fuel gas system. Alternatively, these vapors can be flared or recovered by a flare gas recovery compressor.

Coke Dewatering and Handling System. When a coke drum is being emptied, coke and water must be collected and separated. To accomplish this, the facilities commonly used today include pit dewatering, pad dewatering, dewatering bin, and direct railcar loading. A short description of each follows.

In pit dewatering, the coke and water drop from the coke drum, through a chute, into a large pit which provides several days of storage. Water drains through the coke into a maze where any remaining coke fines settle to the bottom. Coke is removed from the coke pit with an overhead crane. The large storage capacity of the pit makes it especially suited for units with four or more coke drums. Foster Wheeler has implemented this approach in many of the recently built delayed cokers. A typical pit dewatering system is depicted in Figure 3.

Pad type dewatering is similar to pit dewatering. The major difference is that the coke and water drop onto a grade level pad. Traditionally, water drains through coke packed ports in the pad wall and is then clarified of remaining fines in a settling maze. A new coke fines filtering system has been developed by Foster Wheeler which removes coke fines from the decoking water by use of coke filled baskets. Coke is removed from the pad with a front-end loader. Pad dewatering offers a lower capital investment and simpler operation than pit dewatering. The drawbacks of pad dewatering are that the coke storage capacity is limited by plot area. A typical pad dewatering operation is illustrated in Figure 4.

Direct railcar loading allows the coke to drop directly from the coke drum into a railcar. Coke and majority of fines remain in the railcar. Water drains from the railcar to a sump and is

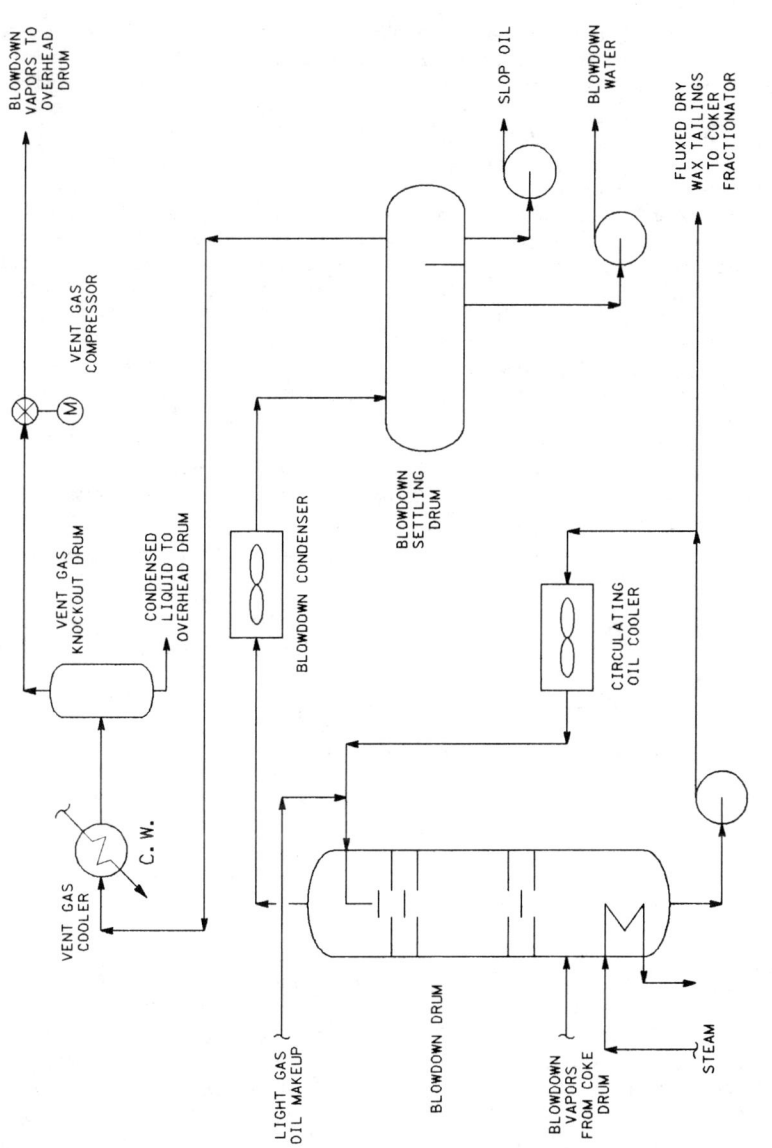

Figure 2. Process Flow Diagram for Typical Blowdown System.

then pumped to a clarifier. This system offers the lowest capital investment, but requires extra time to decoke the coke drum to allow for railcar movement. It is usually limited to units with small coke drums. A direct railcar loading system is shown in Figure 5.

A dewatering bin system is one in which the coke and water fall from the coke drum into a crusher and then, either by gravity or by slurry pump, are directed to a dewatering bin. In a gravity flow system, an innovation developed by Foster Wheeler, the coke drum and crusher are mounted on top of the dewatering bin. In a slurry system the crushed coke is pumped from a sump located directly below the coke drum and crusher to the dewatering bin. The dewatering bin is a large vertical drum where the coke and water are separated by gravity. The coke is allowed to settle and the water is drained from the drum and directed to a water tank for clarification. When dewatering is complete, the coke is discharged from the dewatering bin. Foster Wheeler has developed totally enclosed systems which are especially desirable in areas with strict environmental regulations. The main drawback of the dewatering bin is that it requires the largest capital investment of all dewatering facilities. Of the two systems, gravity and slurry, gravity flow is the more expensive. However, the slurry system requires a substantial amount of water circulated to transport the coke. The slurry flow and gravity flow systems are depicted in Figures 6 and 7, respectively.

Types of Petroleum Coke

The three main types of coke that are produced in a delayed coker are typically categorized as needle coke, sponge coke and shot coke. Needle coke is a premium grade coke, which is considered a specialty coke and is produced from specific aromatic feedstocks. The regular grades of coke are sponge coke and shot coke. A short description of each type of coke follows.

Needle Coke. Usually produced from highly aromatic thermal tar, pyrolysis tar or decanted oil stocks. This coke is typically characterized by a fibrous texture with long, unidirectional "needles" of coke. This form of coke is a premium product, which is sold to the carbon industry for use in the manufacture of large graphite electrodes.

Sponge Coke. This type of coke is considered a form of regular coke and is produced from high resin - asphaltene feeds. It contains small pores with no interconnections. Sponge coke can vary from light "honeycomb" varieties to heavy isotropic types. Sponge coke with low sulfur and ash contents is generally sold as anode coke to be used in the aluminum industry. High sulfur, high metals coke is frequently sold as low value fuel grade coke.

Shot Coke. Another form, generally undesirable, of regular coke is formed as small spheres often held together in a matrix of sponge coke or in large spheres of shot coke alone. The formation is highly dependent upon feedstocks such as Maya, West Texas Sour and

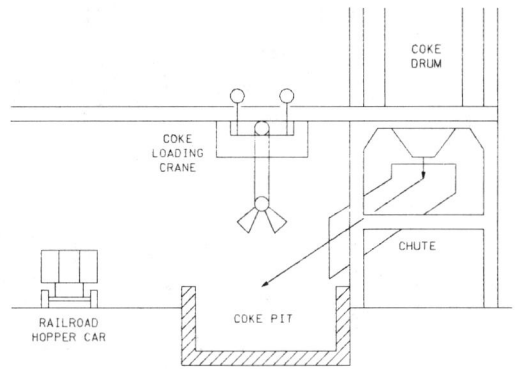

Figure 3. Pit Type Coke Handling System.

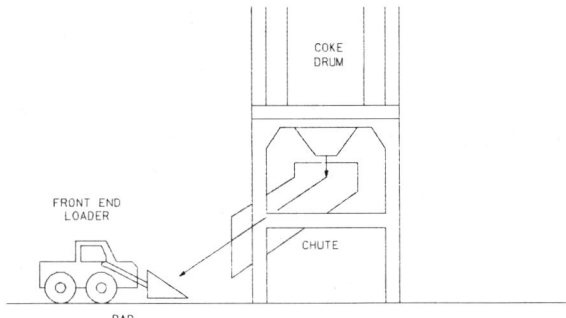

Figure 4. Pad Type Dewatering System.

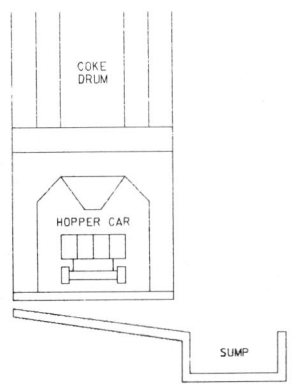

Figure 5. Direct Rail Car Loading.

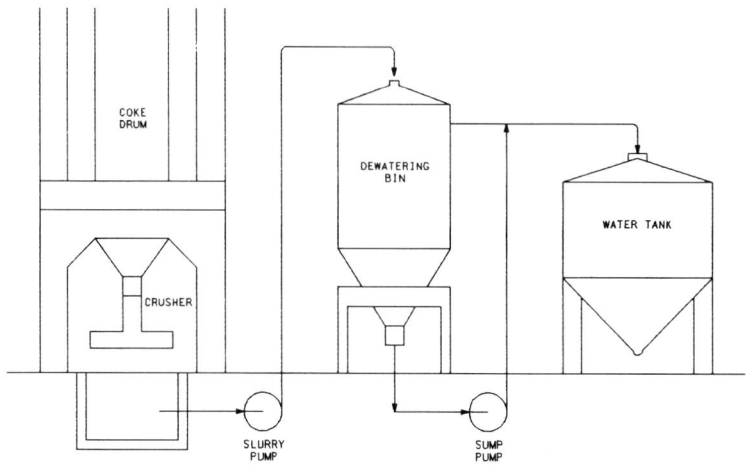

Figure 6. Slurry Dewatering Bin System.

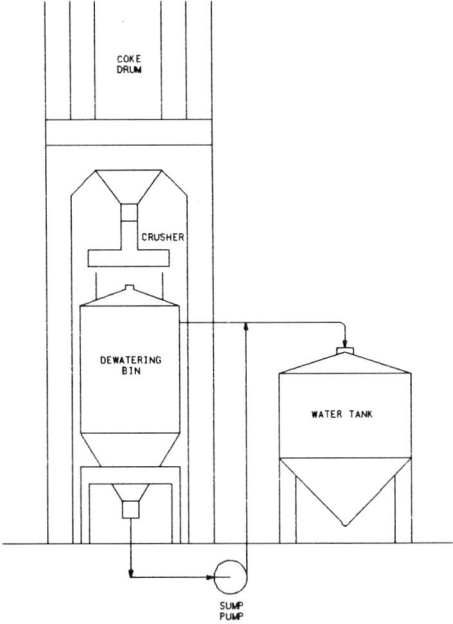

Figure 7. Gravity Flow Dewatering Bin System

some California residues, shale oil and gilsonite. Operating conditions such as temperature, pressure and recycle also affect shot coke formation.

Feedstocks

As crudes become heavier with higher levels of sulfur and metals, it becomes more difficult to produce acceptable marketable coke quality while maximizing desirable liquid product yield. This mandates that refiners and designers scrutinize physical properties, upstream processing and downstream requirements when selecting a feedstock.

The physical properties of a certain feedstock that determine the yields and product qualities include gravity, characterization factor, carbon residue, sulfur content and metals content. The last three properties are of specific importance.

Carbon Residue. The carbon residue is one factor used to determine coke yield as a percentage of fresh feed, and is defined as the carbon residue remaining after evaporation and pyrolysis of the feedstock in a specified procedure. All other operating conditions being the same, as the carbon residue is increased, more coke will be produced. In recent years, as the quality of crudes has diminished, the carbon residue of vacuum residue feedstocks has increased from typical values of 10 to 20 weight % to 20 to 30 weight % and more.

Sulfur Content. Another important feedstock physical property related to delayed coking is the sulfur content. The sulfur present in the feedstock tends to concentrate in the coke, where the sulfur level is usually equal to or higher than that of the feedstock. Sulfur levels as high as 4 weight % in today's feedstocks can cause unacceptably high levels of sulfur in the coke product. The resulting coke may not be acceptable for metallurgical use and may be a problem when burned as fuel.

Metals Content. When producing coke for electrode or anode use, feedstock metals content must be reviewed relative to coke product specifications. As in the case of sulfur, metals tend to concentrate in the coke.

The most common upstream processing methods for producing regular coke feedstocks are atmospheric and vacuum distillation. Another upstream feedstock preparation process is visbreaking. Other alternatives include charging heavy crude oil or asphalt from a solvent deasphalter. Charging whole crude will allow the coker fractionator to operate as both a crude unit, by distilling off the lighter portion of the crude, and a delayed coker by coking and cracking the heavier residual fraction. However, charging whole crude is generally limited to heavy crudes with minimal distillate. Examples of regular grade coke feedstock and product yields are given later.

When producing needle coke, the refiner must be more selective in determining if a feedstock is suitable. Needle coke has a highly crystalline structure which must be produced from an aromatic feedstock with low sulfur and metals content. Needle coke that meets stringent specifications commands a premium price for use in manufacturing graphite electrodes. In general, feedstocks which are to be used for needle coke production should be tested in a pilot plant to assure product quality.

Because of the increased sulfur and impurity levels in crudes currently being processed, refiners in recent years have been considering residue desulfurization units upstream of the delayed coker. In addition to the reduction in sulfur content, residue desulfurization units also lower the metals and carbon residue contents. Due to the reduction in the carbon residue, the liquid product yield is increased and the coke yield reduced. In addition, the coke produced from a desulfurized residue may be suitable for use as anode grade coke. Table I shows the yields and product properties after coking Medium Arabian vacuum residue, with and without upstream residue desulfurization.

Process Variables

Three operating control variables in a delayed coker dictate the product quality and yields for a given feedstock. These variables are the heater outlet temperature, coke drum pressure and the ratio of recycle to fresh feed.

At constant pressure and recycle ratio, the liquid product yield increases with an increase in temperature. This is often a desired effect, but there is only a narrow range over which the temperature can be adjusted. As the temperature is increased, the tendency of the heater and transfer line to coke increases, causing shorter run-lengths. Above a certain temperature, the coke formed can be excessively hard and difficult to remove from the drum with existing hydraulic cutting equipment. Operating the heater at too low an outlet temperature can cause difficulties as well. If the temperature is too low, then the volatile combustible material (VCM) of the coke will be excessively high or, possibly a soft tar or pitch could be produced. Modern units are designed so that an 8 to 12 weight % VCM coke is produced.

A decrease in pressure has the effect of vaporizing more heavy hydrocarbons. As the production of desirable liquid hydrocarbons is increased at low pressures, the coke yield is correspondingly decreased. Thus, most modern delayed cokers have been designed to operate at a low coke drum pressure.

The effect of recycle ratio on coke production is analogous to the effect of pressure. As the recycle ratio is decreased, the production of liquid products is increased. Reduction of recycle also lowers the fuel usage in the furnace because of lowered throughput. Recycle is often reduced to the minimum rate which still produces acceptable product qualities.

Regular Grade Coke Operation - Typical Yields and Product Qualities

Illustrated in Table I are the estimated yields and product qualities of several representative delayed coker feedstocks. The yields were established from generalized correlations using typical operating conditions for the operation noted.

TABLE I. Estimated Yields and Product Properties for Regular Grade Coke Production

Feed	Venezuelan	Venezuelan Visbreaker Tar	Medium Arabian	Desulfurized Medium Arabian	North African
TBP Cut Point, °F	950+	–	1,000+	1,000+	1,000+
Gravity, °API	2.6	1.5	7.0	17.0	15.2
Con. Carbon, Wt%	23.3	28.5	21.0	6.5	16.7
Sulfur, Wt%	4.4	4.0	4.8	0.5	0.7
Operation	------------------------------Maximum Liquid Yield------------------------------				Anode Coke
Products					
Dry Gas, C_4^-, Wt%	8.8	9.1	8.5	6.8	7.7
Naphtha, C_5-380°F, Wt%	14.0	13.6	14.0	13.1	19.9
Gravity, °API	55.0	55.5	58.9	58.6	62.1
Sulfur, Wt%	1.1	0.9	0.6	0.1	0.1
Gas Oil, 380°F+, Wt%	44.7	40.2	46.5	67.5	46.0
Gravity, °API	23.9	23.6	25.7	28.9	34.9
Sulfur, Wt%	2.7	2.0	2.9	0.3	0.5
Coke, Wt%	32.5	37.1	31.0	12.6	26.4
Sulfur, Wt%	5.7	5.2	6.5	1.2	1.2

Needle Coke Operation - Typical Yields and Product Qualities

Table II shows estimated yields and product qualities for three typical needle coker feedstocks. The feedstocks are considered desirable needle coker feedstocks because of their high density, low sulfur content and highly aromatic nature. Note the high production of coke, the result of high pressure and high recycle ratis, which is typical in needle coke production.

Table II. Estimated Yields and Product Properties
For Needle Coke Production

Feed	Thermal Tar	Pyrolysis Tar	Decanted Oil
Gravity, °API	2.4	-3.9	-0.7
Sulfur, Wt%	1.0	0.5	0.5
Products			
Dry Gas, C_4-, Wt%	14.4	10.3	9.8
C_5-380°F, Wt%	16.7	3.5	8.4
Gravity, °API	54.9	41.7	59.8
Sulfur, Wt%	0.04	0.09	0.01
Gas Oil, 380°F+, Wt%	15.7	31.2	41.6
Gravity, °API	23.3	11.5	16.9
Sulfur, Wt%	0.7	0.2	0.3
Coke, Wt%	53.2	55.0	40.2
Sulfur, Wt%	1.0	0.6	0.6

Uses of Petroleum Coke

Petroleum coke is essentially pure carbon and can be utilized wherever one would use a similar carbon product. It may be used as a fuel substitute for coal and can sometimes be used as a feedstock for applications such as partial oxidation. Depending on its properties, petroleum coke has four basic uses: fuel, feedstock for downstream processing, metallurgical applications, and for specialty graphite and carbon products.

As a fuel, the most common uses of petroleum coke are in firing cement kilns and steam generators. In the cement industry, petroleum coke is suitable as fuel in kilns because of its low ash content, high heating value and the process's high sulfur allowances. As much as 50% coke can be burned in combination with bituminous coal or 75% coke when burned in combination with oil and/or gas. The only limitation on coke for cement kiln firing may be its metals content. For steam generation, two options are available. The most common is the burning of petroleum coke in pulverized fuel boilers. This utilization often requires that downstream environmental processing of the flue gas be employed. Another method recently developed by Foster Wheeler for using high sulfur petroleum coke as fuel for steam generation is burning low quality coke in a sulfur capture fluidized bed boiler. The flue gas meets environmental standards for NOx and SOx. The only environmental consideration is the removal of ash from the flue gas in a baghouse and removal of the spent limestone. Fluidized bed boilers can be designed to burn petroleum coke along with the option of burning high sulfur coal or heavy fuel oil. (2)

A potentially attractive use for low quality, regular grade coke is to gasify it to produce ammonia synthesis gas, fuel gas, or hydrogen. Foster Wheeler has investigated promising schemes for air partial oxidation (APO), where the coke is partially combusted with air at elevated pressure to generate a gas consisting essentially of hydrogen, carbon oxides, hydrogen sulfide and nitrogen. After shift conversion, hydrogen sulfide and carbon dioxide are removed by scrubbing. Sulfur may be recovered either as sulfuric acid or elemental sulfur. Depending on the desired end product, nitrogen may be partially removed by cryogenic separation and further removed by pressure swing adsorption (PSA). Residual carbon monoxide is removed by methanation or by PSA.

One of the largest uses of petroleum coke is for anodes employed in the production of aluminum. This usage demands a somewhat premium feedstock to produce sponge coke that is low in metal and sulfur content in order to meet product quality specifications. After production in a delayed coker, anode quality coke must be calcined to remove VCM and moisture.

A specialized application of petroleum coke is the production of electrodes for the steel industry. For this application, it is necessary to use needle coke because its low coefficient of thermal expansion and low resistivity. The needle coke must have low sulfur and low metals content. After production in a delayed coker, needle coke is crushed and calcined in preparation for electrode production.

By 1980, special applications accounted for approximately 11% of the total coke production in the United States. These uses include titanium pigments, carbon raisers and synthetic graphite (3). A specialty use of green coke is as a high purity reactant in the production of calcium and silicon carbide. (4)

Coke Calcining

When petroleum coke is utilized for anode and electrode production and some specialty applications, it is necessary to calcine it to remove moisture and hydrocarbon VCM. Product qualities, along with production rate, are based on feedstock composition, kiln temperature profile, kiln residence time and cooling procedures. The two methods available for calcining coke commercially are the rotary kiln (5) shown in Figure 8 and the rotary hearth (6) shown in Figure 9.

In the rotary kiln process, coke is fed to a rotating cylindrical furnace sloped slightly toward the discharge end. Coke flows down the kiln countercurrent to the hot gas flow. Moisture is liberated from the coke in the feed zone, then the coke passes through the combustion zone where VCM is liberated. As coke leaves the kiln, it is discharged to a cooler where it is quenched with water and then cooled with ambient air. Recent designs have incorporated energy efficient features such as air preheat and steam

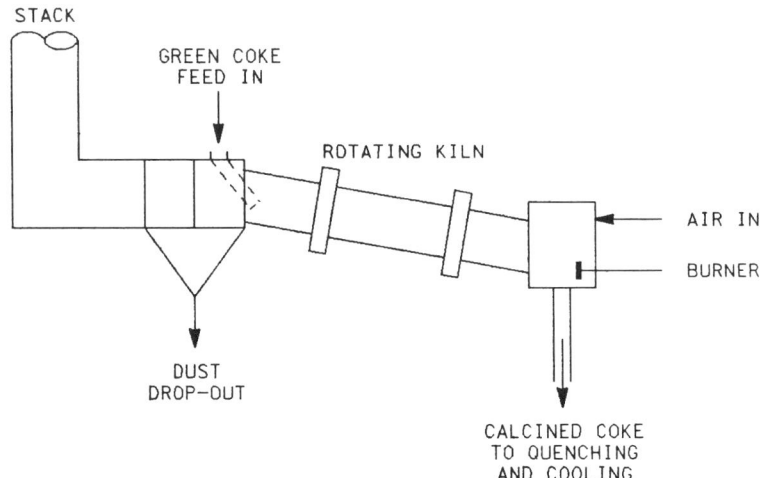

Figure 8. Typical Rotary Kiln Calciner.

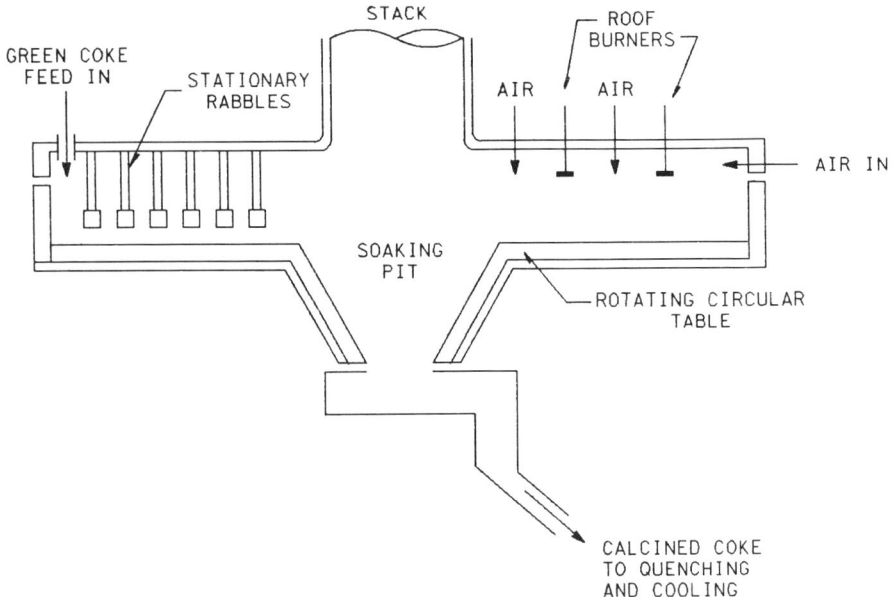

Figure 9. Typical Rotary Hearth Furnace.

generation facilities as well as designs in which the liberated hydrocarbons are used as a fuel in the combustion zone.

In the proprietary rotary hearth method of calcining coke, green coke is fed to the perimeter of the rotating circular table and gently moved toward the center of the hearth in a circular path by stationary rabbles. A combustion zone above the coke bed, formed by liberated volatiles, supplies the heat necessary for calcination. A rotating hearth furnace can also be equipped with energy efficient features, such as steam generation and air preheat.

Design Features and Considerations

Through the years, delayed coking has evolved from a "black art" to a high technology refining process. Both major and minor pieces of equipment have been examined and updated as new technology becomes available, thus assuring a safe, economical and reliable design. Today's designs must combine versatility and state-of-the art technology with low operating and capital investment costs.

To increase heater run lengths with today's heavier crudes while improving heater efficiency, it has become necessary to improve traditional coker heater design. Most of the recent improvements instituted by Foster Wheeler have resulted in a more conservative heater design, giving more flexibility to the refiner to later increase capacity, optimize operating conditions and increase run lengths. Fire boxes have been designed with more liberal dimensions and more space between tubes, which reduces the ratio of peak to average heat flux. Injection steam is used to regulate the oil velocity through the heater coil thereby preventing coking in the tubes. Heaters are designed with each parallel coil havings its own set of burners and independent flow and temperature control. These features permit run lengths that range between 9 to 12 months and longer in normal operation. (1)

Of the new developments that have been incorporated in coke drum design, the most notable is size. The first coke drums used were 10 feet in diameter. With today's improved metallurgy and increased hydraulic cutting capacity, Foster Wheeler has been able to employ coke drums up to 27 feet in diameter and 110 feet in length. (1) Another development by Foster Wheeler has been to determine the effect of shortened decoking cycles on coke drum life by computer analysis. Results indicate that by monitoring and controlling stress on a drum during rapid heating and cooling, coke drum life can be extended.

Decoking equipment and coke handling facilities have undergone significant changes since the advent of delayed coking. Decoking equipment has evolved from mechanical devices employed with the early small diameter drums to the use of high pressure water, delivered through flexible hoses with the ability to cut coke from 27 foot diameter drums. Coke is dewatered by one of the systems described previously. Steam, oil and water removed while cooling coke in the drum are no longer directed to a blowdown pond or settling pool, but are sent to an enclosed blowdown system where the oil, gas and water are separated and recycled.

As with most refinery units, the delayed coker has been updated to be as energy efficient as possible. Modern coker heaters now incorporate boiler feedwater preheat, steam generation or air preheat to raise the heater efficiency. Heaters are no longer designed for operation with 20% excess air but for as little as 5-10% excess air. In addition to the improvements in heater design, cokers have had feed preheat, steam generation, and heat integration with other units added where possible.

Revamps and Retrofits

Refiners with existing delayed cokers often have the option to expand their units at a lower incremental cost than adding new units. For most revamps the main concern is whether the fractionation section has sufficient capacity. Methods of increasing fractionator capacity while lowering operating pressure include using packed beds in the fractionator and full port valves in the vapor lines. Adding upper pumparounds to reduce condenser overhead duty and pressure drop has also proved helpful. The least expensive option for increasing fractionator capacity is to increase the operating pressure. This is accomplished at the expense of lower liquid yield.

Once it has been determined that the capacity of the fractionator section is adequate for the new load, the capacity of the coke drums may be increased. The easiest way to do this is to shorten the coke drum cycle. Refiners have been able to shorten cycles to less than 16 hours with minimum capital investment. If it is not possible to substantially decrease the length of the coking cycle, another alternative successfully employed by Foster Wheeler is to add another coke drum. This will allow cycles for each drum to be shortened to 12 hours or less and still allow adequate decoking time. These methods for increasing the coke capacity are also applicable when processing heavier feedstocks that produce more coke. In this case, the fractionation section will often handle the distillate production without any modification.

Another option open to refiners is to retrofit an existing heavy oil unit such as a visbreaker into delayed coking service. Other than the coking section, the visbreaker is very similar to a delayed coker. If plot area permits, a coking section may be added and the heater modified.

Summary

Since its inception, the basic process of delayed coking has remained basically unchanged, but the feedstocks, process equipment and operating philosophy have changed substantially. Delayed coking has evolved to process a wide range of today's heavy, high sulfur feedstocks, while still producing acceptable product yields and qualities. Processing equipment has been updated to provide more on-stream time and a more energy efficient operation. In years to come, delayed coking is expected to continue to remain an important residual upgrading process.

Literature Cited

1. DeBiase, R., and Elliott, J. D., <u>Oil and Gas Journal</u>, <u>16</u>, 81 (1982).
2. Nagy, R.L., Broeker, R. G., and Gamble, R. L., "Firing Delayed Coke in a Fluidized Bed Steam Boiler", paper presented at the 1983 NPRA Annual Meeting, San Francisco, CA, March 20-22 (1983).
3. Fasullo, P.A., Matson, J., and Tarrillion, T., <u>Oil and Gas Journal</u>, <u>44</u>, 76 (1982).
4. Guthrie, V.B., Ed., <u>Petroleum Products Handbook</u>, McGraw-Hill Book Company, New York, 1960; Chapter 14.
5. Kennedy Van Saun Corporation's Technical Brochure No. COK 1/82(2).
6. Allred, V.D., "Rotary Hearth Calcining of Petroleum Coke", paper presented at the 100th National Meeting of the American Institute of Metallurgical Engineers, New York, NY, March 1-4 (1971).

RECEIVED February 23, 1985

12

Petroleum-Coke Calcining Technology

H. H. Brandt

Cal Carb, Inc., 27 River Ridge Road, Lake Charles, LA 70605

> The oldest and newest petroleum coke calciners employ rotary kilns to thermally upgrade green coke, thus rendering it suitable for use by the amorphous carbon and graphite industries. In response to environmental pressures and rising fuel costs, modern coke calcining facilities are much cleaner and much more thermally efficient. Despite numerous attractive features of the newer rotary hearth calcining technology, most practitioners still favor the well-proven rotary kiln concept.

Basic Calcining Process

Petroleum coke calcining is a process whereby green or raw petroleum coke is thermally upgraded to remove associated moisture and volatile combustible matter (VCM) and to otherwise improve critical physical properties, e.g., electrical conductivity and real density (1). The calcining process is essentially a time-temperature function; the most important variables to control are heating rate, VCM to air ratio and final temperature. To attain the calcined coke properties necessary for its end use by the amorphous carbon or graphite industries, the coke must be heat treated to temperatures of 1200-1350°C (2200-2500°F), or higher, to refine its crystalline structure.

The eventual quality of the calcined coke is directly related to the particular characteristics/quality of the green coke fed to the calciner. While calcination cannot improve upon certain quality limits inherent in the green coke, potential quality can be lost by improper calcining, e.g., by using incorrect heating rates and/or atmospheric conditions.

The oldest and newest coke calciners in North America (which accounts for almost 75% of the free world capacity) are rotary kilns similar to those employed by the cement industry. The original versions were quite unsophisticated, consisting merely of long, inclined, refractory lined, steel cylinders which were attached by a breeching to a stack at the upper end which provided the necessary draft and to a gas or oil fed burner at the lower end which provided the heat (Figure 1). In this countercurrent system, the green coke is fed into the kiln at the elevated end and the process heat enters at the lower end. As the green coke moves downhill due to the rotation of the kiln, its temperature increases. During the journey, moisture evolves in the heat-up zone (up to about 400°C), devolatilization occurs in a second zone between about 400°C and 800-1000°C, and densification takes place in the final zone (up to about 1350°C). The total residence time, which is controlled by the rotational speed of kiln, can be anywhere between 45 and 90 minutes (or even longer in some special cases).

When the hot (1200-1350°C) calcined coke leaves the kiln, it is transferred to a rotary cooler. This consists of a steel cylinder, usually lined with refractory at the feed end, that is slightly inclined to induce downward travel as it is rotated. In the cooler the hot coke is quenched by water sprayed from a number of nozzles; exit temperature is controlled at about 150°C (300°F) to assure a moisture-free product.

This basic process has a very poor energy efficiency, primarily because of the very high temperature of the exhaust gases (including uncombusted volatiles) leaving the system. The stack gas temperature can exceed 1260°C (2300°F). In the days of unlimited supplies of cheap fuel, this thermal inefficiency was largely ignored. Also ignored was the fact that the typical calciner stack emitted several tons of coke particulates per hour, as plants were located in areas where such emissions went unnoticed or were tolerated.

Modern Rotary Kiln Calciners

Changing times have dictated major modifications of the basic calcining process. Skyrocketing energy costs and environmental considerations are undoubtedly the two most important factors providing the impetus for the evolution of petroleum coke calcining technology.

Except for the rotary kiln itself, which now represents only a fraction of the total capital cost, today's modern coke calcining plants bear little resemblance to the original facilities (Figure 2). Even the kiln itself has undergone major changes, not the least of which is increased capacity. A typical modern kiln, measuring approximately 200 ft (length) x 10.5 ft (inside shell diameter), is able to produce in excess of 250,000 short tons annually compared to about 50,000 tons per year for its predecessors.

To improve energy efficiency, refractories with superior K factors are used in lining the kiln, thus reducing radiant heat losses. Moreover, kiln mounted blowers now inject combustion air into the kilns in the zone where the volatiles evolve from the coke, thus permitting utilization of the Btu content in these previously wasted gases. In modern calciners, most of the energy required is obtained by burning the coke volatiles and fine particulate matter in the kiln. In some instances, rotary kilns equipped with kiln mounted blowers actually operate without external fuel (except for start-up). When these units are also equipped with incinerators (to combust the unburned volatiles and emitted coke fines) and waste

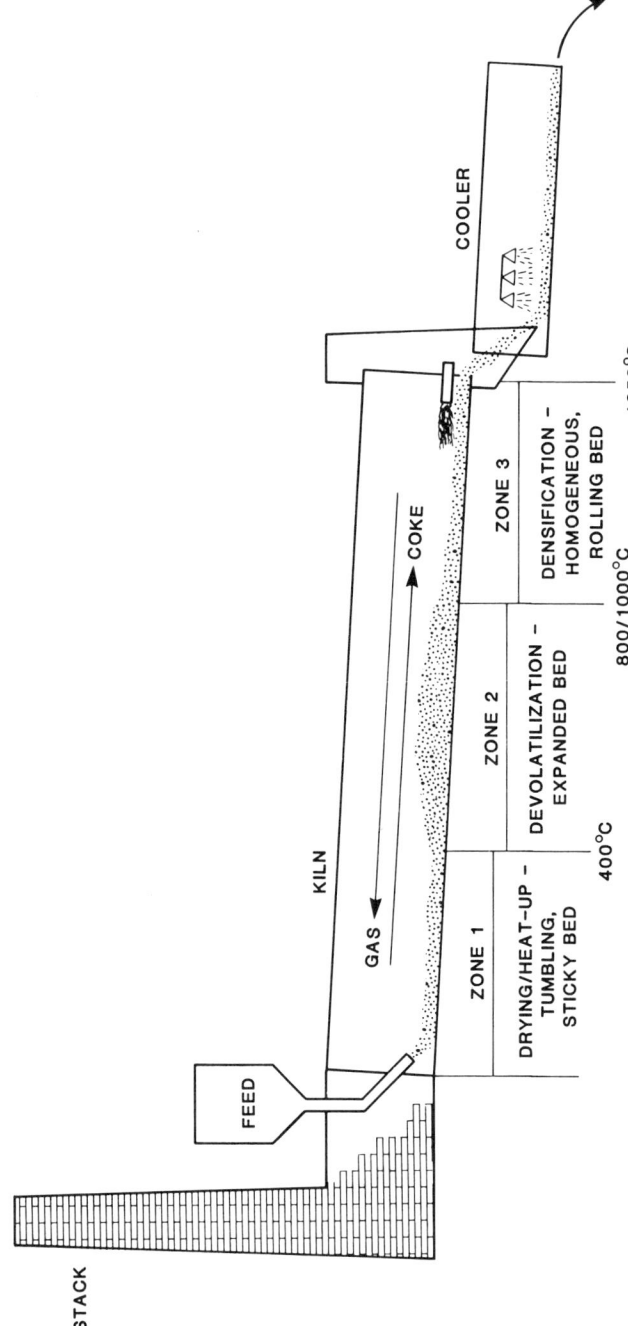

Figure 1. Diagram of original rotary kiln calciner.

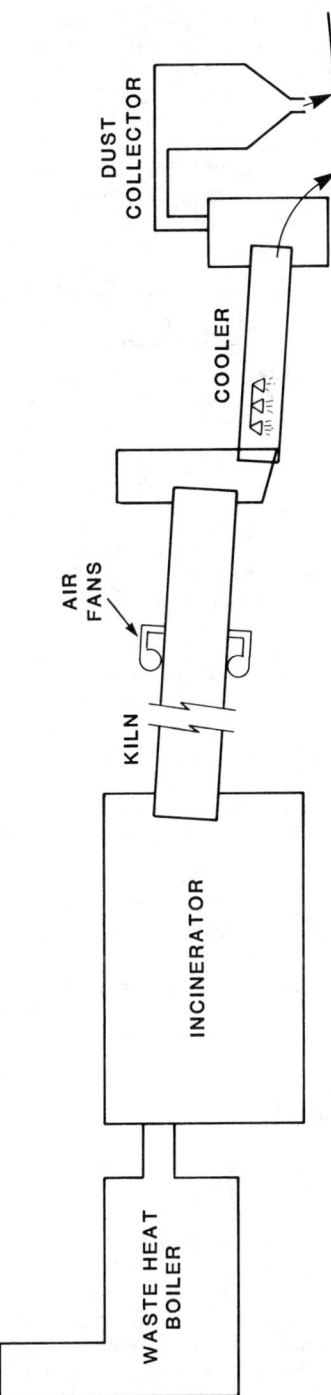

Figure 2. Diagram of modern rotary kiln calciner.

heat boilers, they actually become net energy producers. Some of today's calcining plants even generate power and sell it to utilities.

Other Calciners

Other less common methods of calcining petroleum coke are also or have been practiced. In electric calcining, practiced where an abundant supply of cheap electrical power is available, the required temperature is achieved by electrical arcing and resistance heating; such calciners employ either batch or continuous vertical shaft kilns. Indirect-fired vertical shaft kilns, by-product coking ovens, as well as bee hive coke ovens have also been used. Shaped carbon baking furnaces, when they employ green coke as packing material, can also be considered to be calciners. None of these systems, however, have consistently accounted for significant calcining tonnage and therefore are mentioned only briefly.

Rotary Hearth Calciner

In 1967, a new concept in petroleum coke calcining reached the commercial stage when a rotary hearth calciner was placed on-stream in Burghausen, West Germany. This patented process (2,3), the heart of which is a single, rotating horizontal hearth, was jointly developed by the Marathon Oil Company and the Wise Coal and Coke Company. The green coke is fed in at an outermost position and travels around the hearth in a series of concentric circles, coke movement being achieved by rotation of the hearth (Figure 3). Rabbles, or blades, suspended from the roof force the coke inward and simultaneously turn the bed, thereby uniformly exposing the coke being calcined to the burners located in the roof. The burners use fuel only for start-up; once equilibrium conditions are achieved, the volatiles evolving from the coke are provided with combustion air via the burner ports and the calcining process is self-sustaining.

In the subject process, a soaking pit located at the center of the rotating hearth holds the coke "at temperature" for a period of time necessary to achieve the desired product properties (Figure 4). The soaking pit also acts as a surge bin and a gas seal to prevent air leakage into the hearth section. From the soaking pit, the hot coke is discharged into a cooler for either direct or indirect water quench. The entire process system is tied into a waste heat boiler to further improve its thermal efficiency. Among the advantages claimed for this technology are:

- Particulate-free exhaust gas
- Excellent thermal efficiency
- Simplicity of operation
- Low refractory maintenance
- Improved carbon recovery

The last claim has certainly been verified, but it has some negative aspects. The higher carbon yields are realized because the green coke fines introduced to the calciner exit as calcined product.

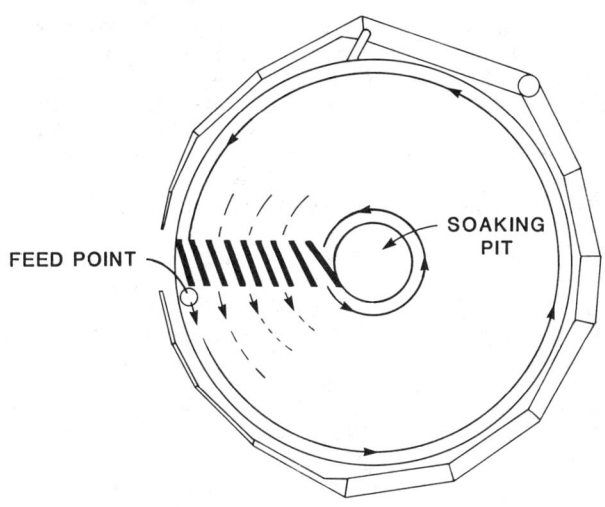

Figure 3. Top view of rotary hearth calciner.

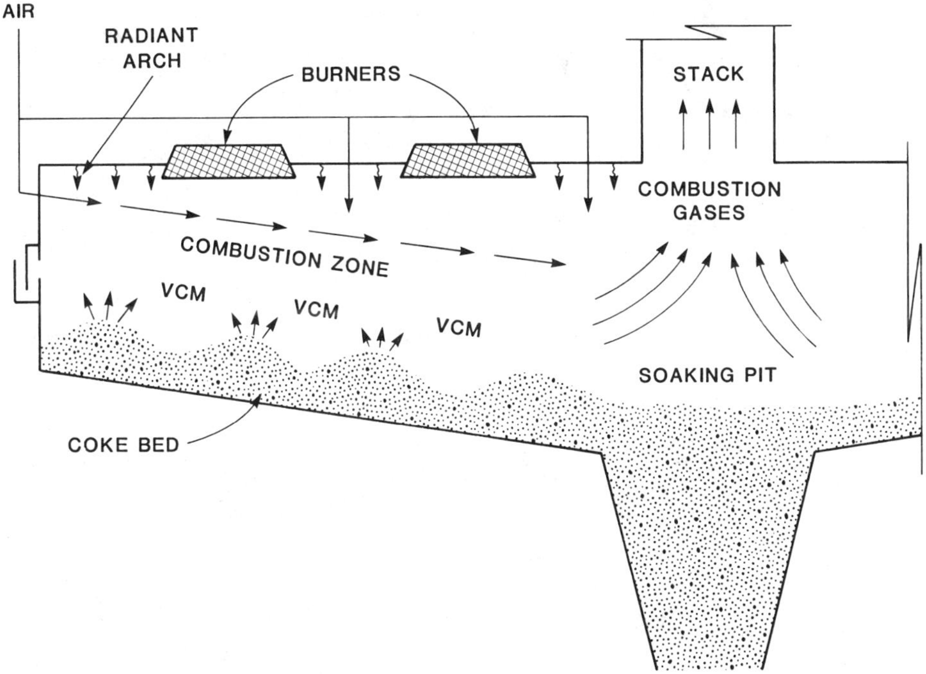

Figure 4. Cross section of rotary hearth calciner.

Since most calcined coke customers have rather stringent limits on fines content, this "advantage" becomes a distinct disadvantage. The solution to this problem is to feed fewer fines to the unit; this can be achieved by very careful coke drum cutting and green coke handling. Alternatively, the green coke or the calcined coke can be screened to remove the fines either prior to or after calcining; either of these solutions, however, would require careful scrutiny of associated costs and alternative outlets for the separated fines.

The other advantages claimed are not unanimously accepted by the industry. It is significant to note that since 1967, a dozen or more new conventional kilns for coke calcining have been placed on-stream in North America and only two rotary hearths. The overseas record is slightly more favorable to the newcomer. Moreover, the only companies adopting this new technology are new entrants to the coke calcining industry; none of the existing calciners which employ conventional kilns have converted to the new technology. Is this because the coke calcining industry is overly conservative or has the evolution of the conventional system met the challenge of this newcomer? It will probably be a long time before this question is answered.

Literature Cited

1. Martin, S. W. "Petroleum Coke" in "Petroleum Products Handbook"; Guthrie, V. B., Ed.; McGraw-Hill Book Co.: New York, 1960.
2. Allred, V. Dean "Rotary Hearth Calcining of Petroleum Coke"; presented at the meeting of the American Institute of Mining and Metallurgical Engineers (AIME), Paper No. A71-26.
3. Reis, T. "About Coke -- And Where the Sulfur Went," Chemtech, June 1977, pp. 366-373.

RECEIVED March 21, 1985

New Calcining Technology of Petroleum Coke

M. Kakuta[1], H. Yamasaki[1], H. Tanaka[1], J. Sato[2], and K. Noguchi[2]

[1] Osaka Research Laboratory, Koa Oil Company, 2-1, Takasago, Takaishi-shi, Osaka 592, Japan
[2] Koa Oil Company, 6-2, Ohte-Machi 2-chome, Chiyoda-ku, Tokyo 100, Japan

>   This technology concerns a new calcining method for
>   reducing the thermal expansion coefficient of the coke
>   at the calcining stage. The low thermal expansion
>   coefficient is an important factor in determining the
>   quality of calcined coke for the production of graphite
>   electrodes. Whereas the traditional calcining method
>   adopts a one stage process, the new calcining method
>   adopts a two stage process. The experimental results
>   indicate that the development of unique microcracks
>   appears in the coke after the new calcination, regard-
>   less of the coke type, and these microcracks contribute
>   to the effective reduction in the thermal expansion
>   coefficient of the coke and the improvement of the
>   puffing characteristics. The optimum process system
>   of this new calcining technology has been studied by
>   using a model calciner pilot plant.

High grade graphite electrodes are required to accomodate the adoption of high-power (HP) and ultra-high power (UHP) operations in the electro-arc steel industry. Calcined coke for artificial graphite electrodes require a high quality standard in terms of the various properties, especially a low thermal expansion coefficient. The process of calcining was normally considered as a heat treatment step at a temperature ranging from 1,300°C to 1,400°C to ensure that the calcined coke had the properties appropriate for production of graphite electrodes. Furthermore, many believe (1-5) that the properties of calcined coke are heavily dependent upon the manufacturing conditions of the green coke. Consequently, various manufacturing methods developed to gain a low thermal expansion coefficient centered on producing a needle coke at the coker. Thus, new technologies for the calcining process were primarily aimed at improving the efficiency and economy of the operation. Our attention was focused on the calcining process and upon investigating the structural changes of green coke as they related to the calcining conditions and characteristics of calcined coke (6,7). Then a new calcining method was developed which reduced the thermal expansion coefficient of calcined coke,

0097-6156/86/0303-0179$06.00/0
© 1986 American Chemical Society

irrespective of the types of green coke to be processed, by appropriately controlling the heating pattern during calcination (1,8). The new calcining process has been named the two-stage calcining process based on the peculiar method adopted. Experimental results indicate that unique microcracks appear in the coke after the new calcination and that these microcracks contribute to the effective reduction in the thermal expansion coefficient of the coke and to the improvement of the puffing characteristics. This paper includes an outline of this new calcining method, results of observations related to the major causes of a reduction of the thermal expansion coefficient and the study of the optimum process system of this new calcining technology.

Outline of New Calcining Technology

To evaluate the effects of the new calcining method, calcining tests were conducted in two different ways, namely, the new and traditional methods, by utilizing the rotary kiln-type electric furnace. As shown in Figure 1, the traditional calcining process involved a one stage process in the calciner with a peak temperature of 1,300°C to 1,400°C and, after maintaining this temperature for a while, cooling to the room temperature. The new calcining process is a two-stage processing in which green coke is calcined initially at a temperature from 600°C to 900°C, cooled, and re-calcined at a temperature of about 1,300°C to 1,400°C - the same level as the traditional method.

Comparison of Calcining Process: New vs. Traditional Methods

Materials

To identify the differences arising from the new method, six different types of green coke (samples A, B, C, D, E and F) were tested. Properties of these green cokes are given in Table I. From these properties and also by microscopic observation results, Coke A was defined as low sulfur standard needle petroleum coke, and Coke B as petroleum premium needle coke mainly of fibrous texture, Coke C of high sulfur content, Coke D of high ash content, and coke E as coke derived from coal. Cokes C, D and E had mixed fibrous and mosaic textures. Coke F had higher density, an extremely low ratio of impurities, and a good orientation of the crystallite. Coke F is now produced by the laboratory coking apparatus and has been given the name of "Supreme Coke" (9-11).

Table I. Properties of Green Cokes

Sample Name		Coke A	Coke B	Coke C	Coke D	Coke E	Coke F
Volatile Matter(wt%)		8.2	7.6	10.6	16.1	7.8	4.0
Elemental Analysis							
C	(wt%)	93.5	93.8	89.0	92.0	94.6	95.2
H	(wt%)	3.4	3.3	3.2	4.2	2.8	3.3
N	(wt%)	1.6	1.1	1.1	1.3	0.8	0.2
S	(wt%)	0.48	0.73	5.7	0.36	0.29	0.14
Real Density(g/cm^3)		1.39	1.39	1.38	1.35	1.41	1.42
Ash	(wt%)	0.08	0.05	0.08	1.22	0.11	0.00

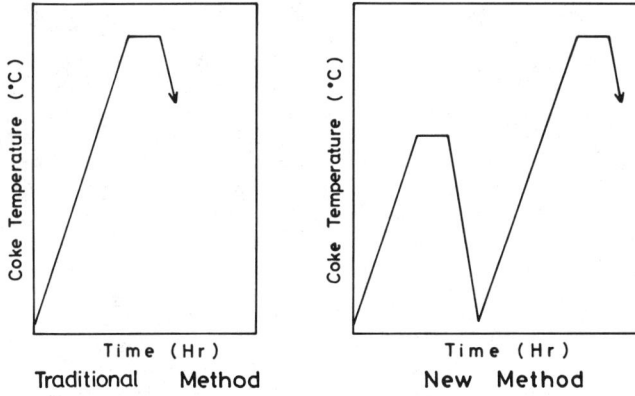

Figure 1. Simplified calcination profile.
Reproduced with permission from reference 8b.
Copyright 1981 Pergamon Press, Inc.

Experimental Techniques. The volatile matter and ash of the green cokes were measured by JIS (Japanese Industrial Standard) M 8812, and the sulfur was obtained by JIS M 8813. The properties of the calcined coke samples were determined by measuring the thermal expansion coefficient, apparent density (test sample of about 30 grams '3.5-4 Tyler mesh' was measured by pycnometer method by displacing water), and real density (about 10 grams '200 Tyler mesh under' of test sample was measured by pycnometer method by displacing n-butyl alcohol). Porosity was calculated from values of the apparent and real densities. Texture and structure of coke samples were observed by polarized light microscopy and by scanning electron microscopy.

Green carbon bodies were prepared by molding (rod; 2x2x8 cm^3) and by extrusion (rod; 2 cm dia. and 11 cm length) from a mixture of appropriately sized calcined coke and coal tar binder pitch. The rods were then baked at a slow rate to 1,000°C in an electric furnace and kept at this temperature for an hour. The extruded rods were further graphitized at 2,800°C for 0.5 hr. under an argon flow.

The thermal expansion coefficient was measured in two directions - perpendicular to the molding direction and parallel to the extruding direction, and thus parallel to the coke particle alignment. Test pieces were obtained on cylinders of 5 mm dia. and 50 mm length cut from rod after heating.

Bending strength and Young's modulus of the graphitized rods were determined by JIS R 7202.

The puffing characteristic was measured in the direction parallel to the molding direction, and thus perpendicular to the coke particle alignment. Baked test pieces (1 inch dia. and 1 inch length) were used for the puffing test in the heat treatment ranging from 100°C to 2,800°C.

Properties of Calcined Coke. Table II shows the properties of calcined cokes A, B, C, D and F by the new and traditional methods. Regardless of the green coke type, the thermal expansion coefficient of calcined coke by the new method was lower than that of traditional calcination method. Porosity values for calcined cokes by the new method have a higher value, regardless of the coke type, than

Table II. Properties of Calcined Cokes

Sample Name	Coke A		Coke B		Coke C	
Calcining Method	New	T	New	T	New	T
CTE (x10^{-6}/°C)	1.5	1.9	1.2	1.6	2.1	2.5
Real Density (g/cm^3)	2.094	2.088	2.095	2.092	2.051	2.040
Porosity (%)	40.3	35.8	37.0	33.1	39.1	36.3

Sample Name	Coke D		Coke E	
Calcining Method	New	T	New	T
CTE (x10^{-6}/°C)	2.0	2.4	0.8	1.0
Real Density (g/cm^3)	2.092	2.082	2.146	2.142
Porosity (%)	43.6	41.4	33.8	31.8

T: Traditional
CTE: Thermal expansion coefficient (30°C-100°C).
  Test pieces: Molded pieces (1,000°CHT, 5 mm dia. x 50 mm length).

those processed under the traditional method. (This is due to the development of unique microcracks within coke processed under the new method, as following mention.)

The properties of extruded graphite rods obtained from calcined cokes A, B and F by two different calcining methods are shown in Table III. The thermal expansion coefficient of graphitized pieces by the new method are lower. Test results on strengths, showed no significant difference between the calcined coke experimentally manufactured by the two different methods. Thus the new method showed no ill effects on the mechanical strengths of electrodes. Puffing characteristics during graphitization were improved by the new calcining method (Table III). Table IV shows the properties of large diameter (20 inches) actual electrodes obtained from the cokes by two different calcining methods. This indicates the effect of the new method, even after the manufacture of actual graphite electrodes.

Table III. Properties of Extruded Graphite Rods

Sample Name		Coke A		Coke B		Coke F	
Calcining Method		New	Traditional	New	Traditional	New	Traditional
CTE	$(\times 10^{-6}/°C)$	1.0	1.2	0.8	1.0	0.6	0.7
Bulk Density	(g/cm^3)	1.52	1.53	1.52	1.52	1.54	1.56
Bending Strength	(kg/cm^2)	115	110	115	105	90	90
Young's Modulus	(kg/mm^2)	790	740	810	740	740	710
Dynamic Puffing $(\%\Delta L)$	Fe_2O_3 0%	0.45	0.81	0.49	1.13	--	--
	Fe_2O_3 1%	0.22	0.50	Shrink	0.16	--	--

CTE: Thermal expansion coefficient (30°C - 300°C). Test pieces; Extruded pieces (2,800°CHT 5 mm dia. x 50 mm length).

Table IV. Properties of Actual Graphite Electrode (20 inches dia.)

Sample Name		Coke B	
Calcining Method		New	Traditional
CTE	$(\times 10^{-6}/°C)$	0.2	0.4
Bulk Density	(g/cm^3)	1.64	1.65
Bending Strength	(kg/cm^2)	140	145
Young's Modulus	(kg/mm^2)	1120	1020

CTE: Thermal expansion coefficient (30°C-100°C). Test pieces; 5 mm dia. x 50 mm length.

Theory on the New Calcining Technology

The mechanisms of the effect of the new calcining technology have been clearly elucidated through the following experimental study.

Materials
As starting materials, four different types of green coke (A, B, C and D) were used for the testing. Properties of these test samples are shown in Table I.

Experimental Techniques.
Distribution of pore size was

determined by the mercury porosimeter. Interlayer spacings and apparent crystalline sizes were obtained by X-ray diffraction method (6). Test pieces were prepared from green cokes A and B by cutting them into specimens of 5 mm x 5 mm x 50 mm. Dimensional changes during calcination were obtained by a differential dilatometer using quartz as the standard material. The heating rates were 10°C/min up to designated temperature under an argon gas atmosphere.

Theoretical Explanation. The major factors of the thermal expansion coefficient of calcined coke are the degree of preferred orientation of the crystallites and void structure (12-14). For example, the thermal expansion coefficient is low for needle coke because it is strongly affected by the preferred orientation of its crystallites. We found the latter factor-voids to be important. Experimental results showed that when green coke was calcined under the new methods, and the derived calcined coke was observed by scanning electron microscopy (Figure 2) and its pore size distribution was measured by mercury porosimetry (Figure 3), microcracks of significant sizes (1 to 60 microns) were developed. This was an important contribution to the reduction of the thermal expansion coefficients of the calcined coke processed under the new method.

Dimensional changes of green coke during calcination were measured between each heat treatment, utilizing heating patterns of the new and traditional calcining processes. An example of such measurements on Coke A and B is given in Figures 4 and 5. While coke material continued to expand until it reached about 600°C, quick contractions started at temperatures above 600°C (15). Specifically, under the new calcining method a considerable contraction was observed during the peak temperature period of the primary processing, followed by some contraction during the cooling stage of the primary processing. During the secondary processing, a slight expansion was observed during heating, followed by a contraction up to the final calcining temperature. On the other hand, dimensional changes observed on coke processed under the traditional calcining method indicated contraction from 600°C to the final calcining temperature, showing that the process of dimensional changes differs between the new and traditional calcining processes. These facts indicate that development and increase of microcracks under the new calcining method was derived from special distribution of stress within the coke caused by the expansion and contraction during this heat treatment pattern.

Table V shows the X-ray parameters obtained from the cokes by two different calcining methods. X-ray diffractometry of interlayer spacing (d_{002}) and apparent crystalline size (L_c) indicated no particular difference between cokes processed under the two methods. Furthermore, no specific differences were found on these samples after graphitization. These facts show that the new calcining method would not lead to adverse effects on the development and rearrangement of coke crystallines.

From the study of the new calcining method, the following facts were revealed (16):
1. During the primary processing, the thermal expansion coefficient of calcined coke changes, and that treatment at the range of 700°C to 900°C is effective.

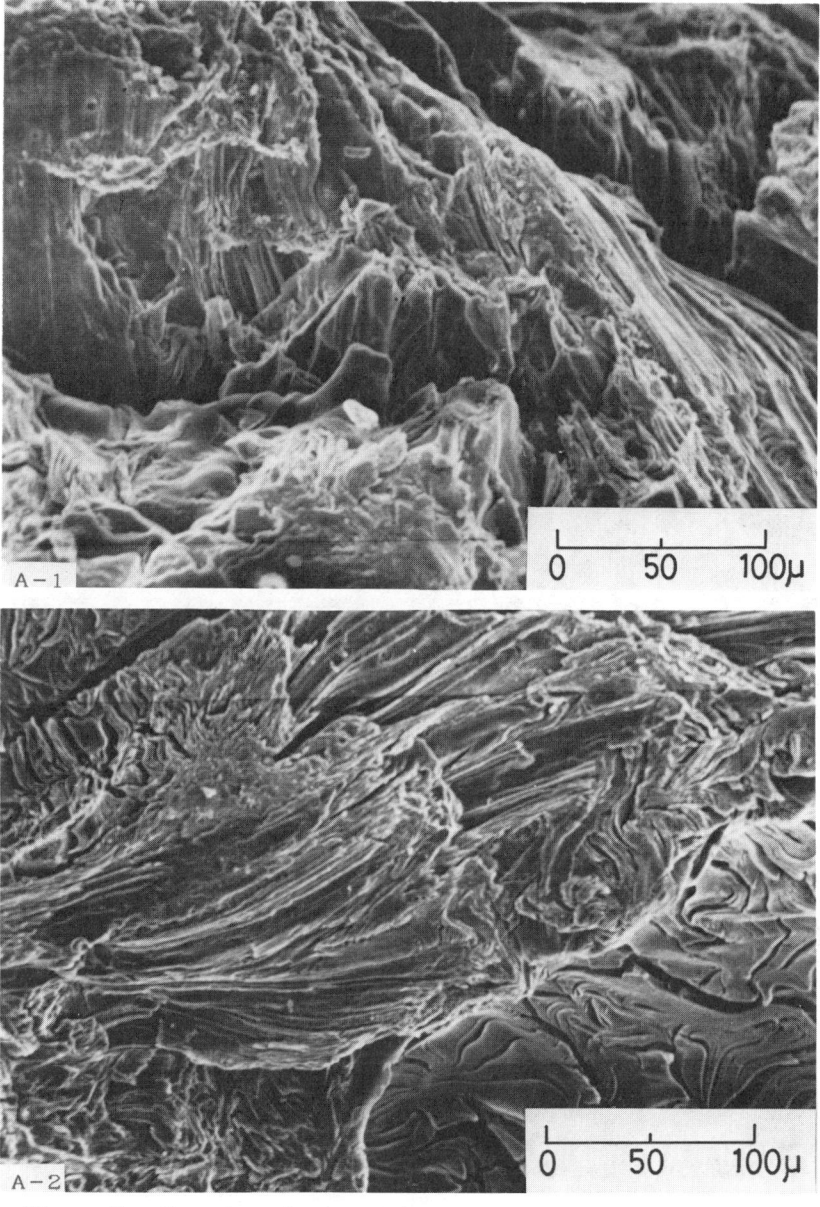

Figure 2. Scanning electron micrographs of the calcined cokes.
A-1; Coke A (traditional method).
A-2; Coke A (new method).

Reproduced with permission from reference 8b. Copyright 1981 Pergamon Press, Inc.

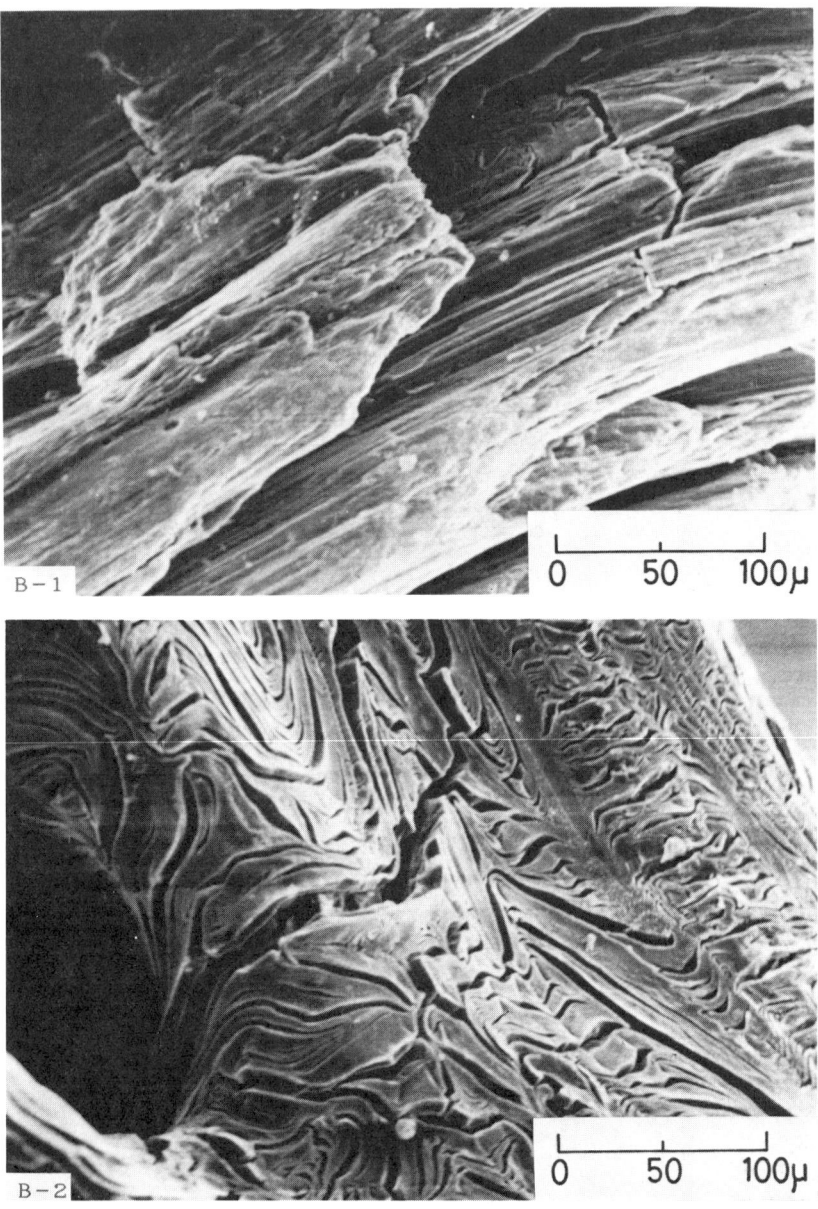

Figure 2. Scanning electron micrographs of the calcined cokes.
B-1; Coke B (traditional method).
B-2; Coke B (new method).
Reproduced with permission from reference 8b.
Copyright 1981 Pergamon Press, Inc.

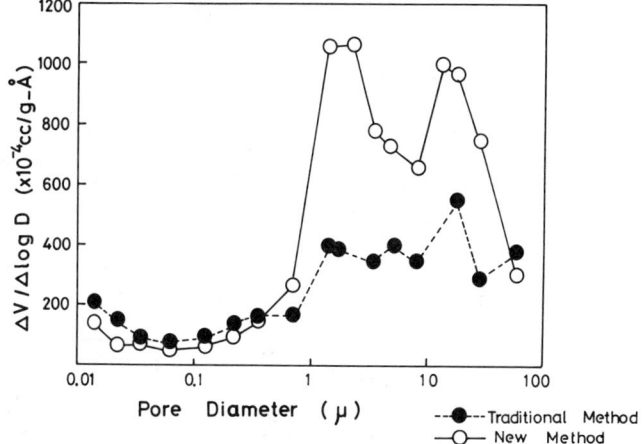

Figure 3. Pore size distribution of mercury penetrable pore volume in calcined Coke A.

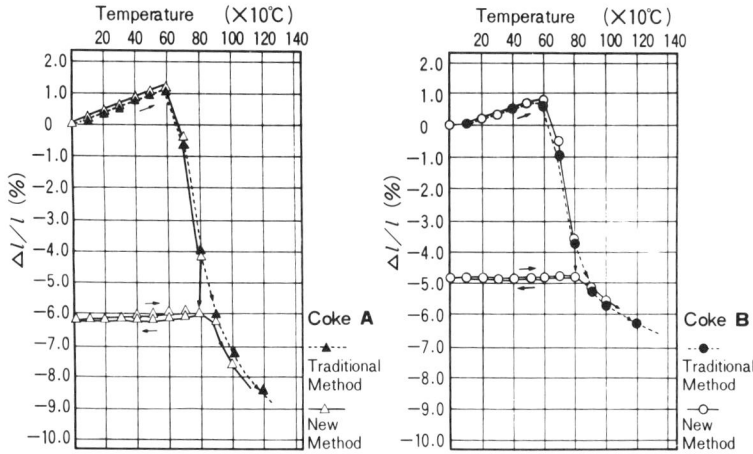

Figure 4. Dimensional changes of coke during calcination.
Reproduced with permission from reference 8b. Copyright 1981 Pergamon Press, Inc.

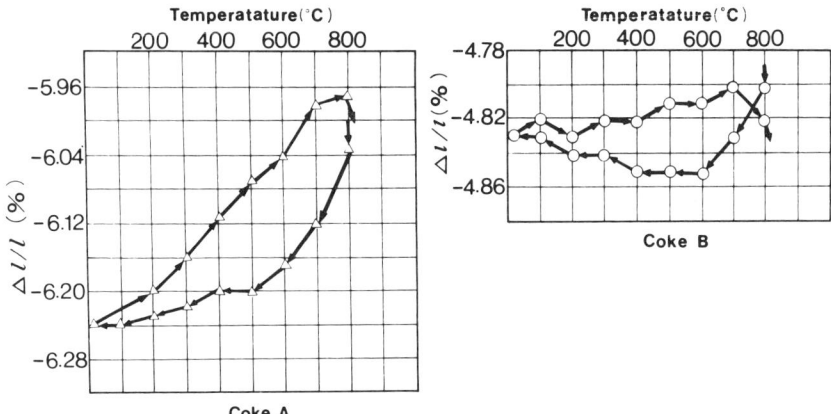

Figure 5. Dimensional changes of coke during cooling process of primary step and heating process of secondary step under new calcining process.
Reproduced with permission from reference 8b. Copyright 1981 Pergamon Press, Inc.

Table V. Comparison of X-Ray Parameters Obtained from the Coke Manufactured by the Two Different Methods

Sample Name	Coke A		Coke B		Coke C		Coke D	
Calcining Method	New	Traditional	New	Traditional	New	Traditional	New	Traditional
Calcined Coke								
d_{002} (Å)	3.444	3.440	3.447	3.446	3.447	3.446	3.442	3.440
Lc (Å)	38	40	38	40	42	41	38	40
Graphitized Coke								
d_{002} (Å)	3.366	3.366	3.365	3.364	3.365	3.365	3.366	3.366
Lc (Å)	840	880	1000	1000	460	460	810	740

Reproduced with permission from reference 8b.

2. After the completion of the primary processing, the coke must be cooled to room temperature. No significant effects result from continued calcining process from the primary to secondary processing without this cooling. Thermal expansion is not affected by a change of cooling rate in the new calcining process.

3. Thermal expansion coefficient of calcining coke will remain almost unchanged even if the primary processing is repeated. Although the effects of the atmosphere and heating rate during calcination on the density of calcined coke can be recognized, the effects on the thermal expansion coefficient were minimal (1).

Study on the Optimum Process System

A commercial scale two-stage continuous calcining process has been conceptually designed and a model calciner pilot plant has been installed. This pilot plant consists of first stage and second stage rotary kilns. Figure 6 shows the pilot plant (300 mm inside dia. x 3500 mm length, and coke feed rate of 20 kg/H). This model calciner is designed to establish the optimum calcining conditions and equipment configuration for economic continuous production of two-stage calcined coke and to obtain basic design and engineering data for a commercial calciner (17,18).

Particular attention, in the design of this model calciner, was given to the utilization of combustion heat of volatile matter emitted from the coke feed and to obtaining an accurate heat balance data for the two-stage calcining system.

Test results obtained through the operation of this model calciner pilot plant are summarized as follows.

The exhaust gas from first stage kiln contains combustible gases (H_2, CH_4 and CO) and it has an approximate calorific value of 640 kcal/m^3 (0°C, 1 atm).

Figure 7 shows a schematic diagram of the model calciner used for studying an economic continuous two-stage calcining system. The exhaust gas from first stage kiln is fed as fuel for second stage kiln.

The operation of this two-stage calcining system indicated that the exhaust gas from first stage kiln, with auxiliary fuel burning, could be utilized as fuel for second stage kiln and the combustion of the exhaust gas in second stage kiln did not cause any difficulties in maintaining a stable operation of second stage calcination.

Figure 6. Photograph of model calciner.

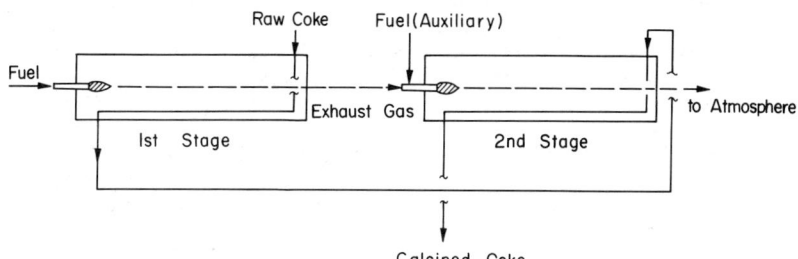

Figure 7. Schematic diagram of two-stage calcining system.

Table VI summarizes a typical heat balance in second stage of the above calcining system (Case A) in comparison with that of an alternative system (Case B) in which the exhaust gas from first stage kiln is not utilized for second stage calcination. The amount, composition and temperature of the exhaust gas from first stage kiln determine the auxiliary fuel requirements for second stage kiln.

By the utilization of combustion heat of volatile matter emitted from the coke feed, the calcining process capable of economic production of two-stage calcined coke can be designed.

Table VI. Heat Balance of Model Calciner (Second Stage)

(Unit: kcal/hr.-calcined coke of 1 kg)

	Case A	Case B
Input Heat		
Combustion Heat of Fuel	512	1,024
Combustion Heat of 1st. Stage Exhaust Gas	667	-
Combustion Heat of Coke & VM	675	655
Sensible Heat of Coke & Air	18	10
(Total Input Heat)	1,872	1,689
Output Heat		
Sensible Heat of Coke	513	513
Sensible Heat of Flue Gas	786	658
Decomposition & Vaporization Heat of VM	10	10
Heat Loss	563	508
(Total Output Heat)	1,872	1,689

VM: Volatile Matter

Conclusion

Application of the new calcining technology reduces the thermal expansion coefficient, regardless of the types of green coke. Major cause for reduction of the thermal expansion coefficient and improvement of the puffing characteristics is the development of unique microcracks (1 to 60 μm) within processed coke. These microcracks function as voids to absorb and relax the expansion of crystallites. Development of microcracks under the new calcining process seems to have been caused by specific distribution of stress within the coke arising from significant expansion and contraction following the heat treatment patterns.

These microcracks may have some effect in terms of mechanical strength when these cokes are processed into artificial graphite electrodes. However, no significant difference in strength between graphitized test pieces obtained from actual electrodes under the new and traditional method was found. An optimum commercial new process system has been studied by using a model calciner pilot plant.

By using two commercial plants conventionally, the trial production of two-stage calcined coke has been carried out and calcined coke samples have been distributed to the graphite electrode manufacturing companies for their evaluation.

Acknowledgments

The authors wish to thank people of Koa Oil Company Marifu Refinery and Osaka Research Laboratory for their assistance.

Literature Cited

1. Yoshimura, K. U.S. Patent 4 100 265, 1978.
2. Thomas, R. Hydrocarbon Proc. 1975, June, P97.
3. Mantell, C.L. "Carbon and Graphite Handbook" Interscience Publ.
4. Kurami, E. J. Japan Petrol. Inst. 1973, 16, (5), 366.
5. Shea, F.L. U.S. Patent 2 775 549, 1956. and Hackley, R.C. U.S. Patent 2 922 755, 1960.
6. Kakuta, M.; Kohriki, M.; Tano, T.; Sanada, Y. Tanso 1979, No. 95, 135.
7. Kakuta, M.; Tanaka, H. 3rd International Carbon Conf. Baden-Baden, Preprints, 1980, P406.
8. a. Noguchi, K.; Kakuta, M.; Tanaka, H.; Sato, J. 3rd International Carbon Conf. Baden-Baden, Preprints, 1980, P401.
   b. Kakuta, M.; Tanaka, H.; Sato, J.; Noguchi, K. Carbon 1981, 19, (5), 347-52.
9. Kakuta, M.; Kohriki, M.; Sanada, Y. J. Materials Sci. 1980, 15, 1671.
10. Kakuta, M. International Symp. on Carbon, Toyohashi, Japan, Preprints 1982, P527.
11. Kakuta, M.; Tsuchiya, N.; Tanaka, H.; Noguchi, K. Sixth London International Carbon and Graphite Conf., London, Preprints, 1982, P10.
12. Mrozowski, S. Proc. 1st and 2nd Carbon Conf., Pergamon Press 1956, P31.
13. Collins, F.M. Proc. 1st and 2nd Carbon Conf., Pergamon Press 1956, P177.
14. Okada, J. Proc. 4th Carbon Conf., Pergamon Press 1960, P547 and P553.
15. Wallouch, R.W.; Fair, F.V. Carbon 1980, 18, (2), 147-53.
16. Kakuta, M. 117 Committee of the Japan Sec. for Promotion of Science, 1979, No. 153.
17. Noguchi, K.; Komi, N. U.S. Patent 4 169 767, 1979.
18. Noguchi, K.; Komi, N. U.S. Patent 4 265 710, 1981.

RECEIVED December 18, 1985

14

Petroleum-Coke Desulfurization
An Improved Thermal-Chemical Process

H. H. Brandt[1] and R. S. Kapner[2]

[1] Diamond West Energy Corp., 27 River Ridge Road, Lake Charles, LA 70605
[2] Chemical Engineering Department, The Cooper Union, 51 Astor Place, New York, NY 10003

Current and foreseeable future supplies of anode grade petroleum coke for the primary aluminum industry are subject to coker feedstock quality limitations, i.e., to higher levels of metallic impurities (V, Ni, Fe and Si) and increasing sulfur content. The presence of metals in coke directly affects refined aluminum quality, while sulfur, though tolerated by producers, causes environmental problems that appear to be on the edge of stringent federal regulations. Steadily rising coke sulfur levels have stimulated interest in removal methods.

This paper briefly reviews the technology of petroleum coke desulfurization and the effects of sulfur removal processes on coke physical properties. A new thermal-chemical process is described which uses hydrogen sulfide as a gaseous reactant to remove coke sulfur after calcining. The process appears able to maintain high bulk and real coke densities by balancing reaction temperature and contact time and is suitable for new calciner construction as well as a relatively easy retrofit to existing calciners.

Impurities in Coke - Nature of the Problem

Free world petroleum coke production has increased dramatically in the past decade. The United States continues to be a dominant factor in the petroleum coke industry, accounting for more than 75% of the total production.

In spite of the phenomenal growth in production, there is a real and justified concern about a coke shortage by the aluminum industry. The shortage is in the supply of anode grade coke. This is the coke that is used by the producers of primary aluminum (about one-half pound per pound of aluminum). This industry consumes approximately 10 million tons of raw petroleum coke annually or about 35% of the total free world output. Most of the recently built cokers, as well as many of the older ones, are being charged with feedstocks that are high in sulfur and metals contents; the resultant cokes are correspondingly high in sulfur and metals contents. While experience indicates that both of these contaminants can be tolerated by the aluminum producers, they do create serious problems. In the case of metallic impurities (i.e., V, Ni, Fe and Si), the problem can be reflected in reduced anode efficiency (e.g., V often acts catalytically to increase oxidation). Furthermore, all metals contaminate the aluminum; the metallurgy required to remove such contaminants is expensive. Sulfur presents another problem whose solution is less clearly defined. The problem is air pollution. Two elements in this scenario are on a collision course. First, due to shrinking supplies of low sulfur (less than 2.5 wt%) cokes, concurrent with increased requirements by the aluminum industry, the average sulfur content of so-called anode grade cokes has increased significantly. Ten years ago the average was 2.0 wt%, while today's specification is 3.5 wt%. Second, SO_x emission limits are being tightened, and there are strong indications that we will see a quantum jump in the environmental restrictions on SO_x emissions in the near future.

Technology of Sulfur Removal

Desulfurization of coker feedstocks to produce lower sulfur cokes has proven not to be the solution to the problem. Nor is there cause for optimism in this area. Not only is this technology extremely expensive, it doesn't always achieve the expected results. A 50% reduction in feedstock sulfur content usually yields a lesser reduction of the sulfur content of the coke.

That leaves us with the challenge to desulfurize the coke itself. There has long been an incentive to remove sulfur from coke since low sulfur coke has always commanded a premium price over its higher sulfur counterpart. There have been dozens of patents filed on petroleum coke desulfurization technology (1-18). Although far from exhaustive, the list of 18 patents cited are those most pertinent to this discussion.

It has been demonstrated in the laboratory that sulfur can be removed from petroleum coke thermally, chemically, or thermal-chemically. Thermal sulfur removal is, as might be expected, a time-temperature function. A long soak at relatively low temperatures, e.g., several days at 2000°F (1095°C), can remove 50% or more of the contained sulfur. While no specific supporting literature reference can be sited, this effect is commonly experienced when baking anodes in ring furnaces where temperatures are held from about 950-1050°C for periods as long as one week. Such treatment applied to petroleum coke per se, however, is impractical because of excessive energy costs and its inability to process large tonnages. Higher temperatures and shorter times, e.g., 2600-2900°F (1425-1595°C) for 45-60 minutes, can also remove 50% or more of the sulfur, but create a high degree of porosity and reduce real density (3,11,12,14,18,20-27). Both of these properties are critical in anode fabrication.

Chemical desulfurization techniques have a common drawback. They require pulverizing the coke to expose maximum surface area to the reactant(s). Thus, the end product is low sulfur coke with no lump content. Since anodes cannot be made without a coarse fraction, agglomeration of the fines would be required. This in turn would require a coal tar binder pitch and a baking step. Even so, the final product becomes a relatively weak lump with prohibitive cost. Some chemical treatments employ various additives, e.g., alkalies, acids, fluorides or organic solvents, which, in addition to being expensive, may leave residues that are unacceptable (1,28-31).

The accumulated data appear to suggest that a thermal-chemical process should have the best chance for success. A considerable amount of research effort has been expended in this area. The reported results, however, are frequently inconclusive and occasionally conflicting. For example, most laboratory results support the contention that H_2S, hydrogen, methane and a variety of other low molecular weight materials such as hydrocarbons can induce desulfurization at elevated temperatures (2-7,11,14-17,24,26,28,34-38); yet several studies conclude (21,32,33) that the sulfur content of petroleum coke actually increases when it is exposed to H_2S at elevated temperatures. Many studies indicate the need to pulverize the coke prior to treatment (9,12,39), while others do not. Still others claim (10,22) that a two-stage temperature treatment is beneficial. Two thermal-chemical systems report (11,18) 90% sulfur removal using a three-step process. In the first step, the volatiles are removed by treatment with an inert gas at 1400°F (760°C). The temperature is then reduced to 700°F (370°C) and the coke partially oxidized (presumably to activate it). The final step consists of reheating the coke to about 1700°F (925°C) and treating it with methane or a similar hydrocarbon.

Some of the inconsistencies in the research are probably due to variation in petroleum cokes. It is generally conceded that the coker is the "garbage can" of the refinery and as such it seldom sees an identical charge from day to day. Cokes with virtually identical chemical analyses can exhibit markedly different behavior when subjected to identical thermal treatments.

A New Thermal-Chemical Desulfurization Process

To date, all of the research effort has not produced a commercial process to desulfurize petroleum coke for use in aluminum cell anodes. However, there is a new and unique process that promises to succeed where so many previous attempts have failed. This new process has been patented (14) by Diamond West Energy Corporation, and the research work, conducted at The Cooper Union in New York City, has been jointly funded by Kennedy Van Saun Corporation and Diamond West.

The process consists of contacting calcined coke with a sulfur bearing gas at elevated temperatures. The concept of treating the calcined coke immediately after it is discharged from the calciner is a key element in minimizing energy costs. The hot coke, at about 2300°F (1260°C), is fed into a reactor directly from the calciner, where it may be further heated up to 2800°F (1540°C), and exposed to a sulfur bearing gas for a period of time up to 90 minutes.

The desulfurization chemistry appears to depend on the presence of free sulfur in the contacting gas and works in conjunction with organic sulfur pyrolytically released from the coke. Gases that decompose to form free sulfur or react to form free sulfur are suitable desulfurization agents and include, for example, refinery sour gas, pure and dilute H_2S, mercaptans, mixtures of CO and SO_2 or COS and H_2O.

Laboratory tests on two-pound samples of green petroleum cokes with sulfur contents of 4 wt% to 6 wt% have successfully reduced the sulfur levels to as low as 0.7 wt%. The degree of desulfurization can readily be controlled by adjusting the temperature (nominally between 2500 and 2800°F), the holding time in the reactor (between 5 and 90 minutes), and the reactant concentrations and flow rates.

The experimental work has been conducted in horizontal furnace tube reactors containing fixed beds of sized (1/2 inch x 4 mesh) coke electrically heated and controlled at precise temperatures over the entire bed length. The coke is first calcined, following a temperature profile similar to that of commercial calcining. The calcining is immediately followed by desulfurization with pure or dilute sulfur gases metered from cylinders.

The data shown in Table I were obtained at reaction temperatures of 2730°F (1500°C) and are typical of measured desulfurization

Table I.

Holding Time (min)*	Thermal Desulfurization % Desulfurization**	With Added Sulfur Gas % Desulfurization**
0	0	0
10	15.2	27.5
20	32.1	48.0
30	55.0	66.8
45	66.8	76.8
60	71.2	83.8

*Holding times are at constant temperature after calcination.
**Percent desulfurization is relative to sulfur content in calcined coke.

results at all temperatures between 2550°F (1400°C) and 2900°F (1595°C). Analysis of these and other data suggests that the desulfurization process is a complex sequence of steps tentatively represented as:

$$(-CS-) \rightleftharpoons (-C-) + S_v \qquad (1)$$

$$S_v + (-CS-) \longrightarrow CS_2 \qquad (2)$$

$$2S_v + (-C-) \longrightarrow CS_2 \qquad (3)$$

Reaction (1) constitutes a purely thermal initiation of the desulfurization process described by (1) and (2). Step (1) would appear to be slow, rate controlling and with a high activation energy. Also it is very responsive to temperature increases. Hence, the generally observed rapid acceleration of thermal desulfurization as the temperature increases from 2400°F (1315°C) to 2900°F (1595°C). By adding sulfur from an external source, the desulfurization described in step (2) is no longer limited by the rate at which sulfur is supplied by the coke itself and desulfurization proceeds more rapidly than by a purely thermal process at every temperature. Experimental data show that the desulfurization with added external sulfur, represented by reaction (2), is dependent on temperature and concentration of sulfur in the reactant gas. It has a significantly lower activation energy than reaction (1).

The observation that the activation energy of reaction (1) is greater than that of reaction (2) is confirmed by experimental data which show that the relative rates of desulfurization by thermal processing and by the addition of a sulfur gas decrease as holding temperature is increased. This points out one of the more valuable features of this technology--a high rate of desulfurization can be supported at temperatures where thermal desulfurization is not generally achieved.

Another valuable feature is the preservation of the critical properties, real density and bulk density. Real density is a specific gravity measurement of the calcined coke (performed on a sample that has been reduced to minus 200 mesh), and bulk density is a porosity determination (performed on a 35 x 65 mesh sample). Both of these properties tend to deteriorate to unacceptable levels during thermal desulfurization at high temperatures due to the creation of micro- and macro-porosity as sulfur is vaporized and leaves the solid coke matrix. The persistence of high real and bulk density values in cokes desulfurized using this new technology is a promising indication that some sort of annealing process occurs during the treatment. Unfortunately, we do not have an explanation for this phenomenon. The data from almost one hundred desulfurization runs strongly support the claim and are summarized in Table II.

Table II. Comparison of Physical Properties of Desulfurized Cokes with Typical Calcined Coke Specifications

Property	Commercial Specification	Sulfur Gas Desulfurized Coke
Real Density (g/cc)	2.05-2.11	2.07-2.16
Bulk Density (lb/cu ft)	49 (minimum)	53-55

Pilot plant work is presently under way at Kennedy Van Saun's facilities in Danville, Pennsylvania. We hope to be able to develop hardware that can be retrofitted to existing calciners.

Literature Cited

1. Aldridge, C. L.; Waghorne, R. H., "Desulfurization of Fluid Petroleum Coke," U.S. Patent 3 600 130, August 17, 1971.
2. Alford, H. E.; Marsh, E. N., "Process for the Desulfurization of Petroleum Coke," U.S. Patent 4 146 434, March 27, 1979.
3. Bauer, W. V.; Isaacs, J. A. C.; LaMont, O. M., "Coke Desulfurization," U.S. Patent 4 203 960, May 20, 1980.
4. Buchmann, F. J.; Hammer, G. P., "Low Sulfur Coke from Virgin Residua," U.S. Patent 3 702 816, November 14, 1972.
5. Ford, F. P.; Nelson, J. F., "Desulfurization of Fluid Coke with Sulfur Dioxide Containing Gas," U.S. Patent 2 739 105, March 20, 1956.
6. Franz, W. F.; Hess, H. V., "Purification of Petroleum Coke," U.S. Patent 3 595 965, July 27, 1971.
7. Franz, W. F.; Hess, H. V., "Purification of Petroleum Coke," U.S. Patent 3 598 528, August 10, 1971.
8. Frischmuth, R. W., "Desulfurization of Carbonaceous Materials," U.S. Patent 4 270 928, June 2, 1981.
9. Gorin, E.; Zielke, C. W., "Desulfurization of Low Temperature Carbonization Char," U.S. Patent 2 717 868, September 13, 1955.
10. Grove, J. H., "Desulfurization of Petroleum Coke," U.S. Patent 2 983 673, May 9, 1961.
11. Hardin, E. E.; Guffy, D. H.; Grindstaff, L. I., "Thermal Desulfurization and Calcination of Petroleum Coke," U.S. Patent 4 160 814, July 10, 1979.
12. Hsu, H. L.; Hardin, E. E.; Grindstaff, L. I., "Process for Calcining and Desulfurizing Petroleum Coke," U.S. Patent 4 298 008, September 22, 1981.
13. Johnson, H. S.; Andersen, A. H., "Process for Preparation of Carbon Disulfide and the Desulfurization of Coke," U.S. Patent 3 009 781, November 21, 1961.
14. Kapner, R. S.; O'Brien, R., "Desulfurization of Delayed Petroleum Coke," U.S. Patent 4 406 872, September 27, 1983.
15. Lutz, I. H., "Purification of Coke," U.S. Patent 4 208 307, September 25, 1978.
16. MacGregor, D., "Process for Desulfurizing Sulfur Bearing Coke," U.S. Patent 4 011 303, March 8, 1977.
17. Nelson, J. F.; Ford, F. P.; Hubbard, A. W., "Desulfurization of Fluid Coke with Hydrogen Above 2400°F," U.S. Patent 2 872 384, February 3, 1959.
18. Yoshimura, K; Hayashi, M., "Process for Preparation of High Quality Coke," U.S. Patent 4 100 265, July 11, 1978.
19. Bopp, A. F.; Goff, G. B.; Howard, B. H., "Influence of Maximum Temperature and Heat-Soak Times on the Properties of Calcined Coke," Light Metals 1984, pp. 869-882.
20. El-Kaddah, N.; Ezz, S. Y., "Thermal Desulfurization of Ultra-High Sulfur Petroleum Coke," Fuel, 1973, 52(2), 128.

21. Gehlbach, R. E.; Grindstaff, L. J.; Whittaker, M. P., "Effect of Calcination Temperature on Real Density of High Sulfur Cokes," presented at 106th AIME Annual Meeting, Atlanta, Georgia, March 6-10, 1977.
22. Jones, S. S.; Hildebrandt, R. D.; Hedlund, M. C., "Influence of High-Sulfur Cokes on Anode Performance in Alumina Reduction," J. Metals, 1979, 31(9), 33-40.
23. Kapner, R. S.; Cianco, A.; Gootzait, E.; Brandt, H. H., "A Thermochemical Process for the Production of Low Sulfur Anode Grade Petroleum Coke," Extended Abstracts, 16th Biennial Conference on Carbon, University of California, San Diego, July 18-23, 1983, pp. 586-587.
24. Miura, K.; Hashimoto, K., "A Model Representing the Change of Pore Structure During the Activation of Carbonaceous Materials," Ind. Eng. Chem. Process Des. Dev., 1984, 23, 138.
25. Rhedy, P., "Structural Changes in Petroleum Coke During Calcination," Trans. Met. Soc. AIME, July 1967, :356239, 1084.
26. Sadilla, F. M.; LaMonte, O. M.; Sze, M. C.; Bauer, W. V., "Thermal Process Is Developed for Petroleum Coke Desulfurization," Oil & Gas J., January 22, 1979, p. 64.
27. Sadilla, F. M.; LaMonte, O. M.; Sze, M. C.; Bauer, W. V., "New Process Desulfurizes Coke," Hydrocarbon Processing, March 1979, p. 97.
28. George, Z. M.; Schneider, L. G.; Tollefson, E. L., "Desulfurization of a Fluid Coke Similar to the Athabasca Oil Sands Coke," Fuel, 1978, 57(8), 497.
29. Mahoud, B. H.; Ayad, S.; Ezz, S. Y., "Desulfurization of Petroleum Coke," Fuel, 1968, 47, 455.
30. Phillips, C. R.; Chao, K. S., "Desulfurization of Athabasca Coke by (a) Chemical Oxidation and (b) Solvent Extraction," Fuel, 1977, 56(1), 70.
31. Sabott, F. K., "A Study of Methods of Removing Sulfur from Petroleum Cokes," Col. School of Mines Quart., July 1952, 47, 1.
32. George, Z. M., "Hydrodesulfurization of Coke from Athabasca Tar Sands Operation," Ind. Eng. Chem. Process Des. Dev., 1975, 14, 298.
33. Mason, R. B., "Hydrodesulfurization of Coke," Ind. Eng. Chem., 1959, 51, 1022.
34. El-Tawil, S. Z.; Morsi, M. B., "Thermal Treatment of Egyptian Petroleum Coke in an Oxygen Atmosphere," J. of Mines, Metals and Fuels, June 1979, p. 185.
35. Hussein, M. H.; El-Tawil, S. Z.; Rabah, M. A., "Desulfurization of High-Sulfur Egyptian Petroleum Coke," J. Institute Fuel, September 1976, 49, 139.
36. Kalinowsky, B.; Wlodarski, R., "The Mechanism of Thermal Desulfurization of Petroleum Cokes with High Sulfur Content," Przemysl Chemiczny, 1968, 47, 351.
37. Schafer, W. C., "Removal of Sulfur from Petroleum Coke," Col. School of Mines Quart., July 1952, 47, 27.
38. Sef, F., "Desulfurization of Petroleum Coke During Calcination," Ind. Eng. Chem., July 1960, 52, 599.
39. Parmer, B. S.; Tollefson, E. L., "Desulfurization of Oil Sands Coke," Can. J. Chem. Eng., 1977, 55, 185.

RECEIVED March 21, 1985

15

Granular Graphitic Carbon

W. M. Goldberger[1], P. R. Carney[1], R. F. Markel[2], and F. J. Deutschle[3]

[1]Superior Graphite Company, 120 S. Riverside Plaza, Chicago, IL 60606
[2]R. F. Markel & Associates, 14 Gallery Center, Taylors, SC 29678
[3]Superior Graphite Company, 4021 Calvin Drive, Hopkinsville, KY 42240

> Desulco is a granular graphitic carbon made by desulfurizing petroleum coke at temperatures approaching 2500C. Desulco meets a sulfur specification of 0.03 percent sulfur maximum as required for use of the product as a recarburiser for production of ductile iron.
> A highly advanced continuous electrothermal furnace process was developed by The Superior Graphite Company for production of Desulco. Commercial operation was first begun in 1977 at Hopkinsville, Kentucky. Plants with total capacity of 30,000 metric tons are in operation in the U.S. and Europe.
> Desulco is characterized by its high purity, relatively low particle density and very high degree of microporosity. In addition to use worldwide as a carbon additive to molten iron, Desulco is finding increasing use in special graphite products such as lubricants, coating agents and electrically conductive fillers.

Commercial cast irons are alloys containing 2-4 percent dissolved carbon. On cooling and solidification, the dissolved carbon precipitates as crystalline graphite. In "gray" iron castings, the graphite has been precipitated in spherical or nodular form. Ductile or nodular cast iron has excellent mechanical working characteristics and is finding increasing markets.

Solid carbons are added to iron melts to adjust the carbon content before casting. A listing and analysis of the various carbon sources used is given in Table I. Sulfur present in the carbon additive will be retained by the iron. This is not critical in making gray irons castings because the sulfur content of the metal is relatively high, in the range 0.04-0.12 percent (1). However, for production of ductile iron, it is essential to maintain sulfur in the range 0.005-0.015 percent. For this reason, synthetic graphite, which is virtually sulfur free, is used to avoid a subsequent need to desulfurize the hot metal before casting. Synthetic graphite has been a preferred "recarburiser" in ductile iron foundries also because it contains no volatiles and almost no ash.

Table I. Typical Analyses of Various Recarburiser Materials

Carbon Type	Analyses, Percent					
	Fixed Carbon	Ash	Volatile Matter	Sulfur	Nitrogen	Hydrogen
Metallurgical Coke	89.0	9.4	1.60	0.93	1.00	-
Brown Coal Char	92.0	2.9	2.90	0.28	0.60	1.10
Calcined Anthracite	89.2	10.0	0.4	0.35	0.10	0.013
Pitch Coke	98.0	0.55	0.50	0.45	0.60	0.20
Natural Graphite	86.3	13.2	0.44	0.35	0.06	0.20
Calcined Petroleum Coke	99.6	0.4	0.4	1.8	0.60	0.15
Synthetic Graphite	99.5	0.5	<0.1	0.02	<0.01	<0.01

Synthetic graphite is made in relatively large shapes, for example, electrodes for steelmaking electric furnaces. The manufacture of shaped graphite includes a series of steps involving the blending of cokes with pitch, shaping, baking and graphitizing at temperatures approaching 3000C in large batch operated resistor furnaces of the Acheson type. As such, primary manufactured graphite is too expensive for use directly as a recarburiser. However, the scrap graphite generated as the machinings and breakage during production of primary manufactured graphites and from the stubs and breakage of graphite electrodes in use has been the main source of recarburiser graphite. As is the case with all secondary materials, the availability and prices fluctuate widely. The Desulco development was intended to provide a stable source of consistent quality graphitic carbon raiser to supplement the supply of scrap graphite.

Development of the Desulco Process

Delayed petroleum coke was considered the primary carbonaceous raw material because it is a graphitizable carbon that was immediately available in tonnages well in excess of foundry usage. As noted in Table II, only about 6 percent of the production of more than six million tons of petroleum coke is for carbon addition in iron and steel. Moreover, petroleum coke would be available in particle sizes suitable for crushing and screening to the size range desired in foundry practice. In addition, except for its high sulfur content, petroleum coke has been a relatively high purity carbon. As the name indicates, the Desulco Process was designed to desulfurize petroleum coke; and more specifically, to desulfurize below the 0.03 percent levels specified by the ductile iron industry for recarburisers.

Table II. Production and Usage of Petroleum Coke—1980 Estimate*

Production	
6,150,000 Tons	Delayed
200,000 Tons	Fluid
6,350,000 Tons	Total

Distribution	Percent
Aluminum Reduction	68
Electrode (Carbon/Graphite)	19
Metallurgical (Recarburisers)	6
Titanium Dioxide	3
Miscellaneous	4
	100

*Courtesy Asbury Graphite, Inc. of California

Review of Process Alternatives. Superior Graphite began in 1968 to investigate possible alternatives for desulfurization of petroleum cokes. The methods considered included various chemical treatments and direct thermal purification processes. The chemical treatment methods included hydrodesulfurization and reactions with various alkali metal compounds. Fine grinding of the coke appeared to be required and reaction conditions generally involved high pressure. The chemical methods reported at the time therefore appeared complex and unlikely to achieve at a reasonable cost the low levels of product sulfur required. Alternatively, thermal desulfurization at about 2500C was known to yield a product meeting sulfur specifications as made evident by the conventional process of graphitizing shaped mixtures of petroleum coke and coal tar pitch. In addition to simplicity, thermal purification offered a significant advantage in not requiring fine grinding of the coke.

A review of available data on thermal desulfurization of various petroleum cokes indicated a generalized desulfurization relationship as shown in Figure 1. To achieve nearly complete desulfurization it was recognized that temperatures in excess of 2400C would be required.

Laboratory studies made by Superior Graphite to better understand the kinetics of thermal desulfurization established that above 2400C, desulfurization is rapid and essentially complete within about 5 minutes regardless of particle size (Figure 2). This was found to be true for both green cokes and calcined cokes. It was of particular interest to also note that despite subjecting cokes to extremely rapid heating, little change occurred in particle size indicating that grains of coke are sufficiently porous to allow a rapid evolution of gas without rupture (Table III); although to some extent, differences in this property will vary with cokes from different origin.

Achieving the required degree of desulfurization dictated electrothermal heating. Fixed-bed electrothermal furnaces of the Acheson type were initially considered for use directly with granular coke; but the test results were disappointing. The product of the Acheson furnace was not uniform in its sulfur content. Adaptation of the Acheson Process, which is a batch process, presented problems in materials handling that were considered very difficult to resolve at the 10,000 tons per year capacity determined to be the

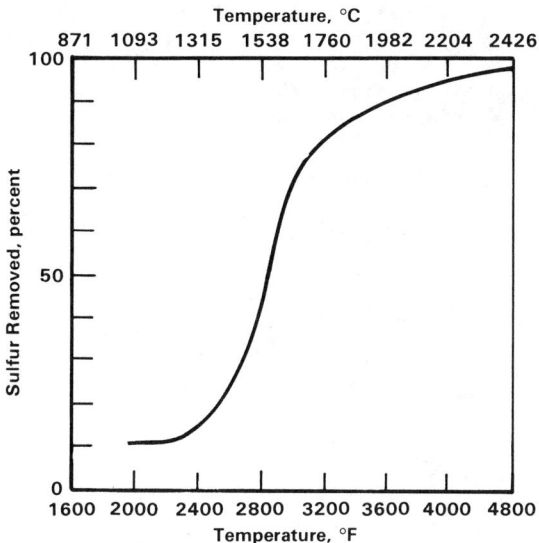

Figure 1. Desulfurization of Petroleum Coke by Thermal Treatment.

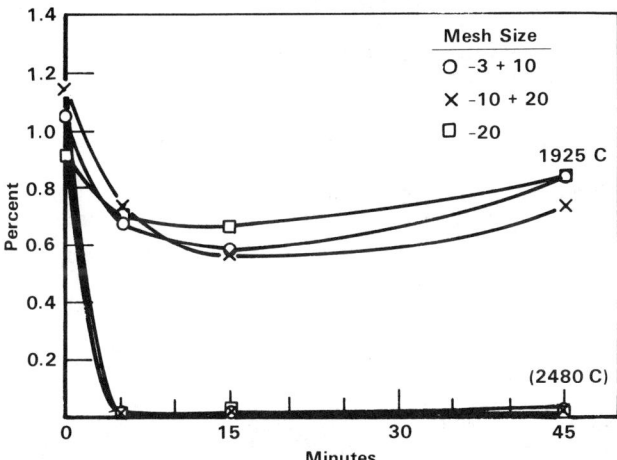

Figure 2. Desulfurization of Calcined Petroleum Coke as a Function of Time at Temperature.

minimum economic production rate. Moreover, a continuous rather than a batch method would be needed to take advantage of the rapid desulfurization kinetics and to best control the gaseous emissions.

Table III. Change in Size Distribution of Calcined Petroleum Coke After Rapid Heating to 2480C

	Weight Percent Retained	
Mesh Size	Coke As Received	Thermally Processed Coke 2400C (4500F)-5 Min.
+12	22.5	16.5
-12 + 20	74.3	78.6
-20 + 28	2.9	4.9
-28 + 35	0.3	TR
-35 + 48	0	TR
-48 + 65	0	TR
-65 + 100	0	TR
	TR=Trace	

Features of the Desulco Process. The Desulco Process subsequently developed by Superior Graphite employs continuous electrothermal charge-resistor heating and is similar in that regard to the Acheson Process but designed specifically for continuous purification of granular carbons. The process has these following features:

- Ability to achieve and sustain a relatively large zone of uniform temperature in excess of 2500C.

- High energy efficiency to enable granular coke to be desulfurized at less than 2 Kw-hr/Kg (0.9 Kw-hr/lb).

- Extremely high rates of energy transfer and high throughput.

- Continuous and uniform movement of coke through the furnace without need for mechanical devices or moving parts within the high temperature zones.

- Steady-state operation with fully automated process control.

- Fully contained gaseous effluent for effective environmental control.

Commercialization. Production of Desulco was begun by Superior Graphite at Hopkinsville, Kentucky in May 1977. The initial production capacity of the Hopkinsville plant was 11,000 short tons per year. This was increased to over 20,000 tons in 1981. To date, more than 80,000 tons of Desulco has been produced at Hopkinsville. A Desulco plant was built at Sundsvall, Sweden by CarboNord, a joint venture company formed by Superior Graphite and the KemaNord Industrial Chemical Company of Sweden. The CarboNord plant, rated at 12,000 metric tons capacity, began production of Desulco in September 1984. The Desulco product made in Sweden will supply the European market.

The Properties of Desulco

Desulco is a partially graphitic carbon grain. Although similar in chemical purity to synthetic scrap graphite, processed at graphitizing temperatures as a grain rather than a formed shape, the crystallinity and other physical properties of Desulco differ from that of scrap graphite grain. These differences in properties are largely attributed to the fact that no pitch was used nor was the grain compacted before heating to graphitizing temperature as is done in the manufacture of shaped synthetic graphite. Also, the time-temperature history in production of Desulco differs substantially from that used in production of formed graphite. Temperature rise rates of particles fed to the Desulco furnaces are estimated to be in the range 50-150C per second with a temperature of 2500C achieved in less than 1 minute. The residence time at temperature in the Desulco process is between 5 and 30 minutes. In the conventional method of making shaped synthetic graphite, the temperature is raised slowly requiring approximately 1 day or longer to reach graphitizing temperature (2). The graphitization cycle of the Acheson process can well exceed 5 days depending on the size of furnace and quantity of material contained.

Chemical Analyses. The production specification and a typical analyses of Desulco is given in Table IV. The trace element analyses of the ash forming elements does vary with the coke. An example analyses of Desulco ash is given in Table V.

Table IV. Specification and Typical Analyses of Desulco

	Weight Percent	
	Specification	Typical
Carbon	99.5 min	99.9
Ash	0.5 max	Less than 0.1
Volatiles	0.2 max	Less than 0.1
Moisture	0.2 max	Trace
Sulfur	0.03 max	Less than 0.02
Hydrogen	-	20-60 ppm
Nitrogen	-	35-70 ppm

Table V. Example of Analysis of Components in Desulco Ash

Component	Wt., Percent
Calcium Oxide	34.3
Silicon Oxide	25.5
Aluminum Oxide	10.3
Vanadium Oxide	8.4
Iron Oxide	3.7
Nickel Oxide	3.5
Titanium Oxide	Nil
Miscellaneous	Balance

The contents of the gases nitrogen and hydrogen are in the less than 100 ppm levels typical of synthetic graphites as shown in the tabulated comparison of Table VI.

Table VI. Comparison of Nitrogen and Hydrogen Content of Desulco with Various Recarburisers

Type	Parts Per Million[1]	
	Nitrogen	Hydrogen
Desulco	35-70	25-60
Synthetic Graphite	30-80	40-70
Mexican Graphite	300-700	1500-2500
Anthracite[2]	760-1,600	130-140
Petroleum Coke[3]	2300-10,600	1600-1650

Notes
1. Range of samples analyzed.
2. Electrically calcined.
3. Calcined.

Particle Size. Run-of-furnace Desulco is a free flowing granular material ranging in size from 0.15 mm to 10 mm (100 mesh to 3/8 in). A general size distribution is given in Table VII. This run-of-furnace product can be sized to various customer preferred grades.

Table VII. Particle Size Distribution of Desulco As Produced

Tyler Mesh Size	Range, mm	Wt. Retained, Percent
Plus 6	Plus 3.36	9
-6 + 10	3.36 x 1.68	26
-10 + 28	1.68 x 0.60	35
-28 + 65	0.60 x 0.21	25
Minus 65	Minus 0.21	5
		100

Bulk Density. As noted, little change occurs in the overall size or shape of the coke particles due to the Desulco furnace treatment; therefore, the loss of volatiles and sulfur reduces the bulk density of the product. The bulk density of Desulco will vary with the coke source and is generally in the range between 0.58 and 0.80 g/cc (35-50 lb./cu. ft.).

Particle Density: Particle density of the as produced Desulco grain determined by helium or liquid pycnometry will also vary with the source of coke in the range 1.40-1.60 g/cc in the run-of-furnace grain. Desulco is therefore substantially more porous than synthetic graphites made by the conventional pitch impregnation and Acheson furnace graphitization process. Particle density varies with particle size; and, the particle density will approach the theoretical density of graphite (2.26 g/cc) as the Desulco particle is reduced to micron sized powder. The particle density of run-of-furnace Desulco grain is compared in Table VIII with calcined delayed petroleum coke before and after graphitization of these coke materials when graphitization is done over a 10 hour heat up period to 2500C and slow cooling cycle under argon.

Table VIII. Comparison of Physical Properties of Desulco with Several Cokes and Graphitic Grains*

Material	Particle[A] Density, g/cm^3	Surface[B] Area, M^2/g	Pore Volume[C] Percent	Pore Radius[D] A°
Desulco[E]	1.46	0.572	35.4	192
CPC-1	1.86	0.769	17.7	2000
CPC-1[F]	2.02	0.362	10.7	420
CPC-2	1.73	0.729	23.5	1080
CPC-2[F]	2.08	0.317	8.0	500
CPC-3	1.86	0.614	17.7	1060
CPC-3[F]	1.98	1.738	12.4	120

*CPC (Calcined Petroleum Coke): All petroleum cokes delayed type graphitized by inductively heating static sample to 2500C over 10 hour period, holding 30 minutes and slow cooling. Argon atmosphere maintained.

Notes

A. Particle density by helium pycnometry.
B. Surface area by BET absorption of argon at liquid nitrogen temperature.
C. Pore volume percent calculated from measured particle density.
D. Determined by mercury porosimetry.
E. Desulco sample in Table VIII prepared from calcined petroleum coke CPC-1.
F. Graphitized.

Electrical. Desulco is isotropic and its electrical resistivity is measured to be between that of calcined petroleum coke and a high carbon content natural vein graphite (Sri Lanka). The measurement of the electrical resistivity of granular materials is not a standardized procedure; Superior Graphite utilizes a high conductivity metal piston to contact and compress slightly a column of the grain or powder within a non-conducting cylinder wall. Table IX shows a comparison of the determined resistivity of Desulco in comparison with other carbon and graphite materials at two size ranges. The resistivity of Desulco will generally be slightly greater than synthetic graphite grain and will vary depending on the precursor petroleum coke used; the values given in Table IX are for comparison only and not intended for design.

Table IX. Comparison of Electrical Resistivity
of Desulco with Various Graphite Materials

Material	Resistivity, ohm-cm Grain Size	
	20x65 Mesh	325 Mesh
Vein Graphite-4712 (97%C)	0.006-0.001	0.020
Vein Graphite-4200 (90%C)	0.011	–
Synthetic Graphite	0.0125	0.030
Desulco	0.0160	0.033
Crystalline Flake Graphite	0.017-0.021	0.043
Microcrystalline (Mexican) Graphite	0.040-0.050	0.090
Calcined Coke	0.1016	–

Structure of Desulco

The crystal structure of Desulco is substantially more ordered than that of the starting petroleum coke; but, not as ordered as that of a synthetic graphite made by the Acheson process.

A comparison of the x-ray peak response of Desulco with that of green and calcined cokes and a typical synthetic graphite made by the Acheson process is shown in Table X. It can be seen that the green coke gave only one peak at the $2\theta = 25.5°$ position. The calcined coke samples give well-defined (0002) graphite peaks at $2\theta = 26°$, indicating a decrease in interlayer spacing from that of the green coke structure. However, the calcined cokes gave broad diffraction maxima at $2\theta = 43°$ with the needle coke (Calcined Coke 1) also showing a very broad response at $2\theta = 53°$. With the Desulco sample, there is some resolution of the peaks in the $2\theta = 42°$ to $46°$ range and considerably better definition of the $2\theta = 54°$ peak. The emergence of the peak at $2\theta = 77°$ is also evident. The Desulco peaks are not as well defined as the Acheson graphite which also shows a peak corresponding to $2\theta = 83.5°$.

Table X. Graphite Diffraction Peak Examinations

Graphite Peaks	d(Å)	2θ°	Green Coke	Calcined Cokes 1	Calcined Cokes 2	Desulco	Synthetic Graphite
0002	3.36	26.5	25.5	26	26	26.4	26.6
10$\bar{1}$0	2.13	42.4	–	{43(b)}	{43(b)}	42.5	42.5
10$\bar{1}$1	2.03	44.6	–			44.3(b)	44.6
10$\bar{1}$2	1.80	50.4	–	–	–	–	–
0004	1.678	54.7	–	53(vb)	–	54.4	54.6
10$\bar{1}$3	1.544	59.9	–	–	–	–	–
11$\bar{2}$0	1.232	77.5	–	–	–	77.5	77.5
11$\bar{2}$2	1.158	83.5	–	–	–	–	83.5

b=broad; vb=very broad

Crystallite size determined from the diffraction pattern measurements are given in Table XI and show the higher degree of three-dimensional ordering of Acheson graphite compared with Desulco.

Table XI. Comparison of Crystallite Sizes
and Interlayer Spacing
(dimension in Angstrom units, A°)

Material	$L_{10\bar{1}0}$	$L_{10\bar{1}1}$	d_{0002}	$2d_{0004}$
Green Coke 1	–	–	3.493	nm
Calcined Coke 1	29	–	3.410	3.409
Calcined Coke 2	22	–	3.418	3.420
Calcined Coke 3	28	–	3.422	3.422
Desulco Product 1	295	45	3.373	3.372
Desulco Product 2	325	40	3.384	3.378
Desulco Product 3	325	35	3.366	3.367
Synthetic Graphite	480	110	3.356	3.359

Notes
nm = not measured
Desulco products 1 and 2 were obtained from Calcined Cokes 1 and 2 respectively.

The ability of Desulco particles to withstand mechanical deformation appears remarkable. Despite the highly porous structure, Desulco grain can be compacted to essentially the theoretical density of graphite without breakage. Upon release of the compaction pressure, the grain will return almost to the full volume it had occupied before compaction. Superior Graphite monitors the degree of "resiliency" of Desulco by measuring the percentage increase in volume of a sample of grain after release of the compaction pressure applied to fully compress the grain. The applied pressure used is 603.3 Kg/cm^2 (10,000 psi). Measurements are made with a closely sized grain. Desulco will exhibit resiliency measurements generally in the range 100-150 percent; that is Desulco grain compacted to the full density of graphite will expand to more than twice that volume upon release of the compaction pressure. Although all graphites and graphitic carbon materials exhibit significant elastic behavior because of the inherent strength of the atomic structure of the graphite basal plane, Desulco exhibits a substantially higher resiliency than other graphitic materials. The relative magnitudes of this difference is given in Table XII.

Table XII. Comparison of Resiliency of Various Materials

Material	Resiliency[A], Percent
Calcined Petroleum Coke	9
Natural Flake Graphite	9-12
Natural Vein Graphite	16
Natural Microcrystalline Graphite	22-25
Synthetic Graphite	45-60
Desulco	100-150

Notes
A. Percentage increase in sample volume over volume under compaction pressure of 603.3 Kg/cm^2 (10,000 psi) when compaction pressure is released.

The degree of resiliency must be related to the porosity of the material; and, Desulco has a very high pore volume as indicated in Table IX. The data in Table VIII also suggest that the porosity is mainly "sealed" as the Desulco sample shown in Table VIII has a significantly greater pore volume than its precursor petroleum coke; but, a lower BET surface area.

This possibility of a significant degree of "sealed" porosity in Desulco is further indicated by the mercury porosimetry results shown in Figure 3. The initial intrusion cycle shows that approximately 70 percent of the pores are smaller in diameter than about 400 Angstroms. However, when the experiment was repeated in the second cycle with the already intruded sample left in place, the displaced volume versus pressure relationship remains the same below the 70 percent point. It is more logical that the observed volume change in this region of calculated pore diameter is due to the compressibility of a material with sealed and inaccessible pores rather than actual mercury intrusion. Otherwise, the mercury would have had to exude from the micropores when the applied pressure was being released to allow the pores to be intruded again during the second cycle as the data would indicate. Thus the Desulco particle appears to be a structure comprising macropores held by a web-like matrix of graphitic carbon containing substantial inaccessible microporosity between layer planes. The series of photomicrographs in Figure 4 shows the Desulco structure.

<u>Factors Influencing the Structure of Desulco</u>. The specific petroleum coke used as the Desulco precursor has a decided effect on the structure obtained. This is most evident in the particle density and bulk density of the product. Temperature and heating rate will also influence Desulco properties as shown in Table XIII. The higher the desulfurization temperature and the more rapid the temperature rise rate of the particles, the higher the degree of microporosity as indicated by the increased electrical resistivity and resiliency; temperature being the more important factor.

Table XIII. Effect of Temperature and Heating Rate on Physical Properties of Thermally Treated Petroleum Coke

Temp. C	Time,[A] Min.	Density,[B] g/cc	Resist., ohm-cm x 10^2	Resil.,[C] Percent
2000	240	1.80	1.751	88
2000	30	1.67	1.869	89
2000	5	1.56	1.896	92
2400	240	1.80	0.870	108
2400	30	1.63	1.007	122
2400	5	1.68	1.013	122

Notes
A. Time to reach temperature.
B. Particle density measured by liquid pycnometry.
C. Percentage increase in sample volume over volume under compaction pressure of 603.3 Kg/cm^2 (10,000 psi) when compaction pressure is released.

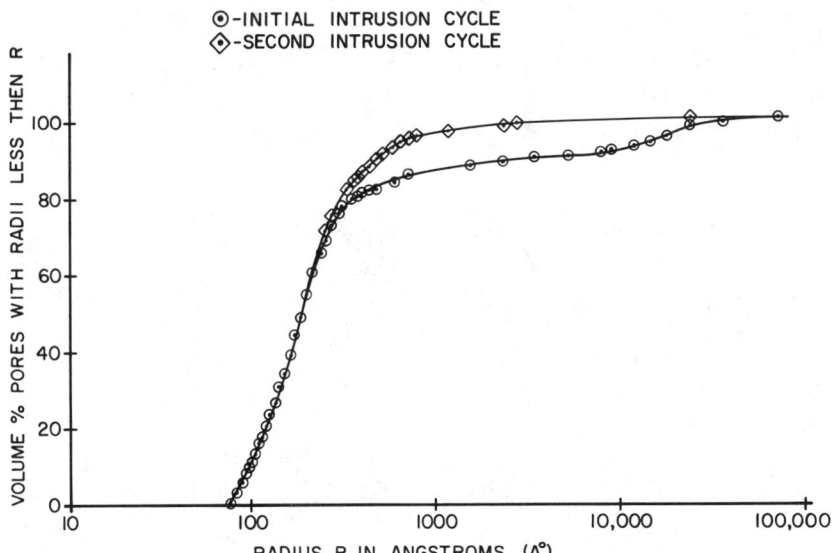

Figure 3. Apparent Pore Size Distribution of Desulco by Mercury Porosimetry.

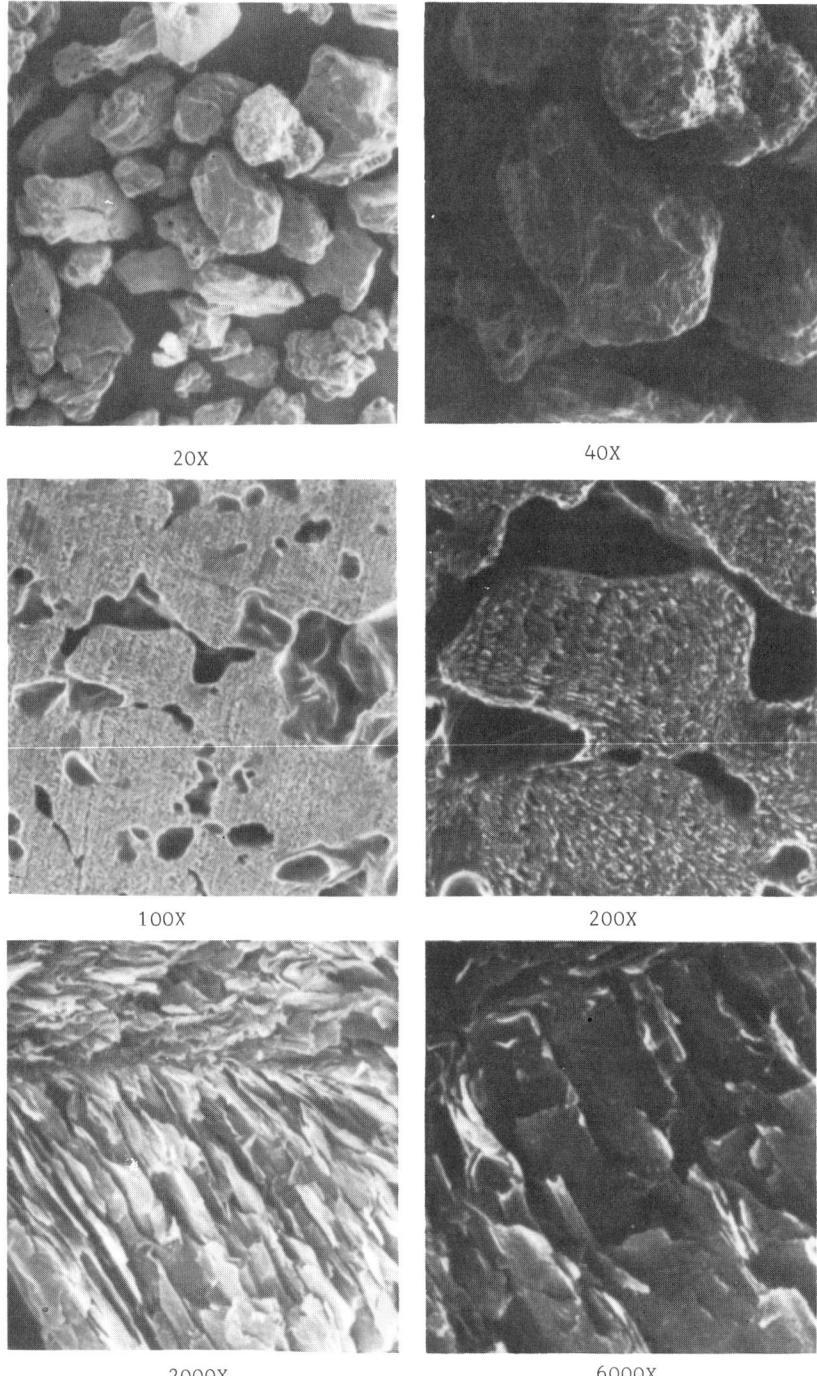

Figure 4. Structure of Desulco—Photomicrographs of Desulco Particles at Various Degrees of Magnification.

Industrial Use of Desulco

The electric furnace melting of steel scrap rather than cupola melting is a trend that will probably continue resulting in increasing need for recarburisers. And as the ductile iron segment of the foundry industry grows, so will the demand for consistent materials that allow improved process control.

The acceptance of Desulco as a premium quality recarburiser in foundry application is due mainly to its rapid dissolution in molten iron. Figure 5 compares the rate of carbon pickup of various recarburiser materials and shows the Desulco rate to be the highest. Ash forming mineral matter and sulfur are known to impede carbon dissolution in molten iron (3,4,5); and, these impurities are removed to extremely low levels in the manufacture of Desulco. And because it is free of mineral matter, slag formers are not added to the molten metal system when Desulco is used.

It is also noteworthy that Desulco is now being used in lubricants and other graphite containing products and therefore this special type of petroleum derived carbon is expected to serve a wider range of industrial needs for many years to come.

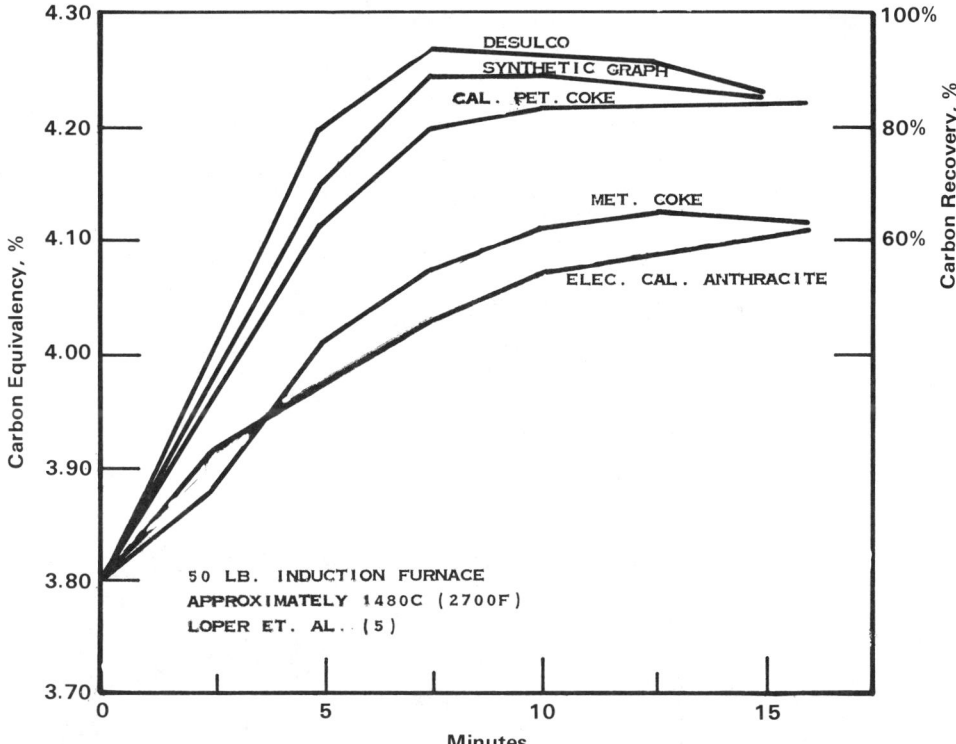

Figure 5. Increase in Carbon Equivalency Versus Solution Time for Various Carbon Raisers.

Literature Cited

1. Coates, R.B., "Types, Selection and Applications of Recarburisers in the U.K. Foundry Industry," The British Foundryman 1979, 72, 178.
2. Liggett, L.M., "Carbon-Baked and Graphitized, Products, Manufacture," Encyclopedia of Chem. Tech. Ed. Kirk-Othmer John Wiley & Son: 1964. Second Edition Vol. 4 158.
3. Coon, P.M., "Some Factors Influencing the Rate of Solution of Carbon in Iron," British Cast Iron Research Association Journal Report No. 1272 July 1977, 391.
4. Coon, P.M., "Carbon Pick-Up in Electric Furnaces," British Cast Iron Research Association Journal Report 1296 March 1978, 133.
5. Loper, C. R. Jr., Shirvani, S. and Liu, S.L. "Carbon Saturation Study of Various Carbon Raisers Added to Cast Iron Melts," Dept. of Met. and Min. Engrg., Univ. Wisconsin May 1983 Report to Superior Graphite Company.

RECEIVED February 12, 1985

Carbonization and Coke Characterization

Harald Tillmanns

Sigri Elektrographit GmbH, Werk Griesheim, 6230 Frankfurt (M) 83, Federal Republic of Germany

> When the carbonization process is divided into its distinct physical and chemical parts and both are considered according to their contributions to the overall process, only then is a description of the mechanism possible. Carbon precursors and the products of their carbonization are characterized by various test methods whose objectives can be the control of coking, a description of the carbon or the determination of its suitability for further application. This paper considers the significance of selected common characterization procedures.

The carbonization of natural or industrial hydrocarbon mixtures annually produces about 2.2 billion tons of solid carbon which, depending on the final heat treatment temperature, is called coke or graphite. A wide range of premium carbons that are not based directly on coal, produced by plastic phase pyrolysis of fusible or liquid isotropic hydrocarbons, constitute about one percent (22 million t/a) of the total solid carbon material produced; included are:

	% of Total Production
• Highly orientated carbons (e.g., needle coke)	5
• Mosaic structured carbons (e.g., regular coke)	75
• Nearly isotropic structured carbon (e.g., binder coke)	13
• Other carbon products (e.g., isotropic coke, glassy carbon and carbon fiber)	7

Three industries produce carbon products based on the carbonization process:

- crude oil and coal tar refining,
- carbon manufacturing, and
- Fe and Al metallurgical industries.

Industries concerned with the carbonization process, the feedstocks, or the products thereof are identified in Figure 1. The areas of activity of the different industries interact, as shown by the overlapping of the carbon products of common interest.

Basically the same testing methods are used for either studying the carbonization process or characterizing carbon materials. In studying carbonization, however, the testing procedures are applied under non-steady state conditions (so called "in situ" measurements) or are applied under steady state conditions with the reaction split into single steps by batch type reaction systems. There are two basic testing strategies for the several objectives of studying carbon materials or the carbonization process:

First: Measure physical and chemical properties of the material (feedstock, intermediate product, or final product).

Second: Measure "empirical data" to describe the material; these are strongly dependent on the testing procedure.

Objectives: Describe carbon material properties or the changes of properties during manufacturing.

Predict manufacturing behavior (coking, calcining, baking, graphitizing).

Predict properties of the product (coke, carbon artifacts, graphite artifacts).

Predict applications behavior.

Measurement of physical or chemical properties result in reproducible data which depend only on the properties of the sample if proper testing methods are used. In comparison, measurement of "empirical data" is determined mainly by sample preparation or testing parameters and no purely material-dependent properties can be evaluated. These principles must be considered if results of different testing methods are compared; reliability of results and conclusions are constrained by the validity of the testing method.

A general description of the carbonization processes using the characterization methods for widely used coke products like binder coke, regular coke, and needle coke is given in this review. The primary aim of the author is to present the following three main aspects of carbonization, based on the schematic model of mesophase- and coke-formation given in Figure 2:

1. differentiating chemical and physical processes;

2. determining the influence of quinoline insoluble constituents in the coking feedstocks; and

3. measuring the carbonization process kinetics.

The different manufacturing techniques for highly specialized carbon products (i.e., isotropic coke, glassy carbon, and carbon fiber) are not covered by this paper. Moreover, a discussion of the multiplicity of testing methods used to study the carbonization behavior of different feedstocks and to characterize different coke qualities is also beyond the scope of this presentation.

Physical and Chemical Pyrolysis Processes

Two aspects of the pyrolysis process must be distinguished:

- distillation as the physical aspect and

- thermal decomposition as the chemical aspect.

The objectives of the testing methods used are to separate the physical and chemical aspects and to define particular states during pyrolysis.

A number of assumptions have to be taken into consideration in evaluating the proposed model and understanding its limitations and range of interpretations. These are:

- The initial material is a pure hydrocarbon containing only hydrogen and carbon or the heteroatom content is so small that it has no effect on the data within the precision of the testing method.

- Pyrolysis is a combination of distillation and thermal decomposition.

- Carbonization is pure thermal decomposition forming a volatile, low molecular weight fraction and a residual high molecular weight fraction by disproportionation reactions.

- Thermal decomposition is dominated by radical reactions.

- Increased temperature decreases the H/C ratio of the hydrocarbons distilled in pyrolysis.

Pure decarbonization is not relevant for hydrocarbon pyrolysis, but may be of interest for high temperature treatment of solid carbon material like coke.

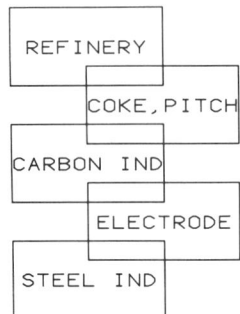

Figure 1. Relation of carbon handling industries.

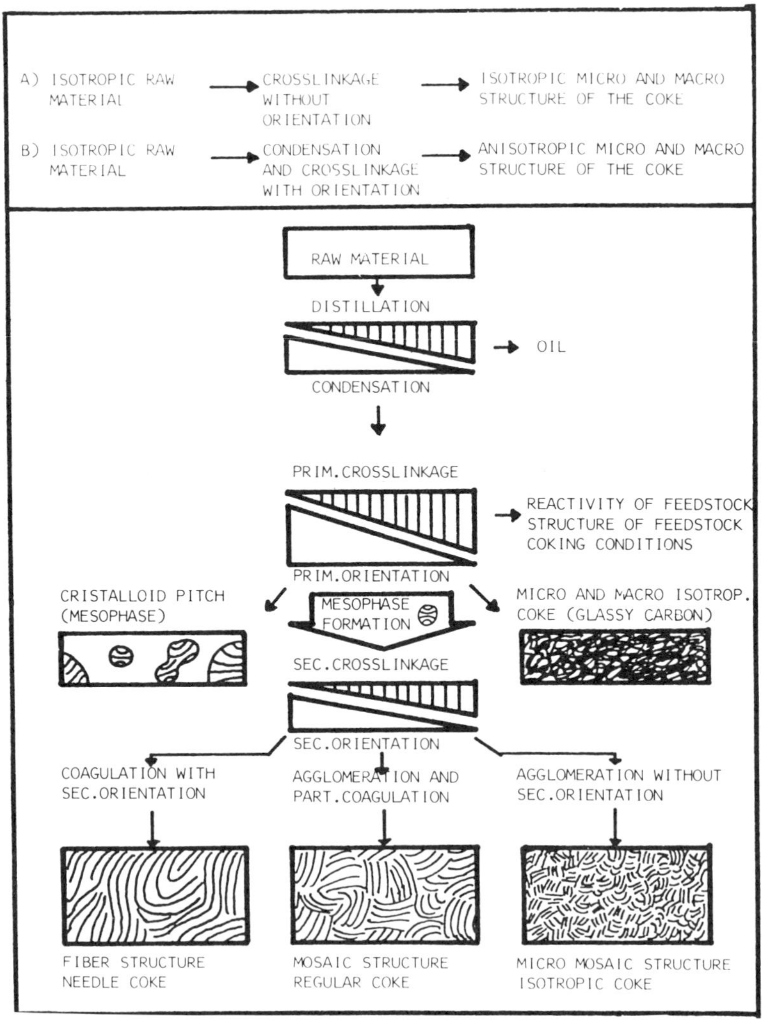

Figure 2. Model of coke formation.

Different processes contributing to pyrolysis are listed in Figure 3. There is a basic difference between the first and the subsequent three processes. Distillation removes molecules unchanged in atomic composition, whereas the other three processes cause a degradation of the molecules. Molecular degradation results in lower molecular weight fractions (which are volatile) and in a residual fraction consisting of non-volatile constituents and recombined radical fragments which form high molecular weight components.

A significant difference is observed in the composition of the volatile products. As illustrated in Figure 4, the hydrogen content of the total volatiles is reduced with increasing distillation temperatures. In contrast, carbonization produces an increase of the hydrogen content of the total volatile products as carbonization progresses. Experimental data shown in Figure 5 demonstrate how spin concentration and the formation of methane and hydrogen indicate the beginning of carbonization and the resultant significant change in the H/C ratio of the volatile matter.

A carbonization model results if it is assumed that, apart from distillation, the carbonization processes of the pyrolysis can be described by three parameters:

1. relative weight loss or residue,

2. relative carbon content, and

3. relative hydrogen content.

Using these three parameters the carbonizing materials or the carbonizing part of the initial material can be characterized in any reaction state by a point laying on the triangle area A-B-C in Figure 6.

The theoretical amount of mesophase (characterized by a hydrogen content of 3.5%) formed is described by the area of D-H-F for decarbonizing materials and the area of H-F-G-C for dehydrogenating materials given in Figure 7.

The intersection of area A-B-C with area D-H-F and the area H-F-G-C represents the state of 100% mesophase defined by the hydrogen content of 3.5%. The state is given in Figure 8 by the dotted line between E-F and F-C.

Pyrolysis follows a reaction path described schematically in Figure 8. The starting material (I') is characterized by its hydrogen and carbon content and its mass is related to the mass of the initial material at the start of carbonization (II). Distillation reduces the mass from I' to the 100% point (II); the reduced carbon and hydrogen content are depicted between I and II. Pure distillation, which is not part of the carbonization, is followed by the thermal degradation II to III; thus weight loss and changes in the hydrogen and carbon content are depicted. At point III, theoretical 100% mesophase is formed, which finally is converted to pure residual carbon (IV).

Distillation

$C_nH_m \longrightarrow C_nH_m\uparrow$

Dehydrogenation

$C_nH_m \longrightarrow n\cdot C + m\cdot H\uparrow$

Decarbonisation

$C_nH_m \longrightarrow C_{n-x}H_m + x\cdot C\uparrow$
$\longrightarrow n\cdot C\uparrow + m\cdot H\uparrow$

Carbonisation

$C_nH_m \longrightarrow C_{n-x}H_{m-y} + C_xH_y\uparrow$
$\longrightarrow C_{n-x} + m-y\cdot H\uparrow + C_xH_y\uparrow$

Figure 3. Possible reactions during the carbonization.

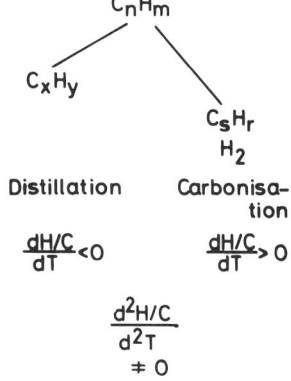

Figure 4. Distillation and carbonization.

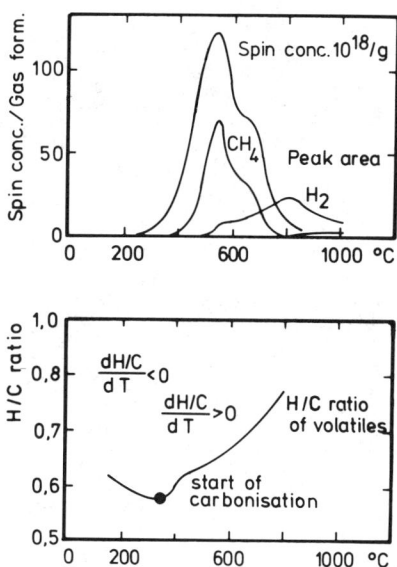

Figure 5. Type and quantity of volatiles.

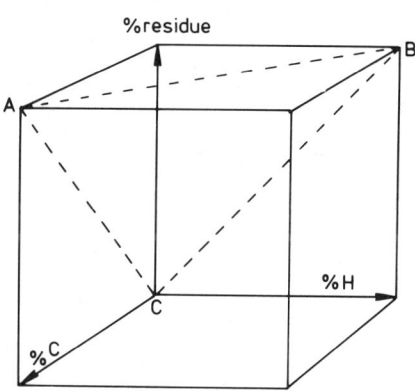

Figure 6. Model of carbonization.

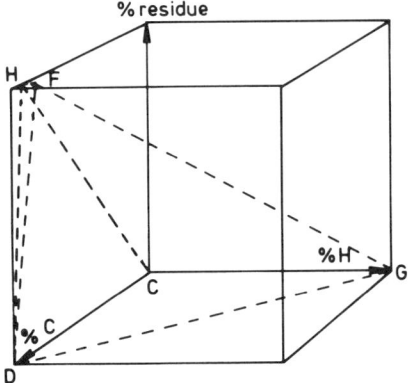

Figure 7. Model of carbonization.

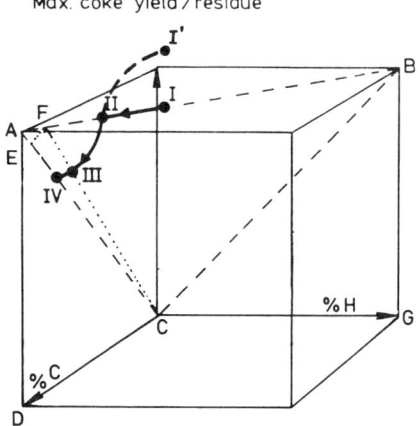

Figure 8. Model of carbonization.

Thus the pyrolysis of hydrocarbons can be generally described by the following information provided by the proposed model:

- H/C ratio of the starting material (I),
- amount of evaporable or distillable material (I', I),
- H/C ratio of the initial material at the start of carbonization (II),
- theoretical yield of mesophase (III) in comparison to residue with hydrogen content of 3.5 %, and
- carbon yield (IV) based on the initial material at the start of carbonization.

This model characterizes the reaction path of the hydrocarbon pyrolysis by four reaction states (I-II-III-IV), with each reaction state being described by three parameters:

- relative hydrogen content,
- relative carbon content, and
- residue related to the start of carbonization.

Influence of Quinoline Insolubles

The model described in the preceding section did not take into account any factor which might influence the structure of the coke formed. As can be seen in Figure 9, quinoline insolubles are formed above the temperature where weight loss has begun. In addition to the reactivity of the isotropic-liquid phase, quinoline insoluble particles within hydrocarbon mixtures are of significant importance in influencing the structure of the coke formed during carbonization. Generally, two different types of quinoline insolubles can be distinguished:

- primary or original type of quinoline insolubles which are only poorly defined optically and have a low hydrogen content of 1.5%; and

- secondary or mesophase type of quinoline insolubles which are spherically shaped, are characterized by well-known optical structure, and have a hydrogen content of 3.5%.

The first question that arises with respect to the carbonization reaction is whether the different types of quinoline insolubles cause differences in the formation rate of quinoline insoluble mesophase. Figure 10 indicates there is no difference in the formation rate of primary and secondary types of quinoline insolubles. The formation rate depends not significantly on the type but strongly on the amount of quinoline insolubles present during carbonization. The data in Figure 10 show that the formation rate at 400°C is increased from 2% QI/h up to 10% QI/h if the quinoline insolubles content in the starting material is increased from 5 to 25%. The formation rate of quinoline insolubles is related nearly linearly to the quinoline insolubles content of the starting material, being (at 400°C in this experiment) 0.4% per hour per % quinoline insolubles of the starting material.

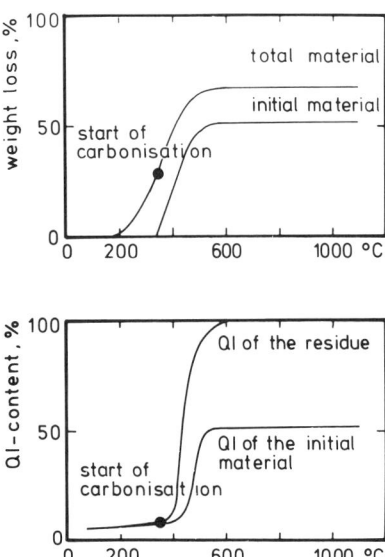

Figure 9. Weight loss and the QI formation.

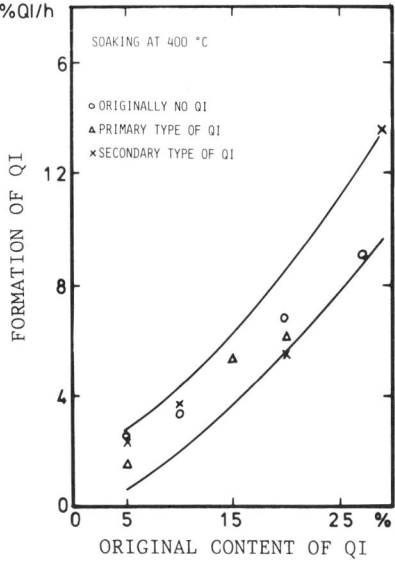

Figure 10. Formation rate of QI as a function of the original QI content.

However, as shown in Figure 11, the type of quinoline insolubles strongly influence the structure of the coke formed. Primary quinoline insolubles lead to an isotropic coke with a very high coefficient of thermal expansion ($16 \times 10^{-6}/K$). In contrast, secondary quinoline insolubles also produce nearly isotropic coke, but they hardly affect the coefficient of thermal expansion (0% sec. QI = $3.5 \times 10^{-6}/K$; 20% sec. QI = $4.1 \times 10^{-6}/K$).

There exists a wide range of other variables which affect coke structure (e.g., soluble reactivity, coking conditions, and additives) which are not discussed here. However, for all technical hydrocarbons, and especially for the coking of coal based tars and pitches, quinoline insolubles have the dominant influence on the structure.

Kinetics of Carbonization

A number of methods are used to evaluate a data set to describe the kinetics of carbonization, such as determining the temperature/time dependence of viscosity, quinoline insolubles formation, or weight loss.

Thermogravimetric analysis can follow weight loss during hydrocarbon pyrolysis. A laboratory thermobalance with a sample size of 100 mg to 1 g gives very similar results for different types of pitches as shown in Figure 12 (KS softening point: 50° and 90°C). But kinetic determinations based on using the Arrhenius equation to meet the weight loss curve must recognize that one basic assumption of this type of calculation is not met. That is the assumption of a homogeneous reaction of the carbonization process from room temperature up to the final state. In reality, it is well known that there is a drastic change in the composition of the residual material which probably modifies the reaction order, the frequency factor, and the activation energy.

In contrast to this, the technique reported herein uses the weight loss of pitch during heat treatment to evaluate a set of apparent kinetic data (reaction order, frequency factor, and activation energy), taking into account that the composition and nature of the reactants are varying.

The experimental way to solve the analytical problem is to heat up the sample to a defined weight loss at different heating rates (assuming that the ultimate composition of the residue is the same) and observe that the point of equivalent weight loss is reached at different temperature levels. At this point the temperature treatment is changed to isothermal conditions and one follows the further weight loss. This technique permits measuring weight loss rate at different reaction temperatures. The experimental procedure is illustrated schematically in Figure 13.

The resulting weight loss curves using this experimental technique are plotted for different isothermal treatment temperatures in Figure 14. The rate of weight loss versus the residue produces a linear relation. The linear lines for the isothermal weight loss curves are shifted along the overall weight loss curve with increasing temperature. In the beginning the slopes of the lines are very similar; above 400°C the slopes of the lines decrease.

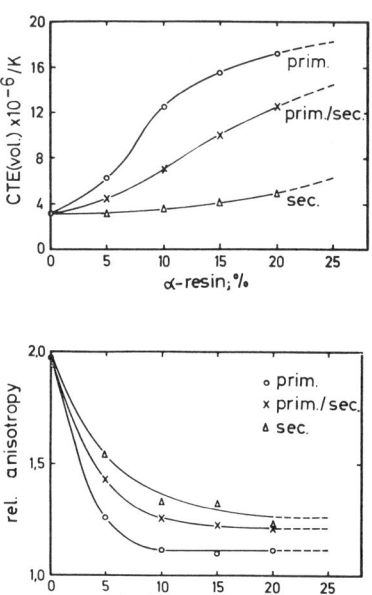

Figure 11. Influence of QI.

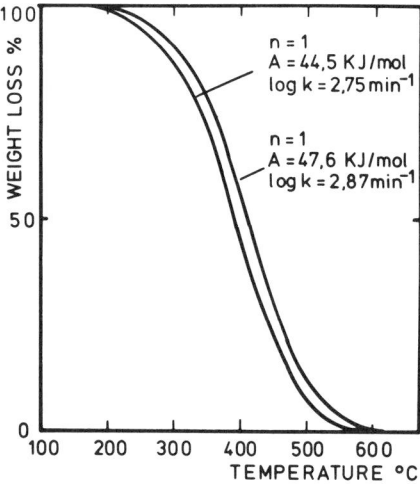

Figure 12. Thermogram of pitch.

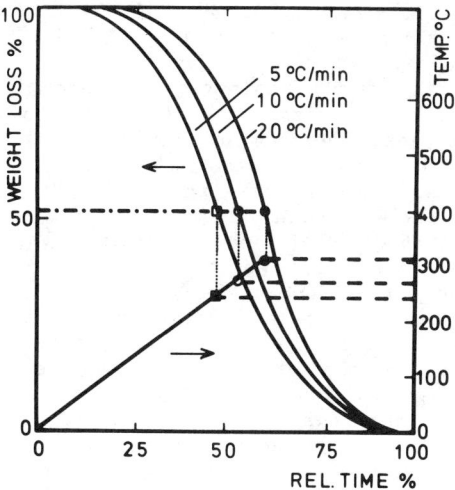

Figure 13. Scheme of evaluation.

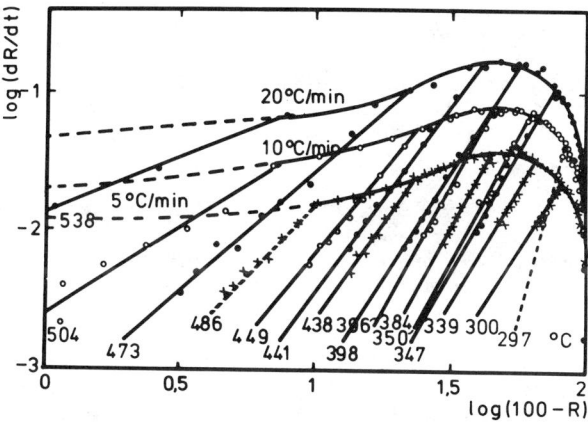

Figure 14. Rate of weight loss vs. residue.

These thermobalance data were used to calculate a set of apparent kinetic data for each state of weight loss. The Arrhenius plots (Figure 15) show an increase of the slope with increasing relative weight loss up to 76.7%; at 91.4% relative weight loss the slope is decreased.

The resulting apparent kinetic data are given in Figure 16. The first conclusion is, as expected, that carbonization is not a homogeneous process described by one set of kinetic data. The reaction order is nearly constant at about 3 up to 400°C and is determined by the experimental procedure. The activation energy increases linearly with weight loss up to 75%. Activation energies range from 30 to 600 KG/Mol. The maximum activation energy is 2 to 2.5 times the evaporation energy of aromatic compounds with a molecular weight of 700-800. The high level of activation energy is certainly caused partly by the strong interaction of the aromatic molecules in the isotropic phase which ultimately orientate and form the mesophase. The frequency factor also increases with weight loss, reaching a peak at about 75% weight loss; the subsequent decrease may be related to solidification of the residue. This drastic change in the phase of the residue must influence the apparent kinetic data significantly.

This mathematical model describing the inhomogeneous pyrolysis reactions by a set of apparent kinetic data (which are changing with the progress of the pyrolysis) should be understood as a first attempt to set up a mechanism to predict pyrolysis. The target of the application of this model would be to evaluate the influence of temperature programs on baking behavior.

Coke Characterization

All the many methods used to characterize coke samples cannot be discussed within the scope of this paper. For brevity and simplicity, this discussion on coke quality focuses on three parameters:

- hydrogen content of calcined coke to characterize the calcination severity,
- mechanical properties of coke grains, and
- irreversible dilatation of carbon artifacts caused by coke puffing.

Hydrogen content of calcined coke is often used to control commercial coke calcining severity. It is well known that overall the hydrogen content is related to final calcining temperature. But detailed analyses show that the hydrogen content alone cannot be used to deduce the final heat treatment for cokes of different quality. Besides the final temperature, there is also an influence of the residence time and the sulfur content of the green coke. Residence time is of lesser significance for technical calcining processes because a state close to equilibrium is reached. The influence of green coke sulfur content is of much greater significance; it can cause misjudgments with respect to the required calcining temperature if different quality cokes are compared.

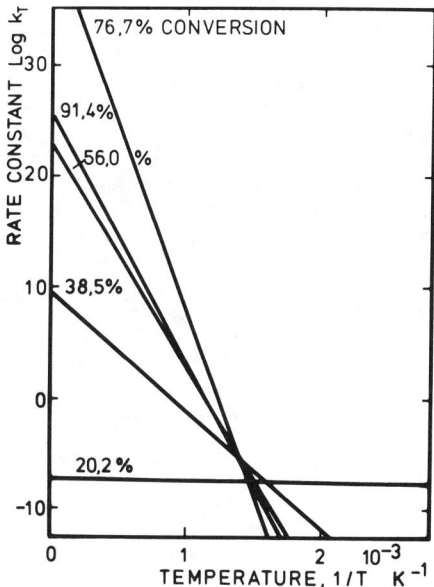

Figure 15. Arrhenius plot.

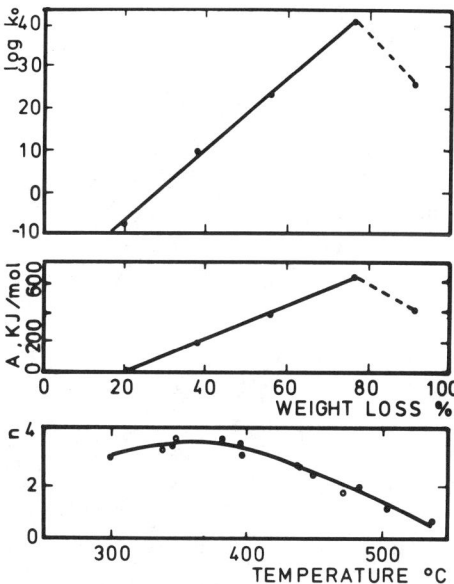

Figure 16. Kinetic data.

As shown in Figure 17, coke with a 1.5% sulfur level has the highest rate of hydrogen release, whereas the coke samples with 0.6% and 0.2% sulfur have a much smaller rate of hydrogen release. Thus, the sulfur present in green coke can result in higher hydrogen release during calcination; if the hydrogen content is solely used as the controlling parameter, over calcination of the coke could occur.

Another important property of commercial coke is the crushing strength of coke grains as an indicator of their handling sensitivity (but not the strength of the final carbon artifact). There is a good correlation between the coefficient of thermal expansion and the crushing strength (Figure 18). A second factor influencing the crushing strength is the bulk density of the coke grains.

The importance of the crushing strength in respect to the coefficient of thermal expansion will increase in the future. Sensitivity of cokes of different quality to handling during their manufacture, their transportation and manufacture of the carbon products therefore may be a criterion for selection.

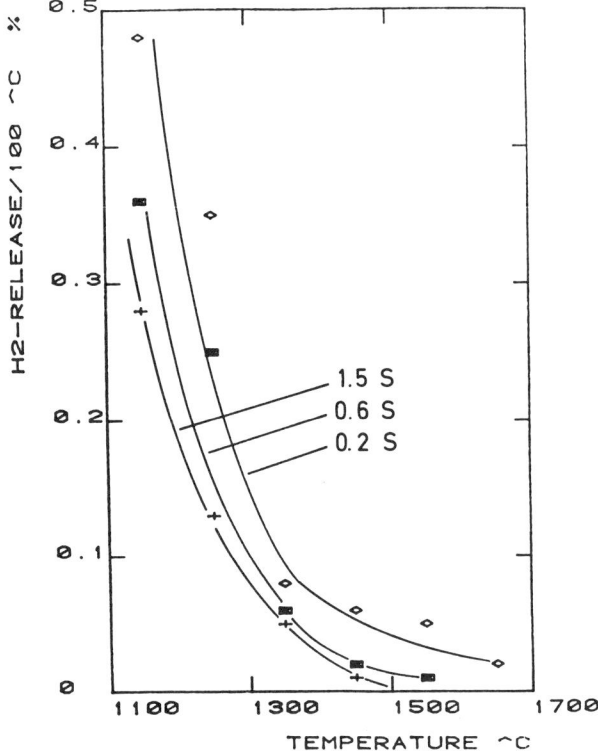

Figure 17. Hydrogen release as function of temperature.

One of the most discussed items in recent times is the puffing behavior of coke, especially of coal tar based needle coke. The dilatation behavior of cokes of different quality, as measured during graphitization of electrodes, are shown in Figure 19. Overall, the curves correspond to the puffing curves measured in laboratory equipment. One facet related to puffing that is seldom discussed is the fact that besides the strong elongation called puffing, a very strong shrinkage in the final state of graphitization also takes place. Insofar as the internal stress/strain situation is concerned, this latter phenomenon may cause problems similar to (or worse than) those caused by puffing.

While there are many applicable methods reported in the literature, this paper was restricted to a subjective selection of testing procedures in the hope that it would stimulate the development of better analytical techniques, recognizing that there is always a limitation of validity for each method and evaluation of results. Analysis of carbon materials and especially of the carbonization process will never lose its attraction as a fascinating application of the art of black magic.

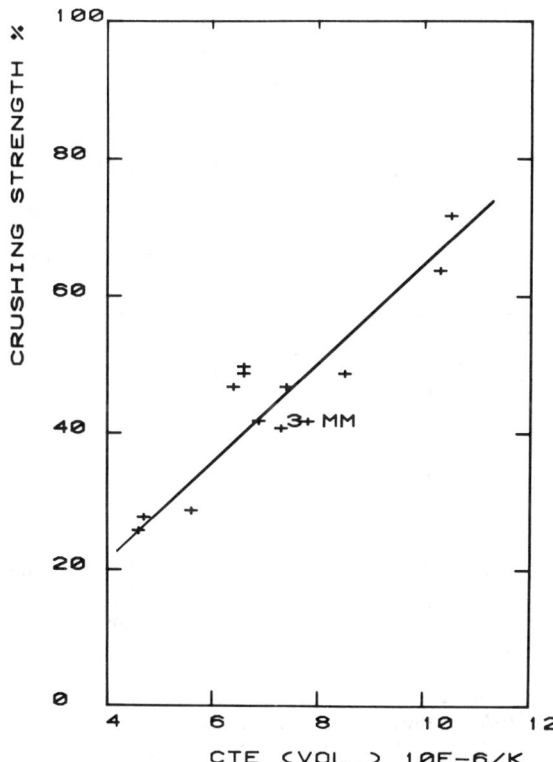

Figure 18. Correlation of crush strength and thermal expansion.

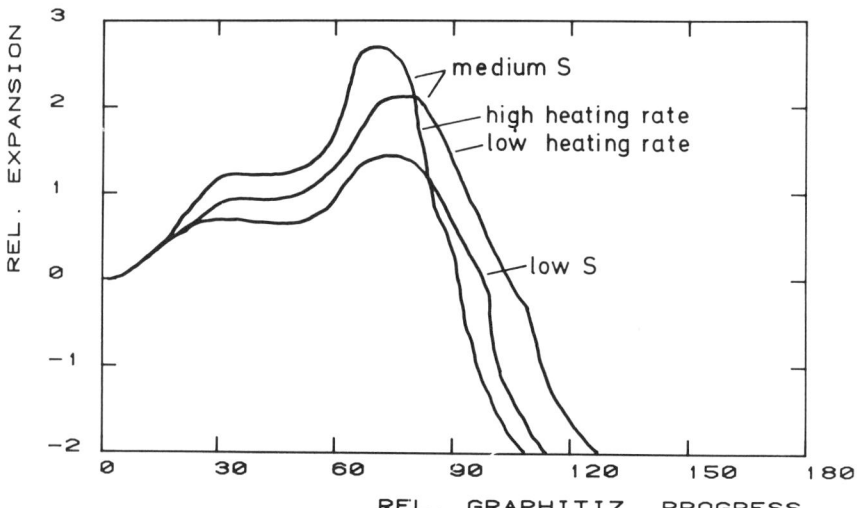

Figure 19. Dilatation curves of electrodes.

BIBLIOGRAPHY

Review Papers:

E. Fitzer, K. Mueller, W. Schaefer, "Chem. and Phys. of Carbon", M. Dekker, Inc. 1970, Vol. 7, p. 237.
K. J. Huettinger, Habilitationsschrift, Universitaet Karlsruhe, Inst. Chem. Tech. 1973.
E. Fitzer, 9th World Petroleum Congress, Tokyo 1975, review paper no. 12.
E. Fitzer, H. Tillmanns, 2nd Iranian Congress on Chemical Engineering, Teheran 1975.
H. Marsh, Ph. L. Walker, Jr., "Chem. and Phys. of Carbon", M. Dekker, Inc. 1980, Vol. 15.
H. A. Kremer, Chemistry and Industry, London No. 18, 1982, p. 702.

Carbonization, Mesophase and Quinoline Insoluble:

J. E. Zimmer, J. L. White, Proc. 12th Bienn. Conf. on Carbon, Pittsburgh 1975, p. 223.
H. Tillmanns, H. Pauls, G. Pietzka, Abstr. Carbon 76, Baden-Baden 1976, p. 374.
H. Tillmanns, G. Pietzka, Abstr. 13th Bienn. Conf. on Carbon, Irvine 1977, p. 332.
G. R. Romovacek, Proc. AIME, 1977, p. 275.
H. Pauls, G. Pietzka, Fuel, 1978, Vol. 57, p. 171.
J. W. Stadelhofer, Fuel, 1980, Vol. 59, p. 360.
A. J. Perotta, R. M. Henry, J. D. Bacha, E. W. Albaugh, High Temp. High Press. 13, 1981, p. 159.
H. Tillmanns, Abstr. 15th Bienn. Conf. on Carbon, Philadelphia 1981, p. 142.

H. A. Kremer, S. Cukier, Proc. Roy. Microscop. Soc., 1981, University of York.
J. L. White, J. F. Tellers, Jour. of Appl. Polym. Sci., Appl. Polym. Symp. No. 33, p. 137.
I. Romey, H. Glaser, R. Marrett, H. Tillmanns, 27. DGMK Haupttagung 1982, Aachen, p. 172.
H. Tillmanns, Proc. 6th Intern. Carbon and Graphite Conf., London 1982, p. 362.

Thermobalance Analysis:

E. S. Freeman, B. Carroll, J. Phys. Chem., 1958, Vol. 62, p. 394.
C. D. Doyle, et al., J. Appl. Polym. Sci., 1961, p. 285.
C. D. Doyle, et al., J. Appl. Polym. Sci., 1962, p. 639.
A. W. Coats, et al., Nature, 1964, Vol. 201, p. 68.
H. L. Friedman, J. Appl. Polym. Sci., 1964, p. 183.
C. D. Doyle, Nature, 1965, Vol. 207, p. 290.
J. Zsako, J. Phys. Chem., 1968, Vol. 72, p. 2406.
K. J. Huettinger, Erdoel und Kohle, 1970, Bd. 9, p. 559.
K. J. Huettinger, Bitumen, Teere und Asphalte, 1970, p. 528.
K. J. Huettinger, Bitumen, Teere und Asphalte, 1970, p. 487.
K. J. Huettinger, Chem. Ing. Tech., 1970, Bd. 42, p. 812.
T. Ozawa, J. of Therm. Analysis, 1976, Vol. 9, p. 217.

RECEIVED December 9, 1985

17

Anode-Carbon Usage in the Aluminum Industry

Samuel S. Jones

Industrial Carbon Consultant, P.O. Box 43698, Tucson, AZ 85733

> The aluminum industry consumes about 0.45 lb. of anode carbon for each pound of aluminum produced. The ideal carbon should have a moderately-isotropic structure with minimum oxidant-accessible surface of low, uniform reactivity, and a maximum ash content, excluding bath salts, of a few tenths of one percent. Industrial anode carbon is a baked composite usually made of calcined petroleum coke filler with a binder of coal-tar pitch coke. While there is no shortage of calcined petroleum coke, the quality is not very good and likely to become worse. Also, coke binders are subject to variability in both quality and supply. In this context, a review is presented of the complex chain of events affecting anode performance, ranging from the properties of precursors for filler cokes and binder pitches, through production of these raw materials and their fabrication into anode carbon, and concluding with anode performance evaluation in full-size prebake and Soderberg cells of different designs.

The aluminum industry consumes much more carbon, as baked anode composites, than the total of all other industrial uses for baked and graphitized carbon products. The free world's total annual aluminum production capacity is approximately 16 million short tons, about one-third being produced in the United States. World aluminum production involves the consumption (oxidation) of about eight million tons of anode carbon. Production occurs by electrolytic deposition from cryolite-alumina melts using a process patented simultaneously, but independently, in 1886 by Hall in America and Heroult in France. While minor process modifications have been made in the intervening years, and productivity greatly increased, substantially the same process is still used. The industrial electrolytic cell consists of a shallow carbon vessel about 10 ft. wide by 30 ft. long, and 1-2 ft. deep, which acts as the cathode and contains the fused salt bath and molten aluminum product. The carbon anodes are supported above the cathode and lowered into the cell at the rate of about one inch per day as the carbon is oxidized. The cell operates at temperatures in the 940-980°C range and produces approximately a ton of aluminum per day with a current of about 150,000 amperes (4-6 amp/in^2 current density) and a cell potential of 4-5 volts.

The anode carbon for the cell is usually a baked composite of calcined petroleum-coke filler bound with coal-tar pitch coke. The carbon composite may either be compacted into blocks which are baked before use in the cell (prebake anode), or be baked in place (as a single block) above the cell as the green paste moves downward toward the anode electrolytic face (Soderberg anode) (1,2,3). For prebake cells, electrical connection is made by inserting a steel conductor rod, or pin, into the top of the anodes. Soderberg anodes may have either vertical (VS) or near-horizontal (HS) conductor rods.

It is the purpose of this paper to review the important factors which affect anode carbon usage in the aluminum industry. Consideration is given to the entire chain of events affecting carbon consumption, from the properties of the precursors for filler cokes and binder pitches, through production of these raw materials and their fabrication into anode carbon, and concluding with anode performance evaluation in full-size prebake and Soderberg cells of different designs.

How Anode Carbon is Consumed

Anode carbon is consumed by four substantially different mechanisms. The relative contribution of each mechanism to total consumption can vary considerably depending on many factors of cell design and operation. However, most carbon is consumed by the basic electrolytic reactions by which aluminum is deposited at the cathode while aggressive nascent oxygen attacks the anode. For example, about 79% of the carbon is removed electrolytically when atomic oxygen, as O^{-2}, is deposited with the primary formation of CO_2 at the normal current density for industrial cells. This oxidation is very aggressive and unselective in attacking the carbon surface microstructure wherever atomic oxygen is deposited. However, reaction overvoltage is lower for more reactive surfaces, and this factor determines where the oxygen will be deposited at low current densities. Overvoltage selectivity decreases with increasing current density used for industrial cells. While the details about various steps in this process remain controversial, probable steps in the process are as follows:

Dissociation of Oxygen-Carrying Complexes

$$AlOF_{x-x}^{1} = AlF_x^{3-x} + O^{-2} \qquad 1)$$

Discharge and Chemisorption of Oxygen

$$O^{-2} + xC = C_xO + 2e \qquad 2)$$

Conversion of C_xO

$$2C_xO = CO_2 \text{ (ads)} + (2x-1)C \qquad 3)$$

Desorption of CO_2
$$CO_2 \text{ (ads)} = CO_2 \text{ (g)} \qquad 4)$$

In the present example, of the 79% carbon consumed by the above mechanism, about 12% represents excess carbon consumed due to current inefficiency, the second consumption mechanism. Present-day cells have current efficiencies in the 85-95% range. Current inefficiency reactions involve back-reactions of electrolytic CO_2 with reduced metal species from the cathode. An example of this reoxidation mechanism, for dissolved aluminum, is given by the following equation:

$$Al\ (dis) + 3/2 CO_2 = 1/2 Al_2O_3 + 3/2 CO \qquad 5)$$

The third mechanism of carbon consumption is airburn of prebake anode tops and the bottom edges of Soderberg anodes during cell operation. This mechanism typically accounts for about 17% of total prebake carbon consumption, but can vary (for different cell designs) from less than 10% to about 40% during severe airburn problems. The following equation represents such airburn reactions:

$$C + O_2\ (air) = CO/CO_2 + surface\ oxides \qquad 6)$$

The fourth consumption mechanism, responsible for about 4% of total carbon consumed, is the reduction of primary CO_2 by carbon in the pores just above the electrolytic face of the anode, and by free carbon in the bath, according to the following equation:

$$CO_2 + C = 2CO \qquad 7)$$

In contrast to the aggressive, unselective attack of the first, and principal carbon removal mechanism, reaction 7 is quite selective. Factors affecting this selectivity are anode gas permeability, temperature, and the presence of impurity oxidation catalysts.

In addition to oxidation losses by the above four mechanisms, mechanical carbon loss (dusting) also occurs due to uneven oxidation of the anode surface. Furthermore, a variety of other factors, related to anode fabrication and use, can affect carbon consumption. The most important of these factors are (1) raw material quality, (2) aggregate sizing and control, (3) pitching level and control, (4) adequate paste mixing, (5) paste compaction, (6) proper baking, (7) rodding procedures, (8) oxidation coatings, and (9) cell operating variables (anode exposure, anode heat balance, bath temperature, gas exhaust rate, housekeeping). The history of anodes, from precursors for raw materials through consumption in the cell, is a complex chain of events which can greatly affect carbon usage. It is difficult to achieve optimum conditions at all points in the chain, and a variety of compromises are possible while still obtaining reasonable carbon economy.

Characteristics of Good Anode Carbon

Maximum-performing anode carbon has minimum oxidant-accessible surface of low, uniform oxidation sensitivity to air (O_2) and CO_2. This ideal carbon should also be as pure as is practical, and have the required electrical conductivity and mechanical strength to function as plant anode material. Anode carbon consists of three elements: (1) graphite crystallites in various sizes, shapes, and orientations, associated with different quantities and arrangements of (2) impurities, and (3) voids. Graphite crystallites are the basic building blocks of anode carbon. They are very anisotropic, with property values for strength, electrical and thermal conductivity, and oxidation resistance which differ markedly in the basal planes of the crystal from those at right angles to the basal planes. For this reason, the properties of bulk anode carbon will reflect the graphite microstructure of which it is composed. The property values of this structure are further affected by its association with impurities and voids. Impurities affect carbon performance by (1) acting as oxidation catalysts, (2) contributing to aluminum impurities, and, for sulfur, (3) increasing carbon consumption while contributing to environmental concerns (SO_2). Voids affect carbon performance by determining the extent of oxidant-accessible surface. In summary, the optimum anode carbon has been found to consist of graphite crystallites in a condition of intermediate disorder, and sized to reflect heat treatment in the 1100-1400°C range. Impurities are present to a maximum of a few tenths of one percent carbon ash content, excluding cell bath salt contributions. Voids are structured to reduce total porosity (usually about 30%) and to produce small closed and blind pores, rather than open connected pores, to reduce oxidant-accessible surfaces. This is usually accomplished by maximizing baked carbon density.

Anode carbon quality is specified by a property value range for certain important physical and chemical properties. The most important single property is baked apparent density (BAD). Satisfactory full-size anode performance has been obtained with BAD values in the 1.40-1.65g/cc range (4,5). However, the most common BAD range is 1.5 - 1.6g/cc. Electrical resistivity should be in the 0.0050-0.0075 ohm-cm range. Because of the low-baked condition, Soderberg carbon typically has a resistivity about 30% higher than prebake carbon. For both carbons, compressive strength should be 350-500kg/cm^2. Regarding other anode carbon mechanical and thermal properties (6), a bending strength (BS) of 60-80kg/cm^2 is considered adequate. Young's modulus (YM) should be 800-1000 times BS to provide acceptable elastic properties. Thermal conductivity (TC) should be in the 3.5-5.5W/m°C range, and the coefficient of thermal expansion (CTE) should be $3.5-5.0 \times 10^{-6}$/°C. Thermal stress resistance (TSR) is determined by the ratio, BS x TC/YM x CTE, and Alcan reports (6) that acceptable TSR values should be ≥1.50. Total carbon porosity should be <30%, with a pore spectrum which minimizes gas permeability and oxidant-accessible surfaces. Even though there is much surface area for submicron porosity, it is usually so diffusion-limited that there is essentially no oxidant-accessible surface (7). Laboratory tests are used to determine oxidation resistance of cylindrical carbon samples to air and CO_2 under conditions which indicate relative chemical reactivity (20-40% weight loss) without oxidant diffusion limitation (8). Airburn values, at 550°C range from ≤20 (excellent) to ≥80 (poor) mg/cm^2/hr. CO_2 oxidation values, at 970°C, vary from ≤10 (excellent) to ≥50 (poor) mg/cm^2/hr. Laboratory electrolytic tests of anode carbon are done under conditions where airburn does not occur. Test results are expressed as a percentage of anode consumption corresponding to CO_2 formation at 100% Faraday efficiency. Good prebake carbon gives test values in the 110-115% range. Soderberg carbon exhibits poorer performance, due to low baking, and acceptable values are 120-125%.

Anode Binder Properties

Anode binder pitches are complex mixtures of highly-aromatic hydrocarbons derived from high-temperature coal-tar and petroleum residues after various distillation, thermal, and sometimes oxidation treatments. Coal tar is a by-product from the high-temperature carbonization of bituminous coal to make metallurgical coke. Coal-tar pitch is the residue remaining after vacuum distillation removal of about half of the more volatile constituents. This pitch has been the binder most frequently used by the aluminum industry for anode fabrication. In principle, it is possible to make a similar binder from various petroleum residues (e.g. vacuum resids, decant oil from fluid catalytic crackers, ethylene steam cracker tar). However, with rare exceptions, most of the petroleum pitches produced have not received enough severe thermal cracking treatment to produce a product with the binding quality of coal-tar pitch.

Pitch binders consist of thousands of different molecular structures with molecular weights in the 100-5000 range. Average binder elemental composition consists of at least 90% carbon, 3-7% hydrogen, and up to several percent each of oxygen, sulfur, and nitrogen. Coal-tar pitches are usually more aromatic and more highly condensed than petroleum pitches. One method which is widely used for binder characterization is the solubility of the pitch in selected solvents, particularly benzene, toluene, or xylene, and quinoline. The more condensed pitches have lower solubility, and a typical coal-tar pitch is ~70% toluene-soluble and ~90% quinoline-soluble. The toluene-soluble (TS) fraction is referred to as oil and acts as a plasticizer, or softener, which affects binder softening point (SP). The fraction which is toluene-insoluble (TI) and quinoline soluble (QS) is referred to as beta resin, and consists of intermediate-size molecules which contribute much to coking value (CV) and bond formation between filler particles. The quinoline-insoluble (QI) fraction contains the largest binder molecules, and has the form of a coarse (micron-size) carbon black with 3-5% volatiles. The QI fraction increases bond-coke quality when present at the 10-25% level (9). When combined to form a pitch binder, oil, beta resin, and QI fractions interact so that whole pitch properties are improved over the properties of the separate fractions.

When pitch binder is pyrolyzed during the carbon bake operation, it is converted from an isotropic liquid, with no structural order, to a liquid-crystal (called mesophase) having a layered structure which is finally converted to layers of carbon atoms in a hexagonal lattice of graphite crystallites. These crystallites of binder coke become more disordered and crosslinked into a more-isotropic coke as the pitch QI content increases. Such moderately-isotropic coke, in contrast to highly-anisotropic microstructure (10), is preferred binder coke because it forms both physical and chemical bonds between filler coke particles which are stronger and more oxidation-resistant (8,9).

Regarding pitch property values, it is desirable to use the highest SP (e.g. 110-120°C) that plant facilities will permit, since

this means fewer volatiles (reducing the carcinogen hazard) and
more binder carbon (and probably better performance) in the baked
anode. The TI content (usually >30%) can be quite low in some
petroleum pitches, but helps indicate the degree of condensation.
QI in coal-tar pitches (e.g. 10-20%) is beneficial, and gives a
better (denser, stronger, more conductive) carbon at levels up
to at least 20% (9). The Conradson CV should be as high as possible
(e.g. 55-60%), and anode properties tend to become marginal as CV
drops below 50%, which can occur with some petroleum pitches (4).
The lower the distillation fraction to 360°C at one atmosphere
pressure (e.g. <5%), the fewer light hydrocarbons (and carcinogens)
are lost during coking, resulting in a higher CV. Specific gravity
(SG) is an indicator of degree of condensation, and should be as
high as possible (>1.25) since this means more bond coke and better
performance for the anode carbon. The C/H atomic ratio also indicates condensation level, and should not be less than ~1.5 (1.7
is common for coal-tar pitches). Finally, binders and particularly
petroleum pitches may have somewhat lower ash levels (e.g. <0.25%)
than available filler cokes. This helps make the binder coke more
oxidation-resistant and adds less impurity to the metal product.
Regarding sulfur content, coal-tar pitches typically have <1%S
(e.g. 0.5-0.7%S), while petroleum pitches may contain much more
(e.g. 3-5%S), which might result in excessive plant SO_2 emission
under some circumstances.

Filler Coke Properties

Coke is produced when organic matter is heated to 400-600°C, essentially in the absence of air. The organic matter used for anode
binder coke has come primarily from coal tar, with minor amounts
from petroleum residues. In contrast, filler coke is produced almost entirely from petroleum, with minor amounts from coal-tar
pitch. Also in recent years, solvent-refined coal (SRC) filler coke
has been found to produce high quality anode carbon.

Coke has an elemental composition of over 80% carbon, with varying amounts of hydrogen, nitrogen, sulfur, oxygen, and a few tenths
of one percent impurities. The specific composition depends on coke
heat treatment, with the carbon content increasing with temperature.
Coke is available in (1) the green or raw state corresponding to a
temperature of ~450°C, and (2) the calcined state corresponding to a
temperature of 1300-1400°C. Green coke is ~85-95% carbon, 3-4.5%
hydrogen, 0.5-2.5% nitrogen, 0.5-6.0% sulfur, and 1.0-8.0% oxygen,
with 6-15% volatile matter (VCM), exclusive of moisture, released
when heated to 950°C. Calcined coke is ~95-98% carbon, 0.03-0.06%
hydrogen, 0.5-1% nitrogen, 0.5-5% sulfur, and 0.1-1% oxygen, with
<0.5% volatile matter. Calcined coke is thus rather pure carbon
(except for sulfur content) as graphite crystallites, associated with
voids, and a few tenths of one percent metallic impurities.

Filler coke is formed by the same general mechanism as that
already described for binder coke. However, the feedstocks used are
various petroleum residual fractions, instead of coal tar. Temperatures of 400-500°C convert these resids into green coke within a day.
A complex series of endothermic pyrolysis reactions produce liquid-
crystal mesophase which is transformed to a carbon polymer of generally graphitic structure. However, there are varying amounts of

structural disorder caused by exothermic crosslinking reactions. The result is coke with a more or less isotropic structure (11), and related property values, depending on the degree of order of the graphite crystallites of which it is composed. If the coke is very isotropic (disordered), it will tend to be more dense, less pure, and less porous (more blind and closed pores), with a high CTE. It will also tend to have the same property values in all directions (i.e. isotropic), reflecting properties of a random array of graphite crystallites. If the coke is very anisotropic (ordered), it will tend to have less impurities, to be more porous (more open, connected pores), with a low CTE. In the extreme case, the product is called needle coke, and has different property values perpendicular and parallel to the graphite basal planes. This coke is sold to the graphite industry at a premium price, where its low CTE is important for producing thermal shock-resistant electrodes for electric steel furnaces. However, such coke is not desired for aluminum industry anodes, but rather a coke with intermediate disorder between the two extreme cases. The important considerations are to obtain (1) high bulk density, (2) low oxidant-accessible surface, (3) moderate thermal shock resistance, and (4) adequate purity, at a reasonable price. Coke impurities tend to increase as the coke becomes more isotropic because one of the principal crosslinking agents in coker feedstocks is asphaltenes, which contain many of the crude oil impurities.

About 90% of all coke is produced by a batch process in which the coker feedstock is heated to about 450°C and then held or "delayed" in large tanks, or drums, for periods up to one day (12,13, 14). The product is called delayed coke (DC), and each coke drum produces 500-1000 tons of coke per day. The remaining 10% of petroleum coke is produced by a continuous fluid-bed process (12, 14). The product, called fluid coke (FC), is produced by coking of liquid feedstock coating tiny coke seeds as they are agitated on a fluid bed, where the temperature is 100-150°C above that for delayed cokers. Relatively little fluid coke is used in the aluminum industry because it is often high in impurities (sulfur and metals), and available only as submillimeter particles which are harder to grind and bind into the anode.

Coke for the aluminum industry must be calcined before use to produce quality anode carbon. This calcined coke should be relatively hard, strong, dense, with low electrical resistivity and oxidation sensitivity, high purity, and available in aggregate sizing from -1 inch particles to cover standard anode filler sizing requirements. The desired range of property values is as as given in Table I.

Rates given in Table I for coke aggregate are measured under conditions which indicate relative reactivity, not limited by oxidant diffusion rates.

Influence of Coking Variables on Coke Quality

Petroleum coke is produced at oil refineries as a method for disposal of very heavy low-value residual fractions of crude oil to

Table I. Desired range property values for calcined coke.

Property	Property Values	
	Delayed Coke	Fluid Coke
Hardgrove Grindability	35-45	28-35
Particle Sizing	-1in/+28mesh	-28/+200mesh
Bulk Density(-8/+14)lbs/ft^3	48-58	70-80
Real Density(kerosene)g/cc	2.04-2.08	1.92-1.98
Volatile Matter:		
moisture,%	<<0.5	<<0.5
VCM,%	<0.5	<0.5
Specific Electrical Resistivity(-35/+65)ohm-in	0.036-0.043	0.036-0.043
Ash,%	<0.5	<0.3
Impurities:		
S,%	<3	<5
Si,%	<0.03	<0.02
Fe,%	<0.03	<0.03
V,%	<0.02	<0.04
Airburn: 550°C(-28/+200)mg/cm^2/hr	5-80	100-150
CO_2 oxidation: 970°C(-28/+200)mg/cm^2/hr	5-20	2-8

obtain lower-boiling cracked products. Coke amounts to 2-3% of the refinery's product yield, and some refineries produce no coke at all. A refinery is a rather complex operation for processing crude petroleum to produce high-profit products, such as gasoline, fuel oil, and lubricants (12). A large refinery can process about 500,000 bbls of crude (i.e. one supertanker load) per day. The crude is first distilled, and some of the distillation fractions are then subjected to other operations, such as alkylation, thermal and catalytic cracking to produce the desired products. The collective process operations result in a series of heavy residual fractions (e.g. distillation resids, decant oil from catalytic cracking, thermal tar from thermal cracking, asphaltics from a lube train (13)) which may be fed to a delayed or fluid coker. In delayed coking, a cyclic batch operation, the coker feed is preheated to about 450-500°C, and then held in a coke drum at pressures of 5-10psi. These conditions cause coke formation within hours, while cracked vapors leave the top of the drum for further processing. Refractory feed may be recycled through the drum to increase the coke yield. When the drum is mostly filled with coke, usually within 24 hours, coker feed flow is switched to another drum. The full drum is steamed out, opened, and the coke removed using hydraulic cutter nozzles with high-pressure water. Fluid coking is a continuous process in which the heated coker feed is sprayed into a fluidized bed of hot coke particles which is maintained at 20-40psi and temperatures well above 500°C. Heat is supplied by recirculating coke particles to a burner vessel. In the coking vessel, or reactor, the feed vaporizes and cracks while forming a liquid film on the particle surfaces. The film thickness must be carefully controlled, by coker feed rate, to prevent excessive sticking of coke particles and loss of fluidization. The product is removed continuously and seed particles are added to maintain fluidization.

The quality of coker feedstock and control of coker operation have an important influence on green coke quality. The purity and density of the feed strongly affects coke purity and yield. Also, the presence of crosslinking agents (e.g. asphaltenes) affects both carbon yield and coke structure. Coke yield and structure are both affected by recycle ratio. As coking severity (time, temperature) increases, coke VCM decreases. Green coke VCM affects the ease of coke removal from the drum, the sizing of the coke removed, and the bulk density of the calcined coke (15). Experience indicates the following VCM ranges for good calcined coke property values: (1) ordinary semi-isotropic delayed, 10-12%, (2) moderately-anisotropic delayed, 8-10%, (3) highly-anisotropic needle delayed, 6-8%, (4) fluid, 5-7%.

There are basic differences in coke sizing and structure for delayed and fluid cokes because of differences in the two coking operations. Sometimes, fluidization occurs in delayed coke drums when the feed has a high asphaltene content and the recycle ratio is low. Such conditions tend to produce millimeter-size "shot" coke, which is very isotropic, and yields a calcined coke which is very hard with low porosity and high density. Such coke is hard to grind and bind into anode carbon, but gives high performance when properly used. For delayed cokers, coke quality also varies with height in the coke drum. Coking severity decreases from bottom to top of the drum, so that bottom coke is more dense with lower VCM than top coke. Since commercial bulk coke is not segregated as it comes from the drum, fluctuations in property values will occur.

How Calcination Affects Coke Quality

It is necessary to calcine green coke, for anode carbon use, to prevent intolerable shrinkage cracks in the baked carbon. The normal calcination process involves exposure of green coke to temperatures up to 1400°C, without using heatup rates high enough to puff the coke and so reduce bulk density. At maximum temperature, calcination continues until specifications for real density and/or resistivity have been met. Most commercial coke calcination is done using rotary kiln calciners (16). These units consist of a heated steel pipe, 8-10ft in diameter and 200-250 ft. long, lined with refractory, which is tilted at a small angle to horizontal and rotates at a few rpm. Green coke enters the upper end and tumbles in a thin cylindrical chord section through the pipe. Some commercial calcination is done using rotary hearth calciners. In this case, green coke is deposited at the outer edge of a slowly-rotating circular refractory table about 40-80 ft. in diameter. The coke is guided by rabbles toward the center of the table where it falls into a soaking pit before removal from the calciner. Both rotary kiln and rotary hearth calciners have been developed into more energy-efficient and environmentally-acceptable equipment.

Much of the energy required for coke calcination is now supplied by burning volatiles from the green coke. Originally, rotary kiln calciners were end-fired with natural gas (17). Then, kiln-mounted air blowers were attached in the central region of the kiln (16), and the energy produced has greatly reduced the amount of natural gas required. The use of lifters for mixing the coke charge is another recent improvement for rotary kiln calciners (18), which appears to have substantially improved the uniformity of coke calcination, and increased coke bulk density by reducing the heatup rate.

During typical rotary kiln calciner operation, coke feed rates are in the range of 20-50 tons per hour. As the green coke enters the low-temperature end of the calciner, residual moisture is flashed to steam and the coke is heated to temperatures last experienced in the coker. At this stage, Zone 1 in the calciner, the coke is in a sticky, plastic state, and the chord section of the coke charge has a relatively high angle of repose on the wall of the rotating calciner. As calciner temperature increases further, maximum coker temperature is exceeded and the coke begins to emit large quantities of hydrogen and light hydrocarbons. This is Zone 2, near the center of the calciner, where vigorous combustion occurs using air supplied by the kiln-mounted blowers. Coke heatup rate appears greatest at the interface between Zones 1 and 2. Since almost all of the energy for kiln operation is generated in Zone 2, heatup rates can easily exceed 100°C/min., particularly if the calciner is operating at maximum temperature and the coke charge is not kept well mixed (15, 17, 19). Such high heatup rates are known to seriously reduce coke bulk density (19). Another characteristic of Zone 2 is reduction of the angle of repose of the coke charge from that in Zone 1. This occurs because the coke loses its sticky, plastic character in Zone 2 and becomes fluidized from volatiles emission. After coke volatiles emission is complete, the charge is no longer fluidized. The coke then arrives at the back section of the calciner, Zone 3, where the angle of repose of the unfluidized charge is greater than that in Zone 2. Heat treatment is completed in Zone 3, and the coke then leaves the calciner after a residence time of about one hour. The coke is finally water-cooled and is usually given a coating of some petroleum fraction to reduce the dust level in later coke processing for anode fabrication.

The calcining operation can have an important influence on coke quality. In principle, cokes with significantly different volatiles contents (quality and quantity), microstructures, and/or impurity levels should be calcined differently. For example, high-VCM cokes require lower heatup rates. Also anisotropic (ordered) cokes require less calcination than isotropic (disordered) cokes (20). In practice, custom calcination is rarely done. Instead, green cokes with substantially different property values are often mixed so as to produce a calcined product with acceptable average property values. It is important that the calcined coke have the property values previously specified, particularly as they apply to bulk density, real density, resistivity, and purity. To produce high-quality calcined coke, it is important to minimize coke puffing from volatiles release, with controlled heatup rate, and to offset actual puffing with shrinkage occurring above 700°C (20-21). With optimum counterbalance of these opposing factors, higher bulk density (and lower oxidant-accessible surface) is achieved. Coke real density and resistivity values are achieved mostly by calcination time near maximum temperature. However, normal calcination has essentially no beneficial influence on coke purity. In fact, calcined coke ash levels tend to be somewhat greater than those for green coke due to volatiles removal. Sulfur is the only important impurity which can be reduced by calcination, if severity is increased by using a maximum temperature near 1500°C (5). Such thermally-desulfurized coke has substantially greater submicron porosity and reduced bulk density, but the oxidant-accessible surface is not increased. It has been determined that thermally-desulfurized coke will produce satisfactory anodes (22). However, the practicality of this method for improving coke quality has not yet been reported.

Important Anode Fabrication Factors

The objective of making good anode carbon, which has the property values already specified, involves four important operations which are applied differently for Soderberg and prebake anode fabrication. These four operations are (1) filler aggregate sizing (2) paste pitching level determination, (3) paste compaction, and (4) compacted composite baking.

The most important objective for anode filler aggregate sizing is to obtain a high vibrated aggregate bulk density (23). For prebake anodes, maximum aggregate bulk density is required to achieve maximum baked carbon density. For Soderberg anodes, this requirement is somewhat modified by paste rheological requirements related to Soderberg cell operation. For most cokes, bulk density increases as particle size is reduced because the large void fraction due to coarse pores is progressively reduced with decreasing particle size. In general, coarse particles act as support pins to hold the composite together, and reduce bake shrinkage while requiring less binder pitch and having less oxidant-accessible surface. However, it is important that these particles be dense as the coarse fraction increases, to prevent reduction in composite compressive strength. Fine coke particles, on the other hand, exhibit more surface, require more binder pitch, and increase composite bake shrinkage. However, in a properly pitched composite, they will increase carbon conductivity and strength. Thus, what is needed is an aggregate balance which emphasizes overall particle sizing skewed toward coarse particles and high vibrated aggregate bulk density to achieve good baked carbon property values and maximum anode performance (24, 25).

Filler aggregate selection is somewhat different for prebake and Soderberg anodes. Prebake plants use four aggregate fractions: (1) coarse butts, and (2) coarse, (3) intermediate, (4) fine coke particles. The butt fraction results from the need to recycle about 25% of the anode carbon near the supporting steel stubs. Butt particles are usually more dense than coke particles due to cell bath impregnation, and tend to increase anode carbon density and conductivity. Maximum butt particle size is about one inch, with finer particle sizing extending through the three coke size fractions (26). The sizing ranges for the coke fractions are approximately as follows: coarse ($-\frac{1}{4}$in/+28 mesh); intermediate (-28/+100), and fine (-100). Soderberg anodes do not have a butt fraction, but the coarse coke fraction extends up to $\sim\frac{1}{2}$in particles, with essentially the same boundaries between intermediate and fine fractions as those for prebake anodes. For both anode types, high vibrated aggregate bulk density is achieved by a series of statistically-designed experiments to obtain the desired result most efficiently.

Once aggregate sizing is determined, the requirements for achieving optimum paste pitching level are different for prebake and Soderberg anodes. Prebake aggregate requires about 15% pitch to produce maximum baked carbon density, strength, and electrical conductivity. This value relates rather directly to the surface area of the aggregate particles coated. The optimum value can be determined by measuring the properties of baked carbon produced from paste batches which are incrementally-pitched about this value. Some plants use a formula based on particle surface area to determine a specific amount of binder. However, since this method does not consider coke particle porosity or the void volume between particles, it is often desirable to do an iteration based on coke and aggregate property values to optimize the pitching level. For Soderberg anodes, about twice as much pitch is used as that for prebake carbon. While the same dense, low-surface carbon is desired for both anode types, Soderberg anode operations require that paste rheology be considered in addition to the prebake factors already discussed. For this reason, maximum aggregate size is somewhat smaller for Soderberg anodes, and more pitch is used to obtain the necessary paste flow characteristics so that paste batches will spread in place and bond to previous batches on the top of operating anodes. To achieve the desired paste rheology, simple paste flow or elongation, is within the required range.

After the optimum pitching level has been determined, anode paste is usually mixed for periods up to an hour, while the paste is at least 60°C above the binder softening point. During this operation, the pitch must coat the filler particles uniformly, to produce well-mixed paste. As a properly mixed condition is approached, compacted paste green density will increase to a constant maximum value. The paste is then ready for compaction into green anode composites. For Soderberg anodes, final paste gravity-compaction occurs on the operating anode top after the paste batch is added, either as loose paste, or as low-pressure ($\leq$1000psi) compacted briquettes. For prebake anodes, the paste is either pressure-molded or vibratory-compacted into blocks which must be baked before use in an operating cell. Pressure-molding techniques usually involve the single (high pressure) or multiple (low-pressure) application of pressures in the 500-7000psi range, to paste which has been cooled to 5-10°C above the binder softening point. Specifically developed compaction methods also include control of loading rate, holding time, and release rate to produce composites with acceptable property values. Vibratory compaction techniques employ much lower pressures ($\sim$100psi), but produce good carbon quality. This method also has the advantage of requiring a smaller plant capital investment than that for pressure-mold compaction. There is some evidence that the two methods give similar results for high-modulus (elastic) cokes, with some advantage for pressure-molding for low-modulus cokes.

The final step in anode fabrication is baking the compacted composite. For Soderberg anodes, baking occurs during cell operation and the maximum finishing temperature can be no higher than bath temperature (940-980°C). This is a relatively low final temperature, which adversely affects performance. Anode heatup rate must also be kept very low ($\sim$1°C/hr.) up to 600°C for high-pitched Soderberg composites to achieve high baked density (27, 28). Such low heatup rates allow for pitch expansion, flow, and volatiles evolution which might otherwise puff the baking composite. For lower-pitched prebaked anodes, corresponding heatup rates can be higher ($\sim$10°C/hr.) because there are more void spaces to accommodate pitch expansion and gas evolution. Above 600°C, the heatup rate can be increased to 50-100°C/hr. without damage to the baking carbon. Prebake anodes should finish the bake operation with about ten hours near a maximum temperature of 1000-1200°C, preferably 1150+50°C. It has been found that increasing bake finishing temperature from 1000 to 1200°C reduces net carbon consumption $\sim$0.21bC/1bAl per 100°C (29). In the 900-1000°C range, the rate is three times greater, and is essentially zero in the 1200-1400°C range. Thus, temperatures above 1200°C do not improve carbon quality, and accelerate furnace deterioration from fluoride attack on the refractory. During this bake operation, normal linear shrinkage for good prebake anodes is 0.2-0.5%.

Anode Performance in the Reduction Cell

There is considerable variety in the design of both prebake and Soderberg cells. For Soderberg cells, the electrical conductor pins, or studs, may either be inserted almost-horizontally (HS) into the side of the anode, or vertically (VS) into the anode top. Since Soderberg anodes emit undesirable (carcinogenic) pitch fumes during baking (30), successful efforts have been made to reduce the anode pitching level from the older "wet top" condition to the more recent "dry top" condition (31). For prebake cells, anode height varies within the 16-26in range, and has marked influence on the temperature profile and performance. The additional oxidation exposure of tall anodes causes greater sensitivity to oxidation catalysts (e.g. V) impurity levels. Another factor affecting anode temperature and performance is the location and manner of crust-breaking for feeding alumina to the cell. In particular, large centerbreak openings which do not close between breaks (possibly due to alumina crusting behavior) can adversely affect anode performance. Cells which operate at low ratio (e.g. 1.1) usually have lower bath temperatures, which increases anode performance in addition to the airburn-reduction characteristics of the added AlF_3 associated with low ratio. Finally, any condition (e.g. poor current distribution, or dusting) which produces a "sick" (hot) cell will put added thermal and oxidation stresses on the anodes, which can reduce performance.

The principal anode performance problem of Soderberg cells is the low-baked anode carbon. This results in preferential attack on binder coke and creates some level of filler dust problem as a standard operating condition. For VS Soderberg anodes, there is additional performance loss for the lower-quality pinhole carbon, which fills the space created when pins are reset. This is due to porosity created when pinhole paste is baked in place by the existing excessive heatup rates. VS Soderberg anodes are also adversely affected if the carbon has a high sulfur content. Conductor pin tips will become coated with an iron sulfide scale, which interferes with electrical conduction in the anode.

Prebake anodes have a clear performance advantage over Soderberg anodes mostly because of the absence of the problems just described. Prebake cells can still have dust problems if the anodes are made with filler coke which is very anisotropic (ordered) and too highly calcined. Also, if dispersion in bake finishing temperatures is too broad (>100°C), or includes too many (>10%) low-baked anodes, bad anodes will tend to reduce the performance of other anodes which would otherwise be of acceptable quality. Prebake anodes also exhibit lower performance when using high-sulfur, high-metals coke. One study (5) indicated an increase in net carbon consumption of 2-3% per 1% increased sulfur content. About half of this appears due to airburn catalyzed by impurities present with the sulfur. Thermally-desulfurized coke anodes may become a practical means to improve anode performance. While prebake anodes generally exhibit better performance than Soderberg anodes, there is some overlap in performance for the improved Soderberg cells (31) and the poorer prebake cells. The net carbon consumption range for prebake anodes is 0.41-0.491bC/1bAl. The corresponding range for Soderberg anodes is 0.45-0.541bC/1bAl.

Summary

The important factors affecting prebake and Soderberg anode carbon usage have been reviewed. The principal variables affecting usage have been emphasized for the complex chain of events from manufacture of binder and filler materials through performance in full-size electrolytic cells. In summary, the key points about anode carbon which have been discussed are (1) porosity, (2) bake level, (3) oxidation sensitivity, (4) impurities, and (5) binder-filler compatibility.

Anode porosity is important because it affects the extent of oxidant-accessible surface. This surface is influenced both by coke microstructure and the fabrication process for converting the raw materials into baked carbon. The prime requirement for good anode carbon is minimum oxidant-accessible surface. It is also desirable that this surface have a low, uniform specific reactivity. Anode surface with pores having diameters in the 1-10 micron range are accessible to oxidation unless blocked in some manner. Submicron porosity, such as that produced by thermal desulfurization of coke, is oxidant diffusion-limited and will not affect carbon consumption significantly. Increasing anode carbon density will usually increase anode performance because the oxidant-accessible surface is reduced.

Anode bake level is important because it affects binder coke reactivity. Increasing binder heat treatment converts it from a high-surface, oxidation-sensitive material to some of the most oxidation-resistant coke in the anode. Soderberg anodes have a low-baked binder disability which usually causes this anode type to exhibit lower performance than prebake anodes. Increasing bake finishing temperature from $900°C$ to $1000°C$ reduces net carbon consumption $\sim 0.061 bC/1bAl$. In the $1000°C$ to $1200°C$ range, carbon consumption is reduced $\sim 0.021 bC/1bAl$ per $100°C$. For best overall results (good anode performance, energy conservation, and low bake furnace deterioration), a finishing temperature of $1150\pm50°C$ is recommended.

Anode carbon is consumed by three different oxidation reactions, each involving a different anode surface, reaction rate, and rate-temperature dependence. Most carbon is consumed by the aggressive electrolytic attack of nascent oxygen. Lesser amounts of carbon are consumed by airburn of prebake anode tops and Soderberg bottom edges, and by primary CO_2 in pores just above the electrolytic face. All these reactions can also produce mechanical carbon loss by dusting. While no carbon can resist the aggressive electrolytic reaction, oxidation sensitivity for airburn and CO_2 oxidation is greatly affected by temperature, accessible carbon surface, and impurity catalysis.

Impurities can have four negative influences on anode performance: (1) catalysis of airburn and CO_2 oxidation reactions, (2) increase anode consumption by electrolytic oxidation, (3) contamination of product aluminum, and (4) environmental contamination. Vanadium, iron and sodium are outstanding examples of selected impurities which catalytically affect carbon oxidation by O_2 and/or CO_2. Sulfur is the most important impurity which both increases anode consumption and produces environmental contamination (SO_2). The principal anode impurities which can reduce aluminum purity are iron, silicon, titanium, vanadium and manganese.

Concerning binder-filler compatibility, the ideal anode carbon should be a pure, homogeneous, moderately-disordered carbon structure. To this end, coal-tar pitch with 10-25% QI produces the most compatible binder coke available. This binder has a high coking value and yields relatively-isotropic coke which gives strong filler-coke bonds and oxidation protection for the filler. Regarding filler coke blends, as well as binder-filler compatibility, experience indicates that coke compatibility decreases as the similarity of physical and chemical properties decreases. Important properties for these considerations are particle shape, density, strength, elastic modulus, pore structure, and impurity content. For example, with filler coke blends, maximum incompatibility has been found for blends of flaky, low-modulus, oxidation-resistant, anisotropic (ordered) delayed coke, and round, high-modulus, oxidation-sensitive, relatively-isotropic (disordered) fluid coke. In this case, anode carbon using such a filler-coke blend exhibited 15% higher carbon consumption than that carbon made with the anisotropic filler alone.

Literature Cited

1. Anderson, W.A.; Haupin, W.E. "Kirk-Othmer Enc. of Chem. Tech. 3rd Ed." 2, 129 (1978).
2. Grjotheim, K.; Krohn, C.; Malinovsky et al "Aluminum Electrolysis, 2nd Ed." 227 (1982).
3. Grjotheim, K.; Welch, B.J. Aluminum Smelter Technology, 43 (1980).
4. Jones, S.S.; Hildebrandt, R.D.; Hedlund, M.C. Paper A77-97, 106th AIME An. Mtg. (1977).
5. Jones, S.S.; Hildebrandt, R.D.; Hedlund, M.C. Light Metals 1979, 553, AIME (1979).
6. Brown, J.A.; Rhedey, P.J. Light Metals 1975, 253, AIME (1975).
7. Turkdogan, E.T.; Olsson, R.G.; Vinters, J.V. Carbon 8, 545 (1970).
8. Jones, S.S.: Hildebrandt, R.D. Light Metals 1974, 901, AIME (1974).
9. Jones, S.S.; Hildebrandt, R.D. Light Metals 1975, 291, AIME (1975).
10. Franklin, R.E. Proc. Roy. Soc. (London), A 209 (1951) 196.
11. Rhedey, P.J.; DuTremblay, D. Light Metals 1977, 301, AIME (1977).
12. Jahnig, C.E. "Kirk-Othmer Enc. of Chem. Tech. 3rd Ed." 17, 183 (1982).
13. Scott, C.B.; Folkins, H.O. Light Metals 1972, Sym. on Anode Raw Mtls., AIME (1972).
14. Kemnitzer, W.J.; Edgerton, Jr., C.D. Bureau of Mines, Info. Cir. 8259 (1965).
15. Jones, S.S.; Hildebrandt, R.D. Light Metals 1981, 423, AIME (1981).
16. Farago, F.J.; Sood, R.R. Light Metals 1976, 351, AIME (1976).
17. Li, K.W.; Friday, J.R. Carbon 12, 225 (1974).
18. Vogt, M.F.; Jones, G.R.; Tyler, G.A. Light Metals 1984, 1697, AIME (1984).
19. Rhedey, R Trans. Met. Soc., AIME, 239, 1084 (1967).
20. Wallouch, R.W.; Fair, F.V. Carbon 18, 147 (1980).
21. Whittaker, M.P.; Miller, F.C.; Fritz, H.C. Ind. Eng. Chem. Pdt. R&D, 9, 187 (1970).

22. Goodnow, W. Panel Remarks, Carb. Mtls. in Al. Ind., 111th AIME An. Mtg. (1982).
23. Belitskus, D. Light Metals 1982, 673, AIME (1982).
24. Belitskus, D. Light Metals 1978, 341, AIME (1978).
25. Rhedey, P. Light Metals 1971, 385, AIME (1971).
26. Peterson, R.W. Light Metals 1982, 691, AIME (1982).
27. Belitskus, D. Light Metals 1983, 741, AIME (1983).
28. Martirena, H. Light Metals 1983, 749, AIME (1983).
29. Jones, S.S.; Hildebrandt, R.D.; Hedlund, M.C. Extd. Abs., 14th Bien. C. Conf. 81 (1979).
30. Augood, D.R.; Jones, S.S.; Seim, H.J. Light Metals 1981, 963, AIME (1981).
31. Hosoi, H.; Sugaya, M.; Tosaka, S. Light Metals 1982, 435, AIME (1982).

RECEIVED September 10, 1985

Utilization of Petroleum Coke in Metallurgical Coke Making

Kenji Matsubara, Hidetoshi Morotomi, and Takashi Miyazu

Technical Research Center, Nippon Kokan K.K., Minamiwatarida-cho 1-1, Kawasaki-ku, Kawasaki 210, Japan

>   In Japan, NKK was the first company to use petroleum
>   coke as the source for metallurgical coke making in
>   1967. Since then, petroleum coke has been utilized
>   to increase carbon content and decrease ash content
>   of coal blends used by Japanese iron and steel
>   companies. This report includes the following
>   subjects:
>   1. The development of the procedure for the use of
>   petroleum coke by Japanese iron and steel companies.
>   2. The discussions about the effect of petroleum coke
>   on coke strength.
>   3. The pitch addition as the advanced utilization.

In order to maintain high productivity in a blast furnace, it is necessary to use high-quality coke having high strength and containing few impurities such as ash and sulfur. Strength is particularly important among the properties of coke. Since coke strength largely depends upon coalification rank and fluidity of coal blends, it is possible to manufacture high-strength coke by keeping these properties at appropriate levels.

Japan imports coking coals from the United States, Canada, Australia and many other foreign countries. Future supplies of coking coals involves some uncertainty relative to procurement of good-quality coals in sufficient quantities. Various measures have been examined to solve this problem in Japan. One such measure is the expansion of the scope of raw materials used for coke-making. Use of petroleum coke as a raw material for coke-making falls into this category. Figure 1 shows annual consumptions of petroleum coke and raw materials for coke-making by the Japanese coking industry.

With a view toward the effective utilization of carbon sources, Nippon Kokan K.K. successfully utilized petroleum coke for the first time in Japan in 1967, taking advantage of the lower ash content relative to that in coals. Now that petroleum coke is used for coke-making in the Japanese coke industry as shown in Figure 1, its application may be considered as an established technology. The ratio of its use relative to coking coals is, however, still at the very low level of 1.0%. Table I shows results of examinations

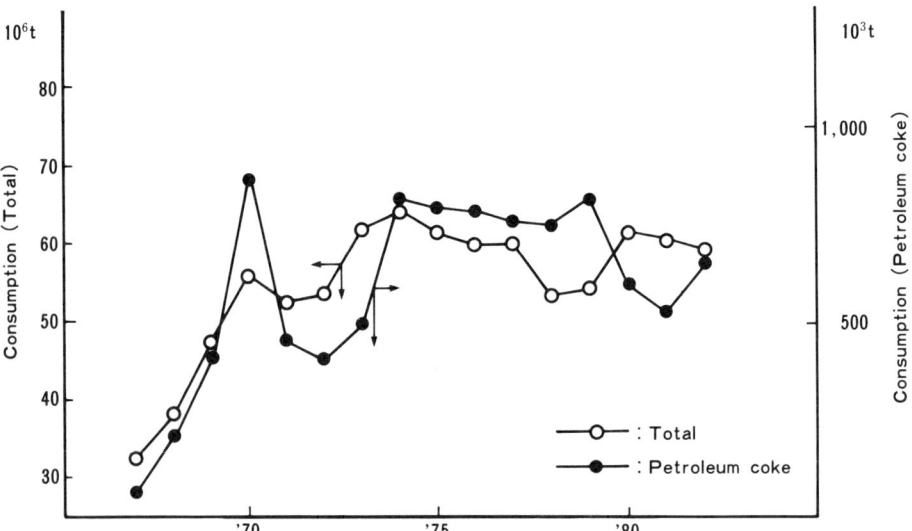

Figure 1. Annual consumption of raw materials for coke making in Japan.

Table I. Views of Japanese steel companies on use of petroleum coke

	Status of petroleum coke blending test	View on blending limit
Company A	Blending ratio: DPC: 0, 3, 5 and 7% FPC: 2, 4, 6 and 8%	1. Blending limit: 5-6%; 2. No marked difference between DPC and FPC, DPC being slightly better. 3. Increased blending of PC decreased coke ash and increased TS, and particularly because of S problem, actual use of PC is limited to about 3%.
Company B	Blending ratio: DPC: 2 and 5% FPC: 5% selective blending to partial briquetting of coal charge is also applied.	1. Blending limit is 3-5%; partial briquetting of coal charge; 2. DPC is superior to FPC.
Company C	Blending ratio: DPC: 0-15%	1. Increasing the amount of added DPC leads to almost linear decrease in DI. 2. Decrease in DI is eliminated by keeping constant Ro and T.In. and maintaining MF at a certain level.
Company D	Blending ratio: DPC: 0, 5, 10 and 15% FPC: 0, 5, 10 and 15%	1. Blending limit is about 10%; 2. No marked difference between FPC and DPC.
NKK	Blending ratio: 1) To ordinary and partial briquetting of coal charge: 6% DPC 2) To partial briquetting of coal charge: 0, 2, 4, 6, 8, 10 and 12% FPC	1. Blending limit: Ordinary coal: 4-5% partial briquetting of coal charge: 8% 2. FPC is a little inferior to DPC.

regarding use of petroleum coke at the five major steel companies in Japan.(5) According to the comprehensive summary of their views, the blending limit of petroleum coke in ordinary blend relative to coking coals is about 5% at the highest, and the quality of delayed coke is better than that of fluid coke.

In this report, the authors evaluate petroleum coke as a raw material for coke-making and examine the above-mentioned blending limit. Besides petroleum coke, examination and evaluation of petroleum residual oil are also presented as a more effective use as a raw material.

Evaluation of Coking Coals

Japan imports coking coals for blast furnaces from many foreign countries. The quality of coke derived from these coals, being subject to operating conditions of the coke oven battery, mainly depends upon properties of the coking coals; coke of an excellent quality is available from appropriate combinations of coals. Blending design of coals forms one of the most remarkable features of the Japanese coke-making technology.

Blending design of coals may be summarized as follows.(1) The most important coke property is the strength, which is commonly expressed by JIS drum index in Japan. Large-capacity blast furnace engineers in Japan assert that coke should have a JIS drum strength, as expressed in DI30/15, of at least 92. Evaluation of a coking coal relative to the coke strength may be expressed with two parameters: the mean maximum reflectance of vitrinite part of coal ($\bar{R}o$) representing the degree of coalification rank of the coal, and the maximum fluidity (MF) indicating the coking property of the coal.

Figure 2 illustrates the relationship between the coke strength (DI30/15) and the maximum fluidity (MF) of blended coals in the coke oven battery at Nippon Kokan's Fukuyama Works with the reflectance ($\bar{R}o$) of blended coals as the parameter.

According to this figure, at an MF over 200 DDPM (Dial Division per Minute) with a constant $\bar{R}o$, the value of DI30/15 is maximized. Within the range of over 200 DDPM, it is necessary to increase $\bar{R}o$ which corresponds to the coalification rank of blended coals, in order to raise the strength. This range is therefore referred to as the coal rank control region. In the region of MF under 200 DDPM, it suffices to increase the fluidity (MF) of blended coals in order to raise the coke strength: this range is therefore called the fluidity control region. With an $\bar{R}o$ of blended coals of 1.15, a DI30/15 of 92 can be ensured by holding MF at 200 DDPM.

A similar test was carried out in a 20-kg test oven at Nippon Kokan's Technical Research Center as shown in Figure 3.(2) Coking conditions included a coking temperature of 880°C and a coking time of 6.5 Hrs. In this case, the coking speed is very high, so that MF corresponding to the saturation of strength is rather low at about 80 DDPM. If $\bar{R}o$ of blended coals is kept at 1.15 under these conditions, a DI30/15 of 92 can be ensured.

Use of this small-capacity test oven can facilitate an experiment, and the experiments presented in this report were conducted in this oven. In this test oven, the value of MF of 80 DDPM forms the boundary between the fluidity control region and the coal rank control region.

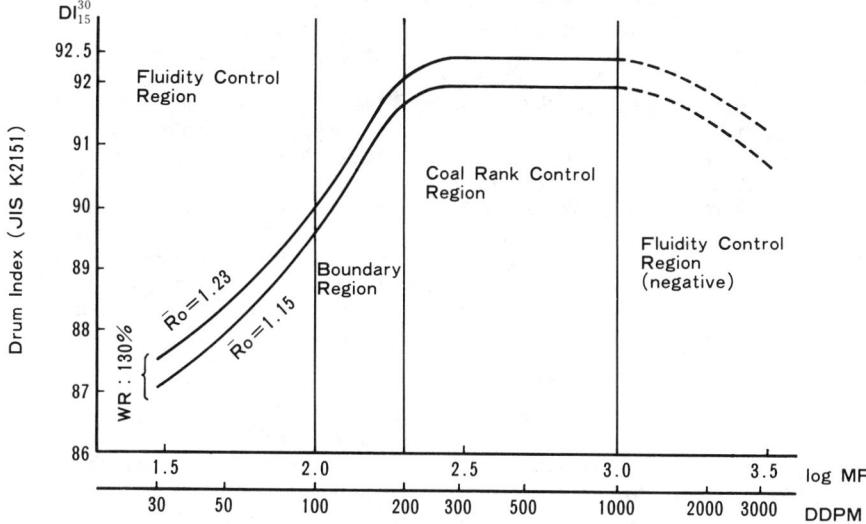

Figure 2. Relation between Gieseler max. fluidity of blends and drum strength of coke yielded in Fukuyama Works.

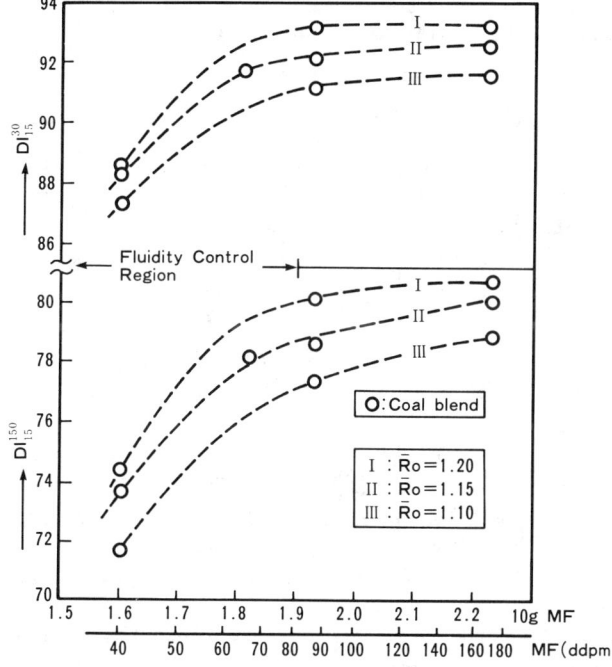

Figure 3. Relation between MF of coal blends and strength of coke made by small test oven.

Coals produced in various parts of the world may be arranged in terms of $\bar{R}o$ and MF as shown in Figure 4, which is referred to as the MOF diagram. Thus, $\bar{R}o$ and MF have significant meanings in the evaluation of a coal in terms of coke strength.

In addition to $\bar{R}o$ and MF, inert matters of coal such as ash and sulfur, which do not soften and melt, are important properties of coal to be evaluated. The relations between FOB prices of coals from various parts of the world and the above-mentioned factors were analyzed by multiple regression analysis.(3) This permits economic evaluation of coals as raw materials for coke-making. Petroleum residual oils and petroleum coke can similarly be evaluated as raw materials. The most difficult problem here is how to evaluate factors corresponding to $\bar{R}o$ and MF in coals.(6) This report presents primarily an estimation of such factors for evaluation.

Test Method

Test Samples. Main properties of the residual oils used in the present test are represented in Table II. It should be noted in this table that: No. 1 is propane deasphalted asphalt; Nos. 2 to 7 are petroleum pitches derived from residual oil heat treated under various conditions; No. 8 is KRP pitch made by Kureha Chemical Industry from crude oil heat treated with hot steam at temperatures over 1,000°C; and Nos. 9 and 10 are both residual oils from coal, No. 9 being solvent refined coal made by NKK and No. 10 being heat treated coal tar pitch. These Nos. 1 to 10 are typical examples of binding material for coke-making in Japan.

Table II   Properties of residual oil

Residual Oil (NO.)	Ash (d.b%)	VM (d.b%)	T.S (d.b%)	C/H (atomic)	Reflectance $\bar{R}o$ (%)	Reflectance $\bar{R}a$ (%)	fa	QS (%)	log (MFc)
1	tr	94.8	0.19	0.56	—	4.26	0.142	100	8.1
2	0.05	85.9	5.30	0.67	—	5.27	0.316	100	15.0
3	tr	75.7	6.75	0.82	0.30	6.25	0.471	100	13.4
4	0.80	55.5	3.75	1.03	0.43	7.30	0.581	80.2	11.1
5	0.96	40.3	6.80	1.23	1.10	9.28	0.637	85.1	11.4
6	0.38	35.1	6.44	1.29	1.05	9.11	0.651	63.7	9.2
7	0.11	36.5	5.93	1.83	0.73	7.95	0.874	83.9	14.2
8	tr	33.0	0.22	1.68	1.87	9.91	0.912	74.9	8.7
9	0.81	60.5	0.76	1.23	0.86	8.78	0.777	99.0	15.9
10	0.50	53.7	0.47	1.78	1.51	9.95	0.939	91.2	13.8

Properties of the petroleum cokes used in the present test are shown in Table III. The symbol MPC represents petroleum coke manufactured with the use of Minas heavy oil by the delayed coking process, and DPC and FPC are, respectively, petroleum cokes provided by a delayed coker and a fluid coker commercially available in Japan.

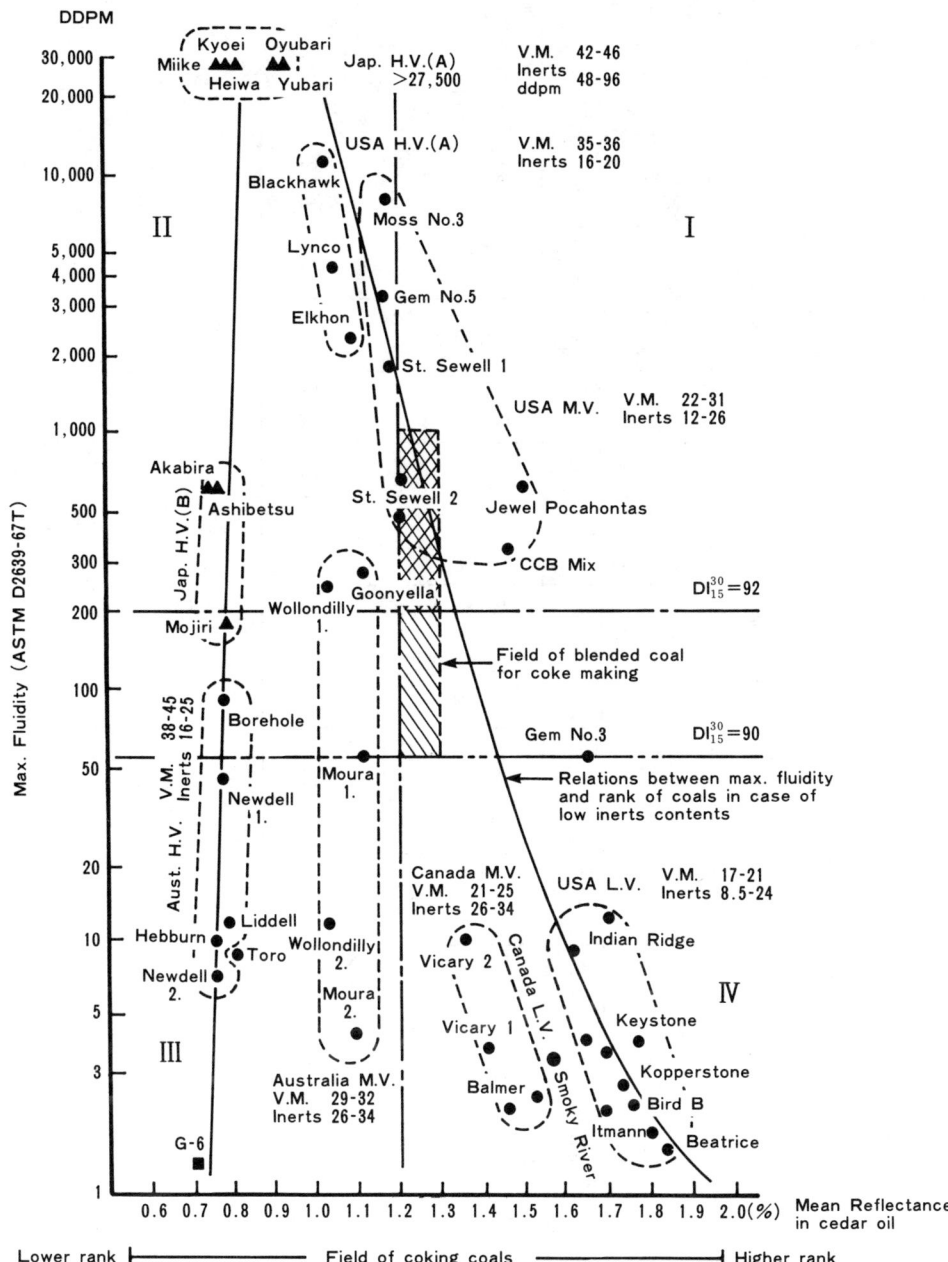

Figure 4. Relation between max. fluidity and rank of coals (MOF diagram).

Table III. Properties of petroleum cokes

Sample	Proximate analysis (d.b.%)			Ultimate analysis (d.a.f.%)					C/H	TS (d.b.%)	Petrographic analysis			Calorific value (kcal/kg)
	Ash	VM	FC	C	H	N	S	O			R_o (%)	R_a (%)	Anis. (vol%)	
MPC	tr	12.47	87.53	93.22	4.33	1.60	0.30	0.49	1.807	0.36	4.67	14.29	100	8,830
DPC	0.35	8.49	91.16	91.62	3.84	2.33	1.74	0.47	2.003	1.73	5.02	14.72	100	8,630
FPC	0.53	6.98	92.49	91.38	1.73	2.62	1.98	2.29	4.433	1.98	7.87	15.68	100	8,220

Residual Oil Addition Test. As previously shown in Figures 2 and 3, the effect of the reflectance and the fluidity of blended coals on the strength of coke that is produced can be expressed in the form of a model as in Figure 5.(3) Using the test oven, the low-volatile coal in the base blends was replaced by the other coals with different coal ranks in the coal rank control region, and the relationship between DI of the coke produced and $\bar{R}o$ of the replacement coal was determined. The regression line is illustrated in Figure 6.

Then, DI of the coke produced by adding residual oil in an amount equal to that of the replacement coal in the manner described above was determined, and the reflectance of the residual oil was determined from the regression line given in Figure 6. The reflectance thus determined is herein defined as the effective reflectance, $\bar{R}o_E$. After determination of $\bar{R}o_E$, DI is determined on the coke produced from a blend of coal and residual coil in the fluidity control region.

Since the values of $\bar{R}o$ of the blended coals are known, it is possible to determine the value of MF for the blend of residual oil and coals from Figure 5, and hence to determine the value of MF for the residual oil because MF for the remaining portion is known. The value of MF of the residual oil thus obtained is defined as the effective fluidity, MF_E.

Petroleum Coke Addition Test. Petroleum coke may be considered as the residual oil which is further coked. Base blends used in this addition test and the blends subjected to the addition test are shown in Table IV, which also gives the properties of the coke carbonized after blending.

When blending petroleum coke, the strength of the coke carbonized after blending is not satisfactory unless the properties of blending coals are good. The degree of contribution of petroleum coke to the coke strength is considered to be much lower than that of residual oil. This means that petroleum coke is considered to exhibit properties considerably different from those of a coking coal. With this fact in view, petroleum coke was blended on the assumption that petroleum coke was the same as the inert matter of coal, and the reactive portion was null, and takes only the fluidity into account. Examination was based on the following combinations with the base blends:
1. Changes in the strength of the coke product were studied by blending petroleum coke simply at 5% and 10% into base blends;
2. Petroleum coke is assumed to be the same as inert matter of coal as described above. Decrease in $\bar{R}o$ of the blend is compensated by coals other than petroleum coke. By considering the amount of blended petroleum coke, the blend is designed so that MF of the blend may be maintained only with that of coals in the base blend, on the assumption, however, that petroleum coke has a log[MF] of 0. Particulars in this case are shown in Table IV.

Results

Results of Residual Oil Test. Values of $\bar{R}o_E$ and MF_E as derived from the results of the test are given in Table V, which suggests that all the residual oils demonstrate excellent properties in many cases. These values are arranged into an MOF diagram in Figure 7.

Table IV. Results of petroleum coke addition test

Blended with	Properties			Blending ratio (%)											
	Ro (%)	MF DDPM	T.In. (%)	II-3	MPC	DPC	FPC	MPC	DPC	FPC	MPC	DPC	FPC	DPC	FPC
Prime (US Lv)	1.60	2	13.8	10	⎫			⎫			9.0	↑	↑	8.5	↑
Pittson (US Mv)	1.10	14,000	18.1	5							4.5	↑	↑	4.2	↑
Balmer (Ca.Lv)	1.36	10	34.5	30	⎬ 95			⎬ 90			31.5	↑	↑	29.8	↑
Fording (Ca.Lv)	1.30	155	35.1	10							4.5	↑	↑	4.2	↑
Wellondilly (Australia Mv)	1.08	1,000	39.2	10	⎭						9.0	↑	↑	8.5	↑
Witbank (S.A.Hv)	0.84	39	36.4	25				⎬			10.1	↑	↑	6.0	↑
Yutoku (Jp.Hv)	0.84	50,000	5.8	10				⎭			21.4	↑	↑	23.8	↑
Petroleum coke		0	100		5	↑	↑	10	↑	↑	10.0	↑	↑	15.0	↑
Properties of coal blend	Ro (%)			1.15	1.15	1.15	1.15	1.15	1.15	1.15	1.15	1.15	1.15	1.15	1.15
	MF (DDPM)			136	107	107	107	83	83	83	107	107	107	107	107
Coke strength	JIS D_{15}^{30}			91.6	91.5	91.3	90.9	90.4	90.0	89.3	91.1	90.6	89.8	90.5	89.4	92.5

Table V Effective reflectance ($\bar{Ro}_E$) and effective fluidity (MF_E) of residual oil

Residual oil No.	Effective reflectance ($\bar{Ro}_E$)	Effective fluidity ($\log MF_E$)
1	0.027	2.55
2	0.208	5.04
3	0.377	5.98
4	0.926	5.01
5	1.144	5.99
6	1.055	5.25
7	1.228	5.38
8	1.605	6.20
9	0.924	5.95
10	1.545	5.82

Results of Petroleum Coke Test. The results of the test are given in Table IV. From the data shown, the relationship between the strength (DI) and the ratio of addition is represented in Figure 8. According to this figure, simple addition to the base blend results in a sharp decrease in DI, and even when $\bar{Ro}$ and MF are compensated, there is still a lower value of DI, suggesting that at least the log[MF] of petroleum coke is lower than 0.

Examination and Evaluation

Examination on Residual Oil. Values of $\bar{Ro}_E$ and MF_E for residual oil have been determined as described above. This measurement, however, requires much labor. Efforts were therefore made to establish a method for estimating $\bar{Ro}_E$ and MF_E from the various parameters of residual oil. Figure 9 shows the relationship between $\bar{Ro}_E$ and faQs(100 - VM); and Figure 10, the relationship between MF_E and faQs. All these figures demonstrate high correlations. These results were arranged and subjected to multiple regression analysis, and the results are given in Table VI. According to this table, $\bar{Ro}_E$ has the closest correlation with faQs(100 - VM), and MF_E' with faQs.

Examination on Petroleum Coke. Evaluation has been made by assuming that petroleum coke was a totally inert content and log[MF] of the fluidity was null, but this is not always the case as shown in the results in the former section.

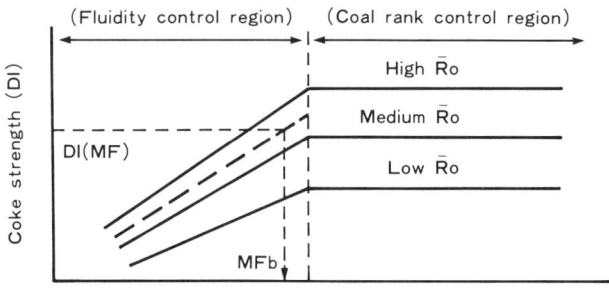

Figure 5. Effect of fluidity and coal rank of coal blend on coke strength.

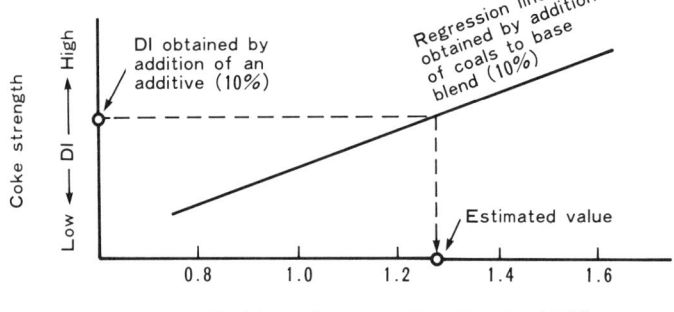

Figure 6. Estimation of effective reflectance ($\bar{R}o_E$) of caking additives.

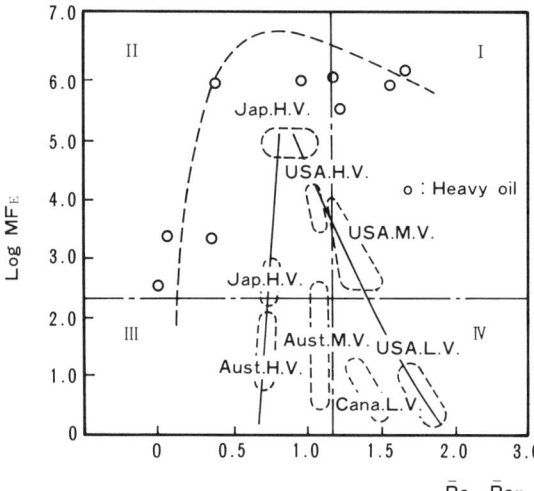

Figure 7. MOF diagram for caking additives (residual oil).

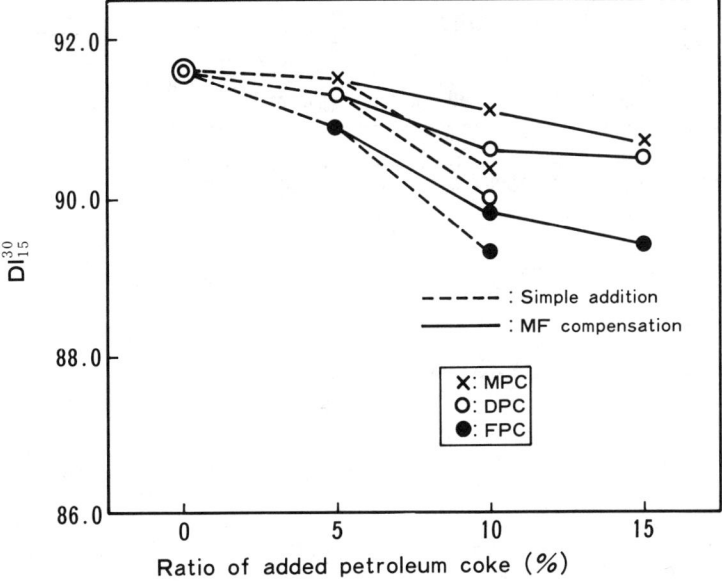

Figure 8. Change in DI30/15 by addition of petroleum coke.

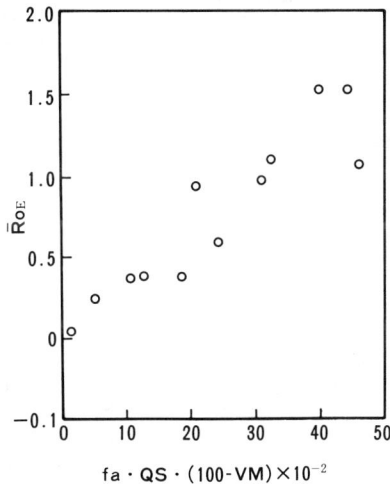

Figure 9. Relation between $\bar{R}o_E$ and $f_aQS(100-VM)$.

Table VI Estimating equations of $\bar{Ro}_E$ and MF_E for residual oil

$\bar{Ro}_E$	$\bar{Ro}_E = -0.02060\,(VM) + 2.021$ (R=-0.8167, N=56)
	$\bar{Ro}_E = 1.1245(C/H) - 0.518$ (R=0.8503, N=56)
	$\boxed{\bar{Ro}_E = 0.03376(FAQS(100 - VM)) + 0.0148 \quad (R=0.9037, N=56)}$
	$\bar{Ro}_E = 0.00285(\bar{R}aQS(100 - VM)) + 0.0287$ (R=0.8724, N=52)
MF_E	$\boxed{\log MF_E = 2.445 + 5.564(FAQS) \quad (R=0.6988, N=50)}$
	$\log MF_E = 1.492 + 0.6024(\bar{R}aQS)$ (R=0.6540, N=43)
	$\log MF_E = 2.429 + 0.218 \log MFc$ (R=0.6689, N=60)
	$\log MF_E = 4.031 + 0.0052(TDc)$ (R=0.5416, N=60)

The values of MF were therefore reviewed while retaining the assumption that petroleum coke was a totally inert content. The results shown in Table VII demonstrate that log MF takes large negative values. The average of these values is defined as MF_E for a particular petroleum coke. The relationship between this MF_E and VM for petroleum coke and residual oil is represented in Figure 11, which shows the possibility of expressing MF_E by VM.

Table VII Effective fluidity (MF_E) of petroleum coke

Sample	(Simple addition) log MF_E		(MF compensation) log MF_E		(Average) log MF_E
MPC	(5%)	-1.17	(10%)	-2.26	-2.20
	(10%)	-2.65	(15%)	-2.17	
DPC	(5%)	-2.75	(10%)	-3.56	-3.13
	(10%)	-3.70	(15%)	-2.52	
FPC	(5%)	-4.83	(10%)	-5.66	-5.12
	(10%)	-5.53	(15%)	-4.44	

Note) Figures in ( ) represent blending (adding) ratios.

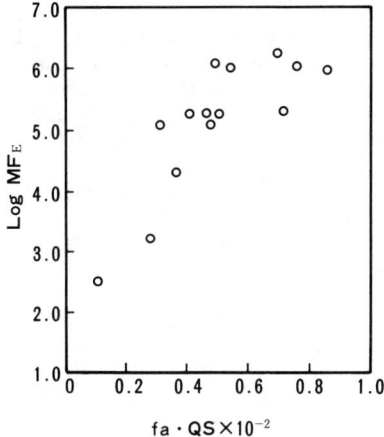

Figure 10. Relation between MF_E and f_aQS.

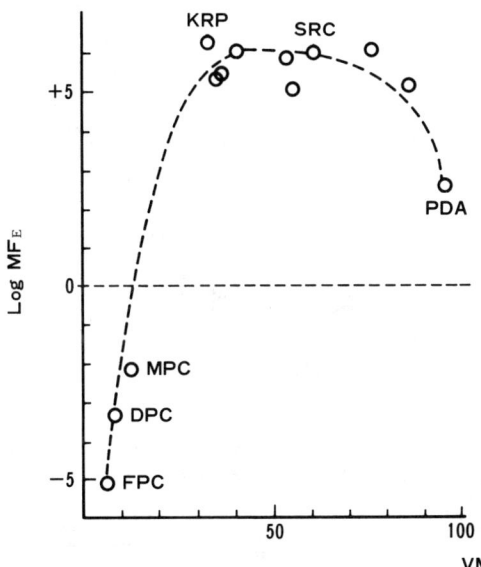

Figure 11. Relation between effective fluidity and VM of heavy residues.

Evaluation

Economic Evaluation. An additive used as a coking raw material can be evaluated from values of $\bar{R}o_E$, MF_E, inert matter, ash, sulfur and so on in the same manner as in a coal, as shown above. The relationships between the individual parameters for coal and the FOB prices have previously been determined at Nippon Kokan. These values are applicable to those of additives as shown in Figure 12.

In this figure, the price is not shown, but only the relative position was estimated by an equation. For heavy residues, the economic evaluation is shown also in this figure. In the estimation formula, C1, C2, C3 and C4 are constant coefficients determined from economic and technical points of view, and u. v. w and s are coefficients determined by multiple regression analysis.

For petroleum coke, the same evaluation is shown also in Figure 12. It will be understood that the value of petroleum heavy oil is better than that of petroleum coke as a raw material for coke making. Petroleum heavy oils can more advantageously be utilized than petroleum coke.

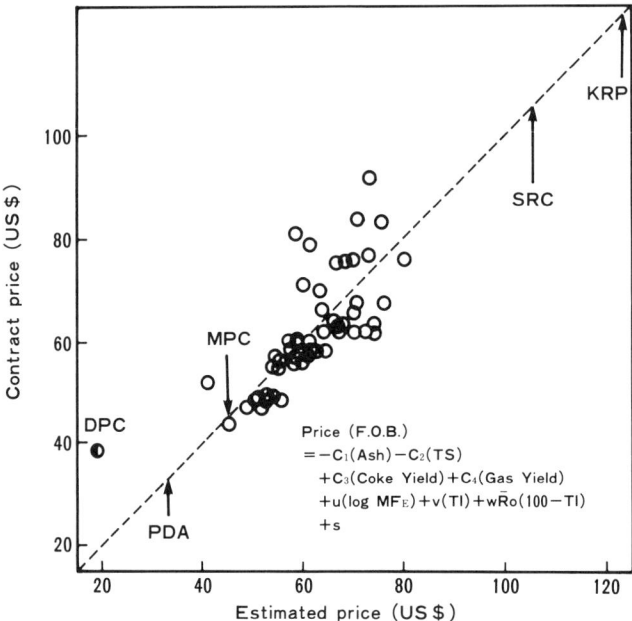

Figure 12. Relation between estimated price and contract price.

Blending Limit of Petroleum Coke. The blending limit of petroleum coke is considered to be about 5% as already mentioned above. This can be explained from the current average value of MF of about 200 to 500 DDPM for coal blends in the Japanese coking industry. Figure 13 shows limit quantities of added petroleum coke for various values of fluidity for the base coals. In the determination of these limit quantities, the quantity of added petroleum coke with which the fluidity of the coal blend decreased to below 200 DDPM under the effect of this blending was deemed as the limit.

According to Figure 13, if DPC is blended in an amount of 5% with a base blend having an MF of 400 DDPM, the resultant coal blend has a sufficient maximum fluidity (MF), whereas a 10% blending results in an MF far inferior to 200 DDPM, showing the impossibility of keeping a satisfactory coke strength. This is also the case with the other petroleum cokes, and the order is: MPC > DPC > FPC. This may be regarded as supporting the information given in Table I.

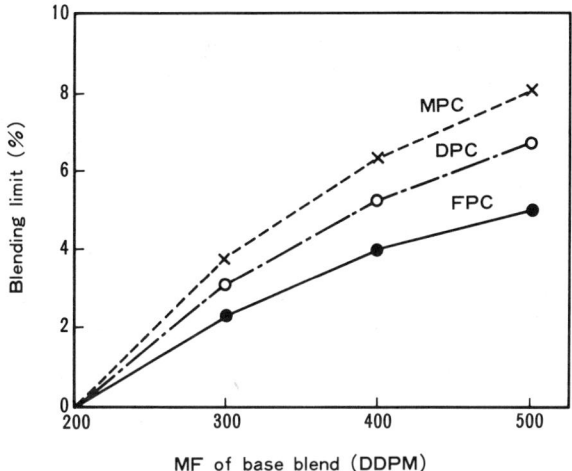

Figure 13. Blending limit of petroleum coke.

Conclusion

Petroleum coke as a coking raw material was evaluated in comparison with residual oil. While residual oil shows excellent properties as a coking raw material, petroleum coke was found to be far inferior to residual oil, and quantitative figures were determined.

Literature Cited

1. Miyazu, T. et al., "The Evaluation and Design of Blend using Many Kinds of Coal for Coke Making", paper presented at IISC Meeting, Dusseldorf, 1974.
2. Miyazu, T. et al., "New Technique for Producing Coke -- Addition of Petroleum Pitch or Solvent Refined Coal", paper presented at IISC & AIME Meeting, Chicago, 1978.
3. Miyazu, T. et al., "Selection of Coal and Additives for the Production of Coke Cost", paper presented at McMaster Symposium, Hamilton, 1980.
4. Miyazu, T. et al., "Evaluation Method of Binding Materials for Coke Making", paper presented at Japan Coal Science Conference, Fukuoka, 1981 (Japanese).
5. The Annual Report in RAROP, 1979 (Japanese).
6. The Annual Report in RAROP, 1982 (Japanese).

   RAROP: Research Association for Residual Oil Processing in Japan

RECEIVED October 21, 1985

Mechanism of Carbon-Black Formation in Relation to Compounded-Rubber Properties

James E. Lewis

Ashland Chemical Company, Columbus, OH 43216

Studies conducted during the last ten years have extended our knowledge of carbon black from the practice of the art to an application of the science. These studies have dealt with materials suitable for fuel and feedstocks, the process for converting these materials to carbon black with predictable properties, the techniques for measuring and defining those properties, a better understanding of the effects of those carbon black properties on processing parameters of compounded goods, and the predictability of the performance properties of finished rubber goods. This presentation briefly considers the status of each of these areas of study relative to today's carbon blacks and tomorrow's rubber products.

"Mechanism" is generally employed as a word to describe the individual rearrangement of atoms as molecules react to form new compounds and products. The mechanism of carbon black formation is used in this context as it appears in the title of this presentation.

In the carbon black industry we talk about carbon blacks in relations to adhesion to steel belts, better abrasion resistance, different traction, changed hysteresis requirements, different flexural properties, different extrusion rates and properties, hardness, and on ad infinitum, and all in new rubber compounds with different synergistic effects or loss of synergism as the case may be with changes in ingredients. Carbon black has traditionally been manufactured by those skilled in the "art", so as soon as a product is defined it can fairly quickly be matched. But, how is the first major product change to be made, and for what specific objectives?

Progress is achieved through change which is understood. So, in this instance, scientific knowledge must surpass the art, and a knowledge of the Mechanism of Carbon Black Formation in Relation to Compounded Rubber Properties prevails.

The carbon black industry in the past has devoted much of its attention to the equipment, its engineering and feedstocks, for the production of carbon black and the empirical study of the effect of various carbon black properties on rubber processing and rubber product properties. As a consequence, most of this published information is in the form of patents covering carbon black reactors, or in the scientific literature reporting on the performance of carbon black in various systems, and the characterization of blacks through a variety of test methods. Being primarily empirical, each area has deficiencies with the resultant cry that we need better methods for testing carbon black, more reliable means for predicting the behavior of any carbon black in any compound, and better carbon blacks.

This last statement typifies chemical products in many industries and is not unique to the rubber or carbon black industry. For that matter, it may be said to be a brief history of our industries, as well as a projection of the future.

In order to continue to progress toward these changing objectives, it has become necessary to learn more about carbon black in the last ten years than through its entire history. In other words, our knowledge has been extended from the practice of the art to an application of the science. These studies have dealt with materials suitable for fuel and feedstocks; the process for converting these materials to carbon black with predictable properties; the techniques for measuring and defining those properties; and a better understanding of the effects of those carbon black properties on processing properties and parameters of compounded goods; and, finally, the predictability of the performance properties of finished rubber goods.

Materials

It has been known for many years that molecular structure of a fuel has a direct bearing on the tendency of that fuel to smoke, i.e., to form carbon or soot in a flame. For example, in 1954 Schalla (41), reporting on a study of diffusion flames, indicated that the rate at which hydrocarbons could be burned smoke free varied in the order: n-paraffins -- mono-olefins -- alkynes -- aromatics. This same phenomena has been reconfirmed by many authors in a variety of systems and always in the same general order (6, 8, 15, 17, 19, 26, 43, 45). Paraffins have the least tendency to smoke, whereas the naphthalene series have the greatest tendency to smoke.

The application of the "tendency to smoke" to the manufacture of carbon blacks is discussed in detail by Austin (4). The relationship of aromaticity, as expressed by the BMCI, is discussed as it correlates with yield of carbon black produced in a furnace reactor.

In our laboratories, we were concerned with yield of black production; however, we were more concerned with the influence of feedstock structure on the quality of product produced. Molecular structures of feedstocks investigated were pure compounds and mixtures of pure and similar compounds including monocyclic aromatics with and without side chains, dicyclic aromatics, tricyclic aromatics, mixtures of higher molecular weight aromatics and high and low molecular weight paraffins. Some examples of the types of compounds studied are:

1. Light paraffins, C12 to C22
2. Heavy paraffins, C16 to C35
3. Benzene, C6
4. Alkyl benzene, C8 to C12
5. Alkyl naphthalenes, C10 to C12
6. Anthracenes, C14
7. Mixed tricyclics, C14 to C18
8. Mixed tetracyclics, C16 to C20
9. Hexane insoluble asphaltenes

These feedstocks were tested in a number of carbon black furnace reactors employing different geometries and different operating conditions. The experiments were designed to develop data for correlation between feedstock composition, yield, production rate, and properties of carbon black produced. The physical properties of the carbon blacks were measured as were the properties of uncured rubber compounds and vulcanizates in standard test recipes.

In general, the conclusions demonstrated that:

1. Structure of carbon black is directly related to the type of molecules in the feedstock employed. (Fig. 1)

2. Yields of carbon black produced are related to the size of aromatic molecules in the feedstock. (Fig. 2)

3. Flow rates and molecular structure effects of the oils are additive on a weight basis for blends of the feedstocks tested.

The total combination of data obtained permitted optimization of process variables relative to reactor geometry employed, and optimization of carbon black quality produced. The array of samples tested enabled us to prepare a mathematical relationship of these principal variables to commercial feedstock properties.

The System

Limiting our discussion to a carbon black manufacturing system for the production of tread grades in a furnace reactor, these systems have been discussed in detail in a number of articles, viz., Smith and Bean ([35](#)), Austin ([4](#)), Burgess ([5](#)), and Stokes ([36](#)), just to name a few.

It is well-known that feedstock, reaction rate, and temperature affect particle size; alkali metal salt additives affect structure; and the use of preheated air, increased reactor size and increased throughput affect yield.

On the other hand, it is also well-known that new blacks or tailor-made blacks for specific applications are produced by selecting various combinations of particle diameter, structure, and surface area ([10](#), [46](#), [9](#), [20](#), [16](#), [30](#)). In fact, the measurement of these physical properties constitutes the principal quality control

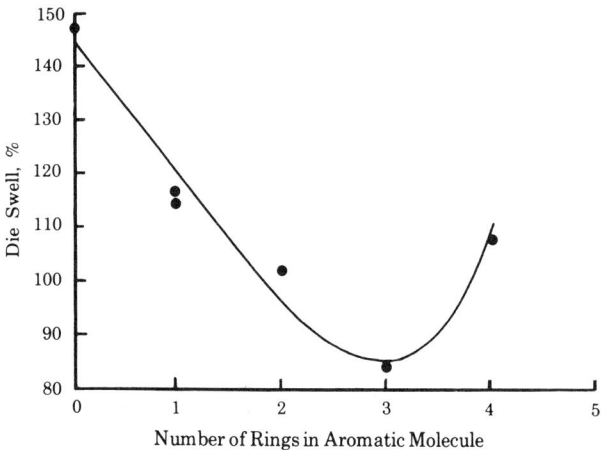

Figure 1. Number of Rings in Aromatic Molecule

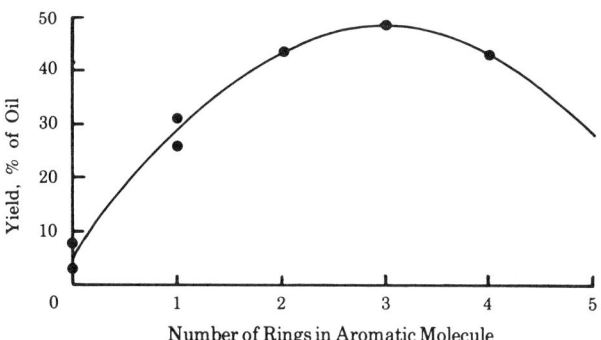

Figure 2. Number of Rings in Aromatic Molecule

tools employed by most carbon black manufacturers (29). (ASTM Methods D3265-73, D1510-70, D2414-72.)

In 1968 Mantell (28) described several processes for the production of carbon black. He also lists 24 distinct grades of blacks identified by their respective particle diameters (average), surface area and oil absorption (structure). These include the then common grades of channel black, gas furnace blacks, oil furnace blacks, thermal blacks, lampblack, and acetylene blacks, in language sufficient for the manufacturer to identify each grade by these three variables.

The Mechanism of Carbon Black Formation and the Manufacturing Process

There are a great many published reports describing studies of carbon black formation in flames. Many of these deal with gaseous fuels in either pre-mixed flames or diffusion flames. The principal objectives are to develop a better knowledge of combustion through an understanding of the kinetics and mechanism of carbon formation. A thorough familiarity of these works and the theories presented is essential for effective studies of the mechanism of carbon black formation; however, a review of these works here is well beyond the scope of this paper. A number of extensive review articles and books are recommended (2, 15, 17, 25, 31, 32, 33, 42).

Experimental

As indicated previously, a number of carbon black reactor systems were employed through the course of these studies as were a number of feed materials. The studies on feedstocks allowed the development of a set of parameters which were applicable to commercial feed materials and, of course, these investigations on feed materials were intimately linked with the studies on reaction kinetics and mechanism.

To simplify this paper, one reactor system and one feedstock were selected for the preparation of two distinct grades of carbon blacks. In the program as conducted six different reactors were studied with all of the reinforcing grades of furnace blacks being produced. A variety of feed materials were studied and a number of different gaseous and liquid fuels were employed.

A reactor (Fig. 3) as described in U.S. Patent 3,060,003 was supplied with a feedstock of BCMI-120 using air and natural gas and a water quench to produce standard HAF and standard ISAF. All carbon black samples were tested substantially by physical tests and standard rubber tests.

The primary techniques employed were based upon establishing a steady state operation of the total reactor system in a mode for producing a standard grade of carbon black meeting existing specifications for rubber use and then sampling the system extensively at a variety of locations.

Fig. 4 gives an example of the sampling locations. Each point designated within the reactor was tested for pressure, gas composition, temperature and materials distribution. These will be discussed separately in order to describe the techniques employed.

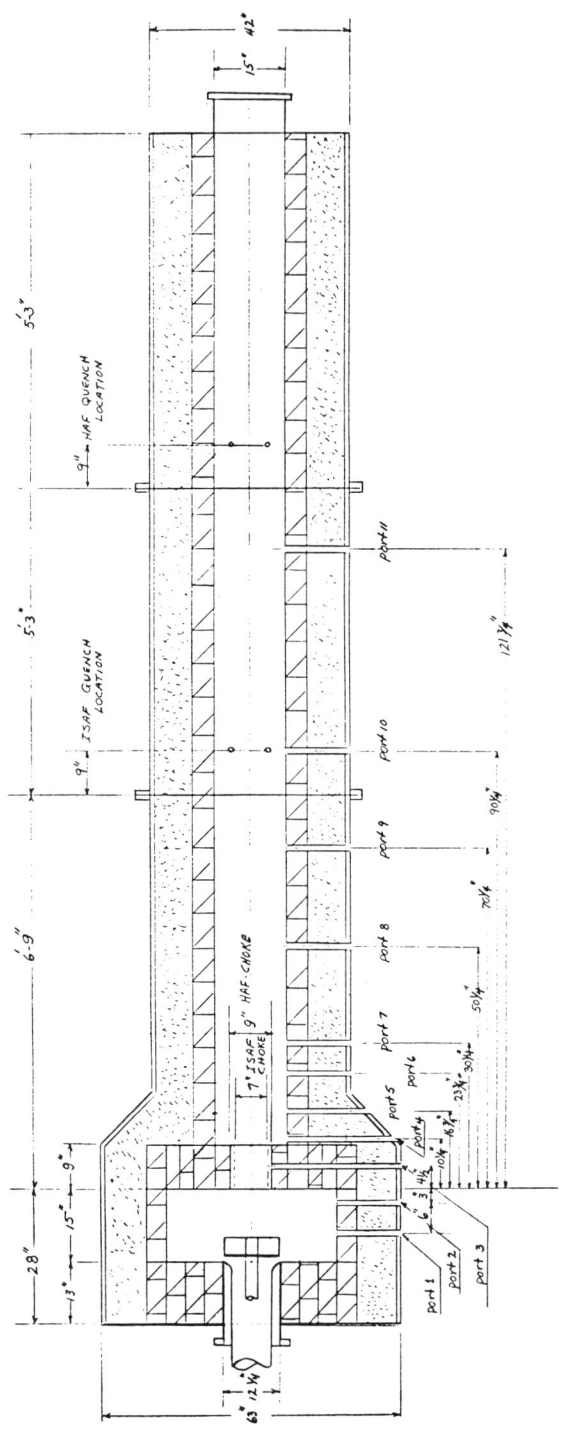

Figure 3. Sampling Ports and Quench Locations on Reactor

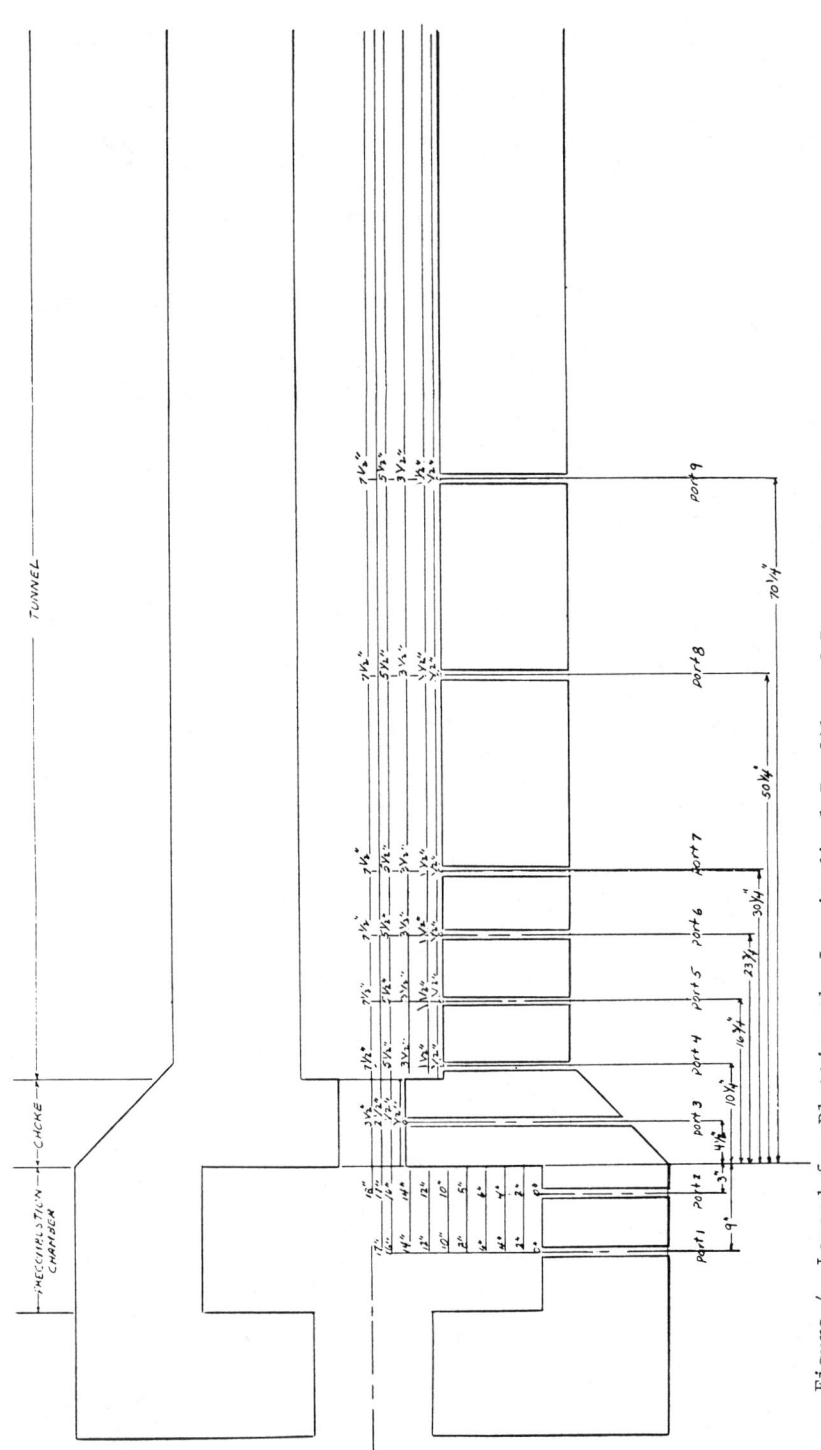

Figure 4. Legend for Plotting the Longitudinal Profiles of Reactor Test Data

First, however, we should discuss sampling as it pertains to a carbon black reactor. A commercial carbon black reactor operating at production conditions may be described as being in a steady state of thermodynamic inequilibrium.

Put in different language, care must be taken that the sampling procedure does not in itself alter the "steady state," and one must recognize that any sample taken from the reaction zone is changed. In fact, the sample is substantially changed from its true existence by the mere act of taking the sample. With a reaction temperature of up to 3600° F, a mass flow rate exceeding a ton an hour in the vicinity of sonic velocities, there are no known sampling techniques that give direct read-out of the total system. Noble efforts have been made such as reported by Wersborg (48), employing a molecular beam sampling system, combined with electrical deflection of the beam. However, the mere disposition of the carbon by the beam makes the sample unsuitable for analysis in terms of its existence before deposition. Gaydon (15) discusses the problems associated with many efforts employing spectroscopic analysis. Using spectroscopy certain species can be identified reliably -- but many cannot -- the simplest of which is carbon vapor. These techniques have not been employed successfully in carbon black reactors although they can be employed in certain simple flames.

For our system we chose the simplest approach -- a fast sample conduit with quick quench into an evacuated sample container. For temperature measurements we used a similar probe outfitted with platinum/6% rhodium-platinum/30% rhodium thermocouple. For pressure measurements the same general type probe mentioned was employed but without extracting samples. This probe had one hole opening perpendicular to the longitudinal axis of the probe such that when inserted into the reactor it could be rotated 360°. In this manner the pressures were read from a precision pressure gauge with the opening facing 0°, 90°, 180°, and 270° relative to the direction of flow in the reactor.

For materials distribution sampling, the same probe used for gas sampling was employed.

In all cases the sample probes were constructed of stainless steel with an internal diameter of ¼" (Fig. 5). This conduit was surrounded by a water jacket to prevent destruction of the probe in the high temperature atmosphere. The probes were accurately measured and inscribed with markings so they could be inserted to a previously determined penetration, i.e., to an exact location in the interior of the reactor. The sampling ports in the reactor were fitted with gate valves so they were easily accessible during the course of sampling. Swagelok fittings on these valves provided tight seals between the wall of the probes and the opening. On the exterior end of the probe another Swagelok fitting was provided for fastening a ¼" I.D. copper tee. On one arm of the tee a compound gauge for reading both vacuum and pressure was attached when taking materials samples. On the other arm was a fitting to permit quick connect and disconnect of the sampling bombs of 104 cubic inch volume. These parts are also illustrated in Fig. 5. The tee was fitted with a valve to allow flushing of the line whether the bomb was in place or not. Each bomb was fitted with a valve and each

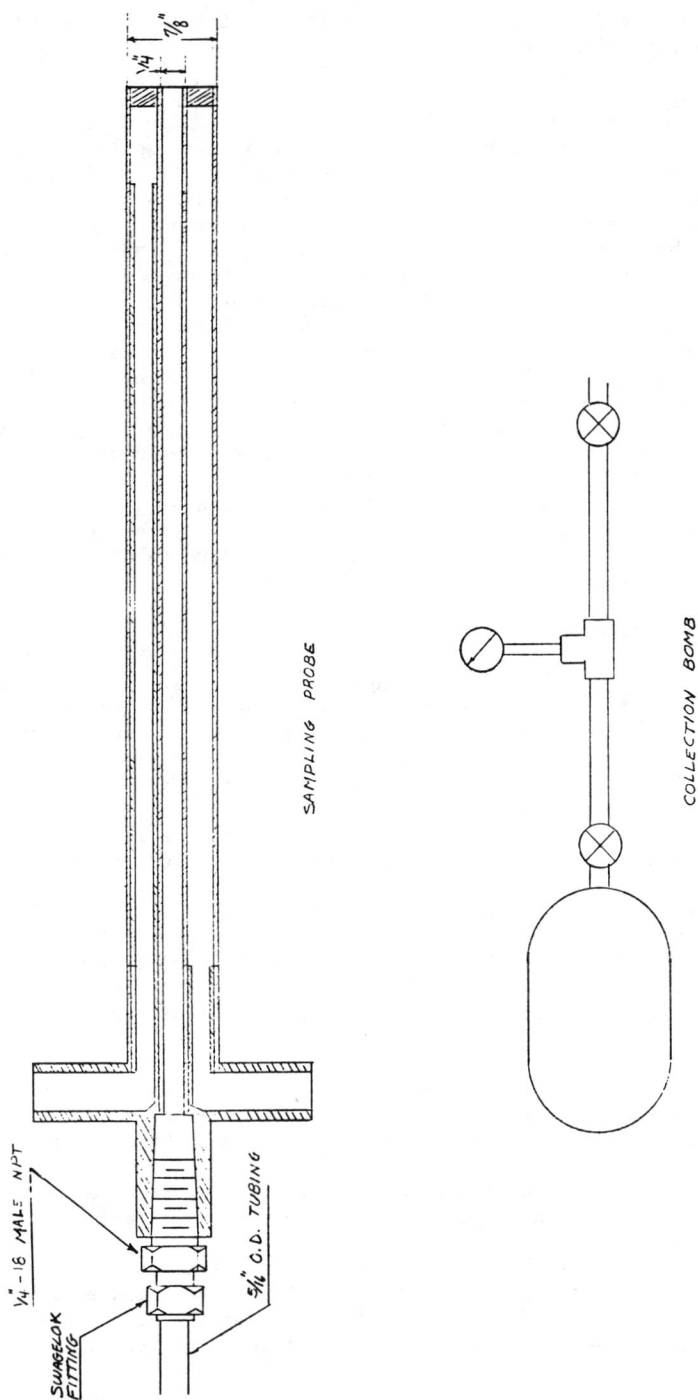

Figure 5. Sampling Probe and Collection Bomb

bomb was evacuated to a few microns pressure prior to use to avoid any sample contamination.

Gas Sampling

With the reactor operating at steady state producing quality product carbon black, the sample probe was inserted in port 1 and positioned to 0" from the reactor wall. The probe line was purged of the gases in the probe, a sample bomb attached and the valve on the bomb was opened to check the vacuum in the bomb. The sample probe was opened allowing gases to flow from the reactor until the pressure in the bomb was in equilibrium with that in the reactor. With the valves closed, the sample bomb was removed, the probe inserted to a new position, purged, a new sample extracted, and the process was repeated at all 11 sample ports at incremental distances from the reactor wall to the exact center of the reactor.

The additional samples for materials distribution, temperature measurements, and pressure measurements followed the identical procedure and employed exactly the same sampling grid.

The samples for gas composition were analyzed by gas chromatography for hydrogen, oxygen, carbon dioxide, carbon monoxide, methane, nitrogen, argon, minor constituents of acetylene, ethane, and ethylene.

Materials Distribution

The principal materials being fed into the reactor are air, gas (or fuel), and the feedstock. The materials coming out of the reactor are, of course, combustion gases, carbon black, water, and the nitrogen which comes into the reactor in the air and passes through unchanged. The question we have then is how is the feedstock converted to carbon black? What are the reactions that occur, where do they occur, and under what variables, what are the reactor rates, and what are the mechanisms of carbon black formation? What are the key variables in the system relative to controlling particle size, surface area, and structure?

Armed with substantial empirical experience regarding the operating of a reactor we know, for example, that the fuel used for heat must be consumed before coming into contact with the feedstock; that the ratio of air to fuel and their rates control the temperature profile in the zones of the reactor, and that the molecular structure of the oil feedstock, the preheat temperature of the oil, i.e., whether it enters the reactor as a liquid or a vapor, the oil rate, and this rate as a ratio to the air/fuel rate all have a bearing on the reactor process, its control, and the properties of the product produced. We also know that the mechanical means employed for injection of the feedstock must be kept constant as well as the reactor geometry and all of the rates and ratios mentioned above.

So, how do we "see" into the reactor when it is operating at 2500 to 3500° F with over a ton of material per hour flow rate? As discussed before, we no longer consider the sampling of reactive species; instead we would like to know what the distribution of materials is at each point in the reactor relative to the temperature, gases and other variables.

A radioactive tracer technique was ultimately chosen as a mapping tool. The isotope Kr-85 was selected because it can represent the gaseous, vaporous and solid materials in the reactor. Being an inert gas it does not interact chemically with any of the species present. It is readily dispersible in the gas phase and it is soluble in hydrocarbons (Stephen and Stephen, Solubilities of Inorganic and Organic Compounds, Vol. 1, MacMillan Co. 1963). Further, this isotope emits energetic radiation that is easy to detect.

Two separate types of experiments were undertaken. In one case Kr-85 was metered from a pressurized container into the natural gas feed line into the reactor; and in the second case, separately from the first, the Kr-85 was metered into the heated feedstock oil line feeding the reactor. In each case the reactor was in steady-state operation producing standard HAF or Standard ISAF as described above.

The sample probes, containers, and procedures were the same as described for gas sampling of the reactor. In each case the sampling gridwork was the same as previously described. Each sampling period through one full grid of the reactor required about one hour so it was easy to tell from the operating charts that no variations occurred in feed rates or reactor operations that would affect the sample series. Each test required approximately 50 millicuries of Kr-85. Before and after each series of tests the reactor was sampled without radioactive isotope injection to insure a "blank," i.e., no radioactivity in the system from any other conceivable or inconceivable source.

The analysis of these samples may be carried out by a variety of techniques, and we chose the technique simplest for us, because we had an internal gas geiger counting system already calibrated for Kr-85 and in operation in our laboratories for other purposes.

Interpretation of the Data

In each case described profiles of the interior of the reactor can be plotted from the collected data. The pressure measurements were the least informative, as was expected. Pressure in a carbon black reactor is employed primarily as a conveyor of materials, i.e., the materials are fed into the reactor with sufficient pressure to move the feed materials at the rates described through the reaction zone and into the collection system after the reaction has been "killed" (quenched).

The temperature measurements were the most difficult to make. The thermocouple used, mentioned before, is a fragile, brittle wire and was subject to frequent breakage as a result of the turbulent bombardment it received in this hostile atmosphere. Sufficient temperatures were determined to verify the combustion gases temperature already known. Most significant, were the measurements of the temperature gradients at the initiation of the carbon black reaction and the disappearance of gradients as the reaction reached its first primary completion. The points within the reactor where these maximum temperatures and their associated gradients occur differ between grades (particle size) being produced. For smaller particle size blacks this primary reaction zone is closer to the point of entry of materials and the reaction path length is shorter.

Gas composition combined with materials distribution were most informative. Hydrogen, oxygen, carbon dioxide, and carbon monoxide were the only constituents of significance. Nitrogen and argon followed the same pattern, with argon nearly constant at 1%, and reaching a slight minimum at the exact center. Nitrogen was essentially constant, also showing a minimum at the center of the combustion chamber and in the vicinity of the principal reaction. Nitrogen on a percentage basis also changed as more gaseous products were produced.

With the particular burner employed in these experiments, the methane was completely consumed at the point of the first sampling point, and was not otherwise present.

Acetylene was not detected in the combustion chamber; however, it began to appear in the reaction zone and reached a maximum of 0.5-0.9% in this zone.

Carbon dioxide concentration appears as a maximum about half-way between the wall and the center of the combustion chamber (Fig 6). This maximum diminishes toward the reaction zone and disappears in the tunnel. The carbon dioxide has practically constant concentration across the reactor tunnel, but this concentration decreases with tunnel length.

Some oxygen in excess of that required for stoichiometric combustion is fed to the reactor which subsequently reacts with feedstock in the principal reaction zone. The conversion of the oil to carbon black is a highly endothermic reaction and the excess oxygen maintains heat for the reaction to go to completion. The oxygen concentration in the reactor is at a minimum at the wall of the combustion chamber, rising to a maximum at the center (Fig. 7). This gradient is reversed in the restriction ring as oxygen is consumed by oil; and then it effectively drops to zero in the reactor tunnel.

Hydrogen is nonexistent in the combustion chamber (Fig. 8). It begins to appear in the reaction zone at a maximum concentration in the center, diminishing to its lowest concentration at the wall. This gradient quickly disappears across the reactor with a constantly increasing hydrogen content with tunnel length.

Carbon monoxide appears at the entrance of the combustion chamber close to the wall but quickly disappears. Carbon monoxide reappears again with a pattern very similar to hydrogen in the vicinity of the reaction zone (Fig. 9). Beyond this zone the carbon monoxide concentration is essentially constant across the reactor, with increasing concentration with tunnel length.

When Kr-85 was injected into the fuel gas (CH4) the concentration patterns followed those of the carbon dioxide (Fig. 10). In the combustion chamber the concentration was at a maximum about half-way between the wall and the center. This doughnut shape persisted into the restriction ring, and the maximum disappeared in the tunnel. This disappearance of the maximum in the tunnel demonstrated total mixing of the combustion gases and carbon black as the oil was converted to carbon black.

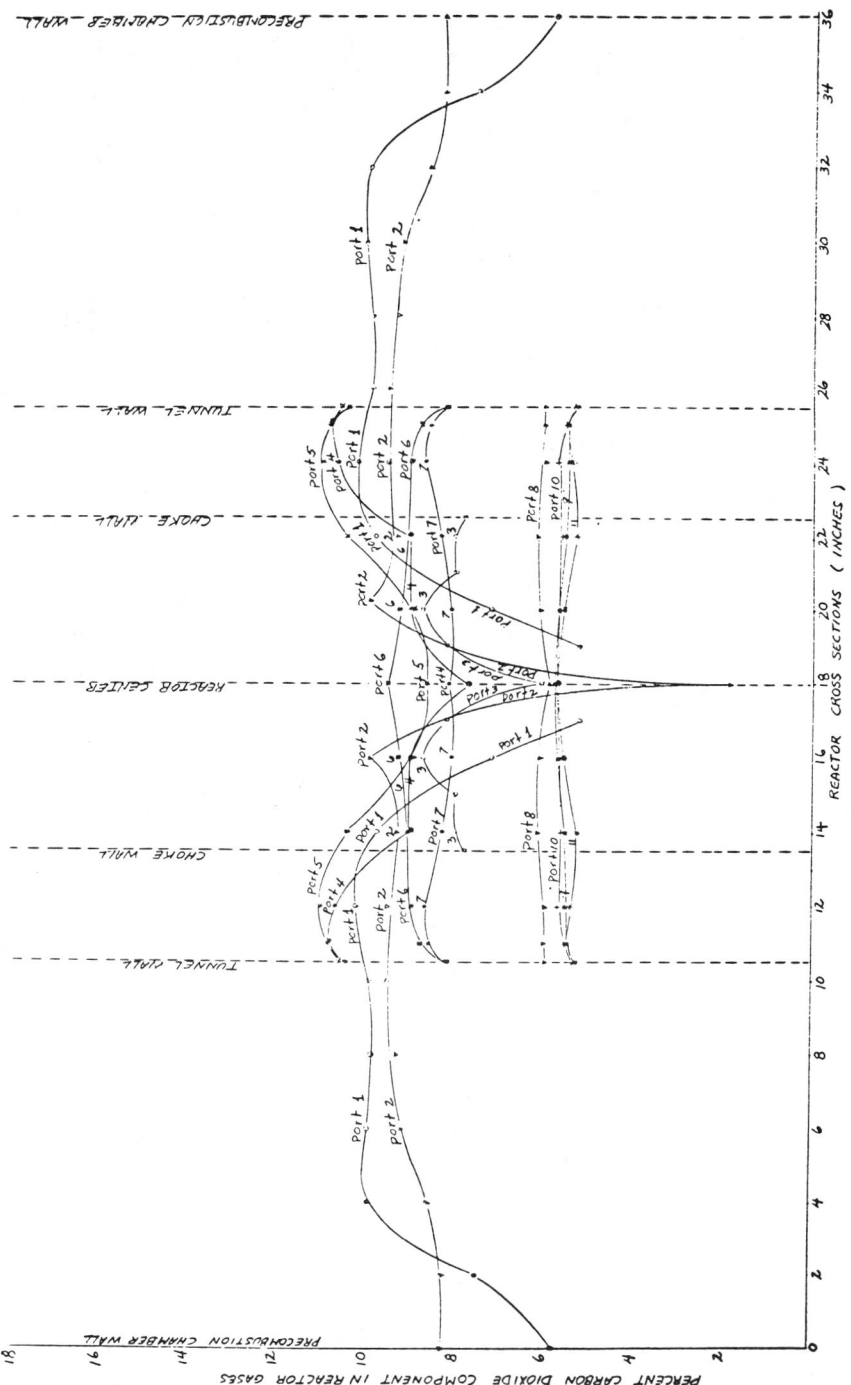

Figure 6. The Cross Sectional Profiles of the Carbon Dioxide Component at Reactor Port Locations (HAF Data)

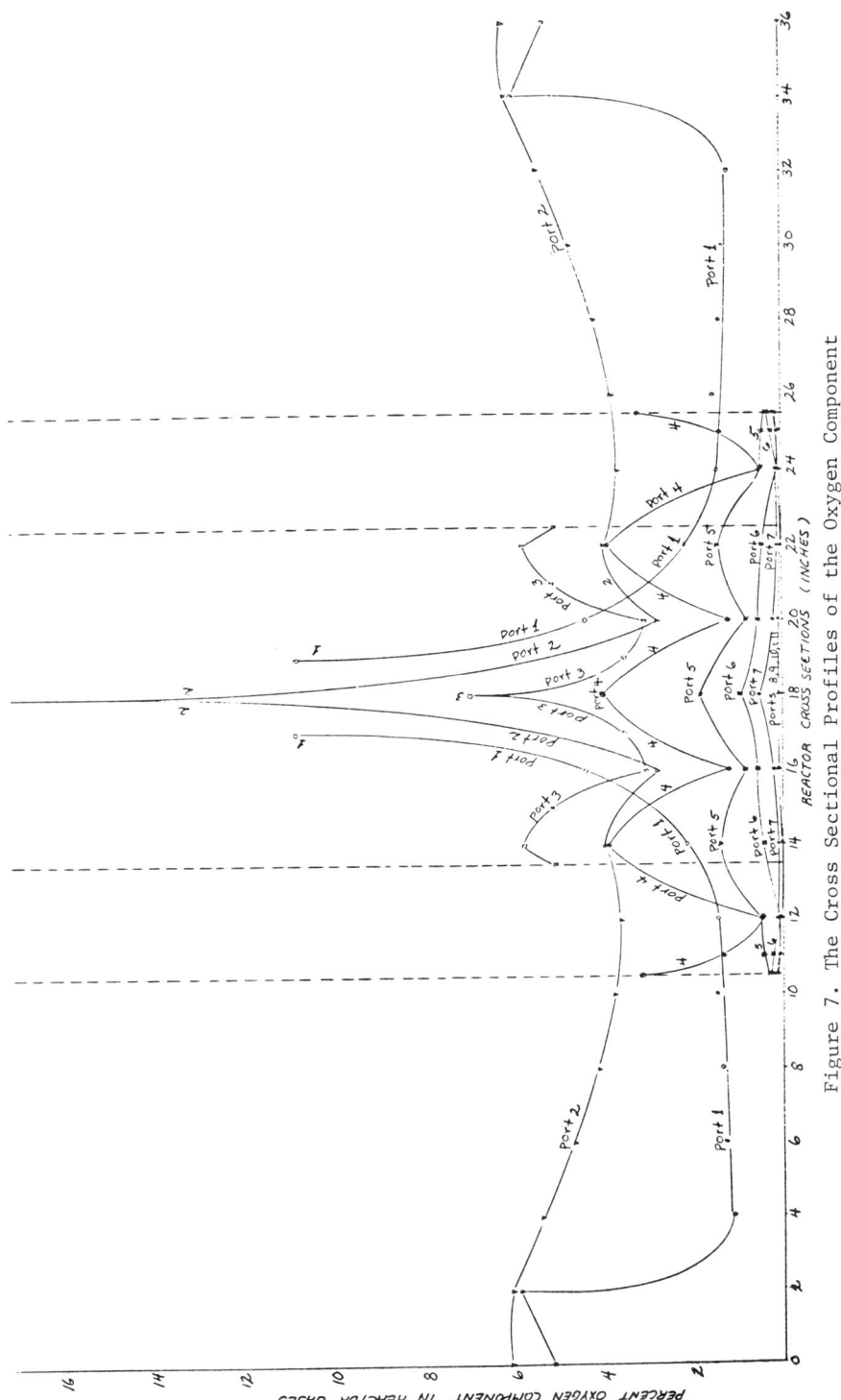

Figure 7. The Cross Sectional Profiles of the Oxygen Component at Reactor Port Locations (HAF Data)

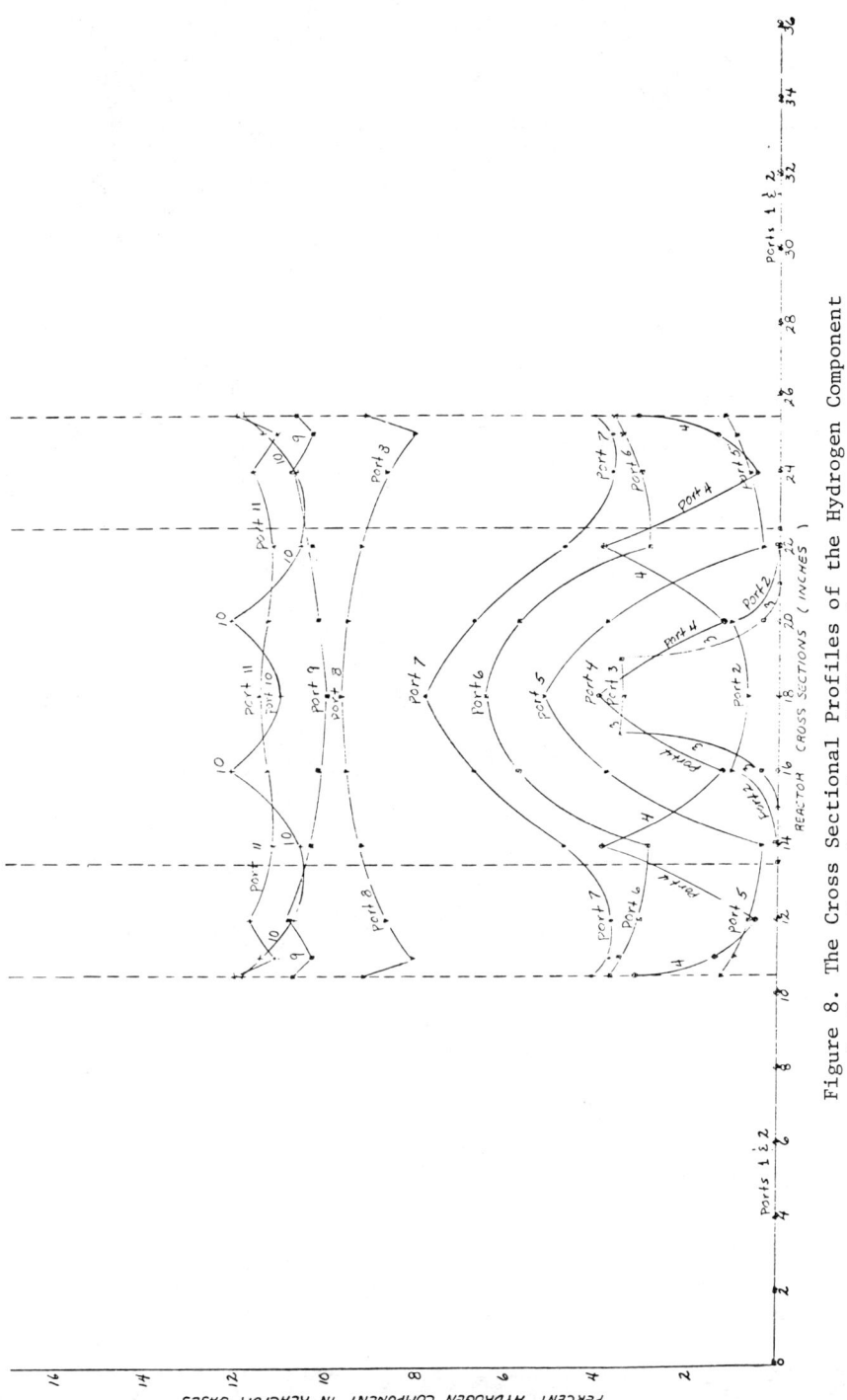

Figure 8. The Cross Sectional Profiles of the Hydrogen Component at the Reactor Port Locations (HAF Data)

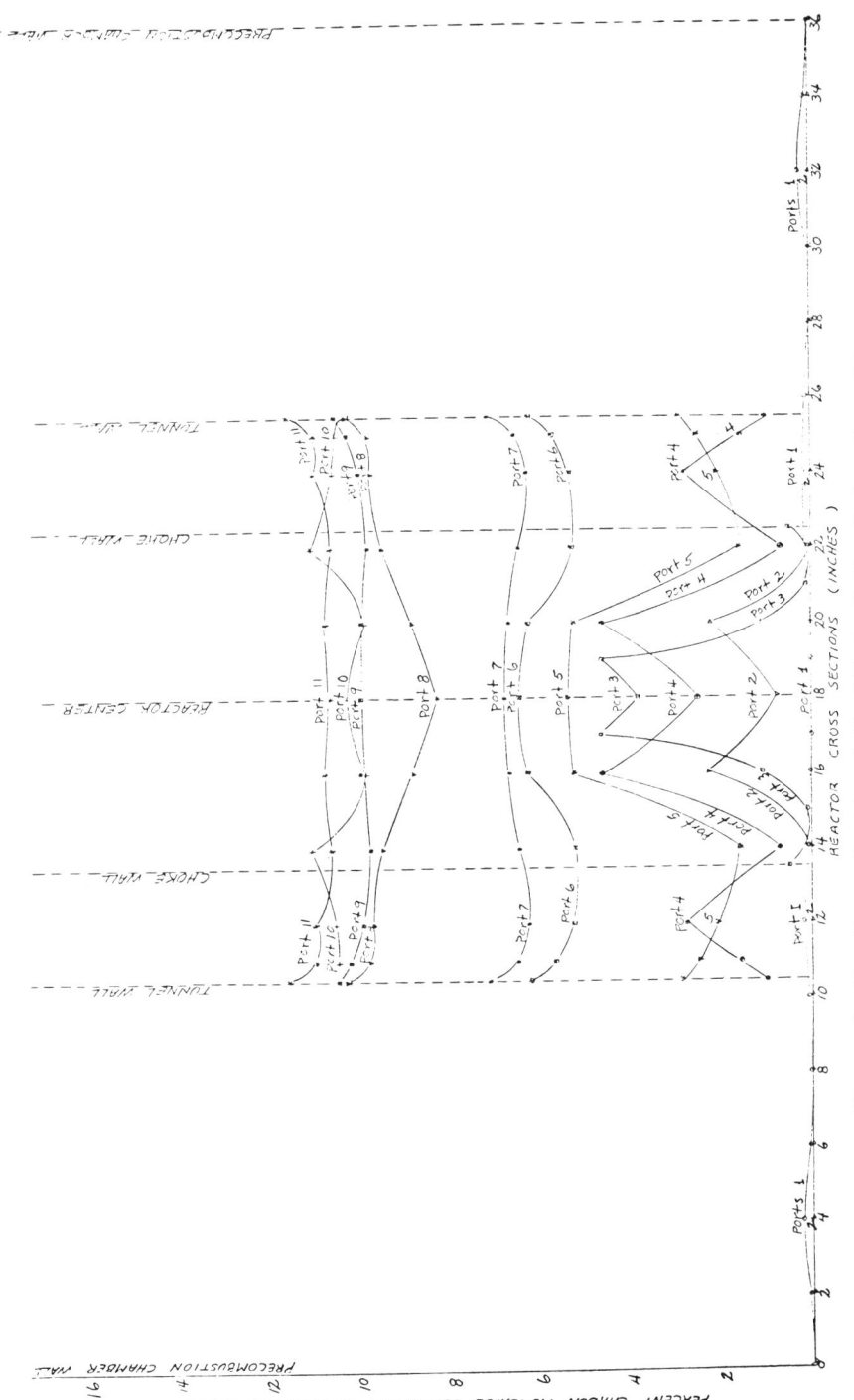

Figure 9. The Cross Sectional Profiles of the Carbon Monoxide Component at Reactor Port Locations (HAF Data)

19. LEWIS *Mechanism of Carbon-Black Formation* 285

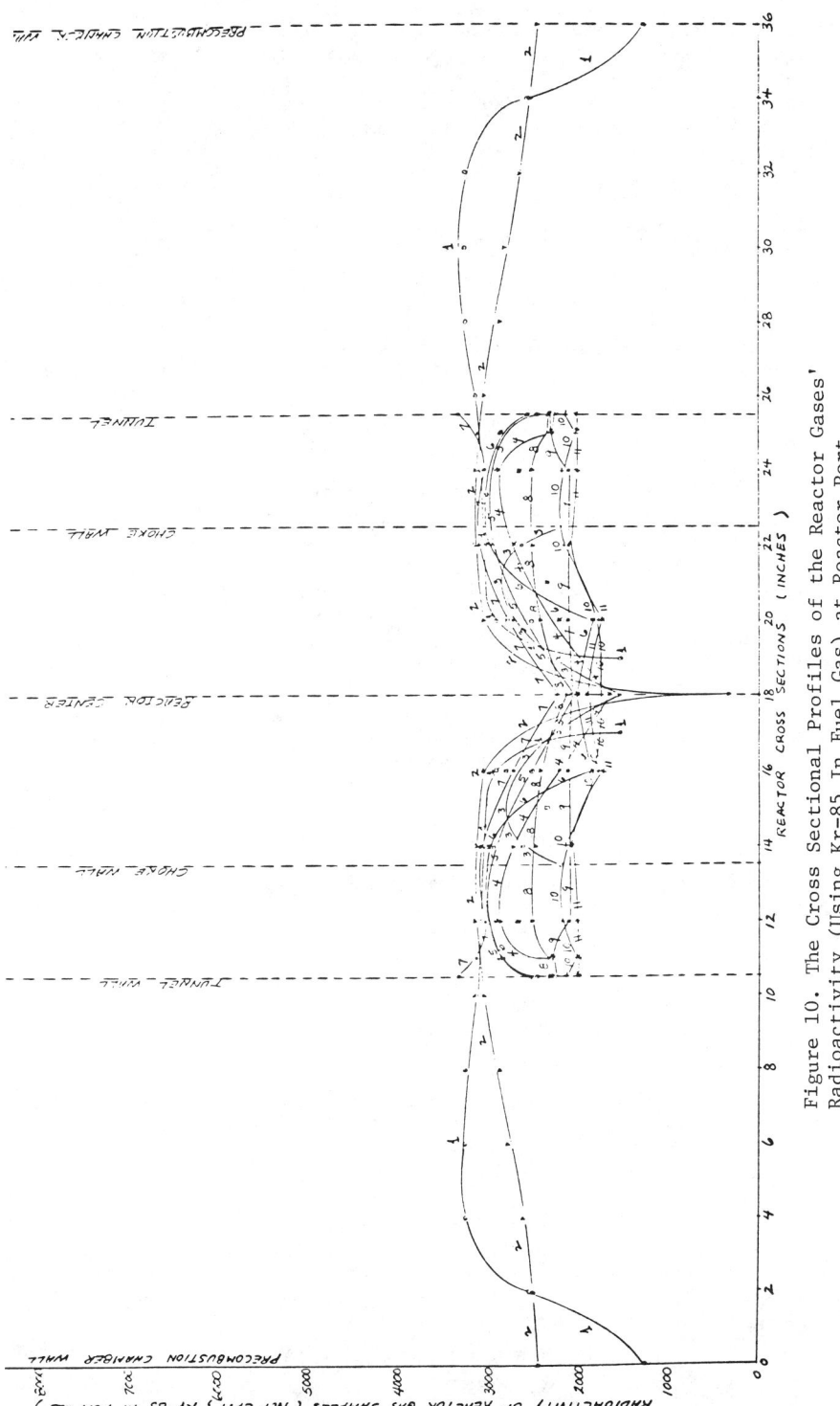

Figure 10. The Cross Sectional Profiles of the Reactor Gases' Radioactivity (Using Kr-85 In Fuel Gas) at Reactor Port Locations (HAF Data)

By contrast, when Kr-85 was injected into the oil feedstock no radioactivity was detected in the combustion chamber to within one inch of the oil stream (injector) (Fig. 11). This pattern is also illustrated by the data in the restriction ring; however, now the oil-black has expanded to where it occupies about half of this cylindrical volume. At the exit of the restriction violent turbulence causes an almost explosive expansion of the oil-black across the diameter of the tunnel. The gradient of the radioactivity from the wall to the center of the reactor smooths out beyond the reaction zone indicating homogeneous distribution of all constituents.

The turbulence causes very rapid dispersion of the oil, creating an intimate mixture with the combustion gases, a very rapid heat transfer as the temperature of the oil increases from about 700° F to 3300° F, and an extremely fast conversion of hydrocarbons to carbon black.

The shape of this reaction zone depends, of course, on a number of physical constraints; however, in the case described it may assume the shape of a teardrop.

The most important observed result is that within this teardrop two key properties of the carbon black are created: particle size and structure.

The reaction from oil to carbon black is almost instantaneous as far as the formation of particle size and structure. The reaction times with an unknown error of measurement were determined to be 1.2 msec. for ISAF and 2.0 msec. for HAF. The error, if any, is plus 0 and minus X. Each of these two properties can be controlled in the reactor independently of the other.

In the particular reactor system employed, structure of the black is more easily controlled by injecting the selected feedstock as a liquid, rather than as a vapor or gas. In this case, the distribution of the oil injected into the combustion chamber can be controlled by mechanical means. The point of initiation of the reaction can also be moved along a line in the reactor to make it short or long at will. Structure then is more dependent upon the feedstock, its boiling range and preheat temperature, and the point within the reactor selected for reaction initiation. Structure can also, of course, be altered downward through the use of feedstock additives.

Particle size, on the other hand, is controlled primarily by feedstock rates and temperature.

Surface area is controlled independently of either particle size or structure. If the reaction were quenched at the end of its primary reaction time (as mentioned, at the point of particle size and structure formation, about 1.2 to 2 msec.), the surface area for this particular black would be at its lowest. If, on the other hand, the reaction mass is allowed to remain at temperature for a period of 8 to 10 times the initial period, i.e., for 12 to 20 msec., additional reactions continue to occur. Although the reaction for the formation of carbon black has been completed, further reactions between carbon, carbon dioxide, and water result in an increase in surface area. These reactions are slow compared to the formation of carbon reactions and, allowing these reactions to proceed for about 20 msec., the effect on particle size and

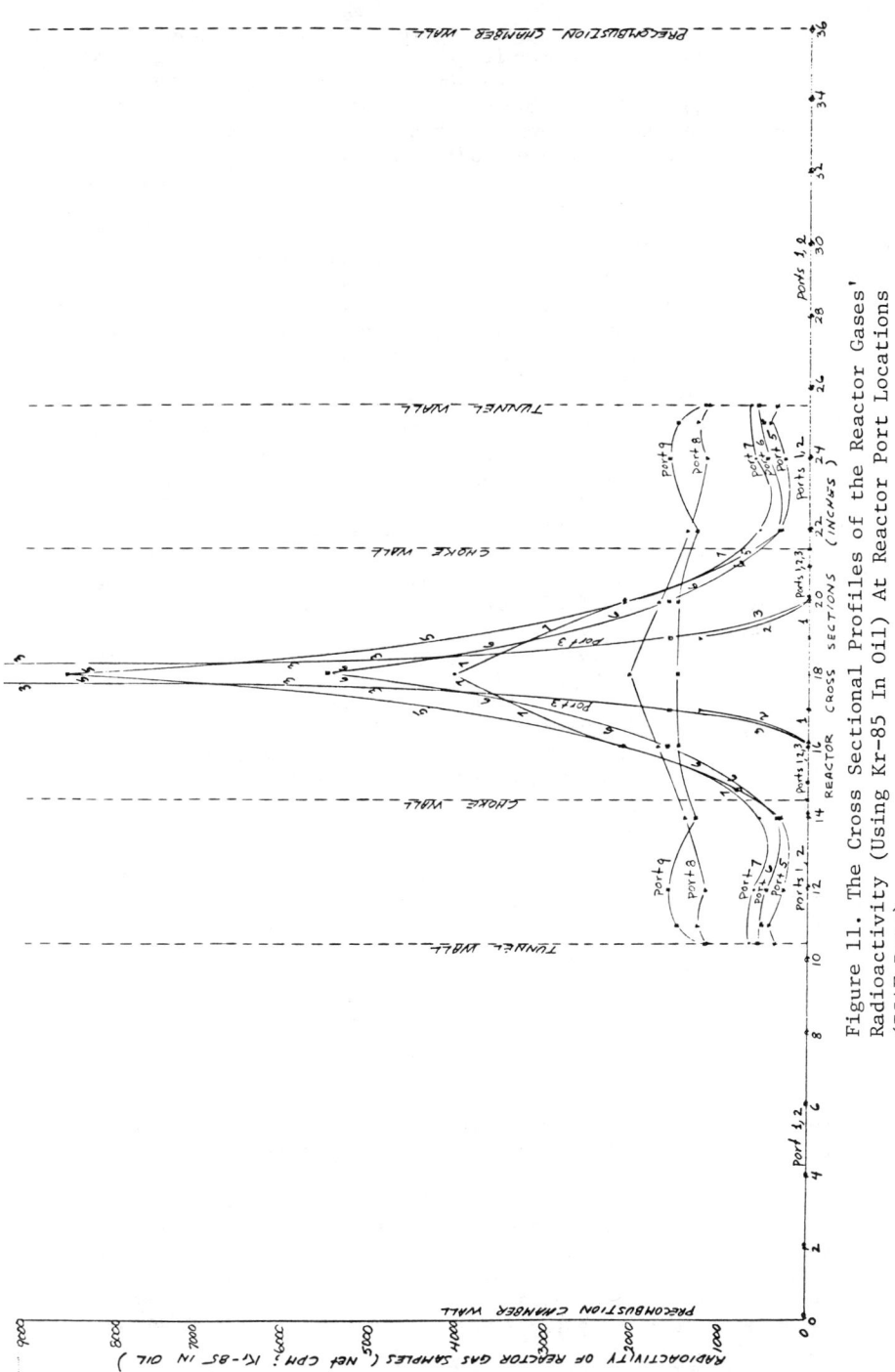

Figure 11. The Cross Sectional Profiles of the Reactor Gases' Radioactivity (Using Kr-85 In Oil) At Reactor Port Locations (ISAF Data)

structure is too small to be detected. If the reactions were
allowed to continue indefinitely in an inordinately long reaction
tunnel both particle size and structure would be affected. In fact,
the carbon particles would eventually disappear, as they are converted to gaseous species. In these tests as conducted, surface
area was controlled from an iodine number of about 60 to 160.

This latter conclusion is in agreement with other published
reports on the mechanism of carbon black formation. For example,
Tesner (42, 43, 44, 45) is most well known for his theories of
nucleation and growth of carbon particles. In a communication with
Gaydon and Wolfhard (15), Tesner revealed that carbon particles in
contact with carbon monoxide in a furnace show no growth whatsoever
at any temperature. A mechanism reporting nucleation followed by
growth in the carbon black reaction was recently reported by Dahmen
(10); however, none of the experimental techniques nor data were
presented.

We have to conclude from our experimental observations that the
carbon black has a direct dependence upon the molecular structure of
the feedstock. This dependence is fairly broad in that, as long as
the feedstock is predominately aromatic, other variables of reactor
geometry, rates and ratios of feed materials, and temperature can be
adjusted to produce various grades of carbon black. Commercial
production of carbon blacks have added incentive of optimizing the
economic variables of efficiency, yield, and throughput.

However, we must set aside the theory of polymerization to
giant molecules followed by graphitization from within as proposed
by Lahaye (27) and Gaydon (14) along with others.

Stehling, Frazee, and Anderson (39) suggested a mechanism of
dehydrogenation followed by condensation. We feel that this may
more closely describe the carbon black reaction. From our
observations it appears that aromatic molecules are stripped of
hydrogen at the point of very rapid temperature rise; this nearly
plasma-like atmosphere does not lose the principle of graphitic-like
nuclear structure. The condensation of these ring structures occurs
almost simultaneously with the dehydrogenation reactions, forming
both particles and particle-to-particle bonding, resulting in the
phenomena we call structure.

The Product

As a result of these studies reactors of different sizes could be
operated with factors of scale that had a high confidence factor.
We are now operating furnace reactors with feedstock feed rates as
low as 5 gal/hour. The larger reactors have feed rates of several
hundred gal/hour, with the largest operating at about 1000 gal/hour.
All of the process parameters are consistent and the products
produced are interchangeable.

Using medium sized commercial reactors 100 different "grades"
of carbon black were produced. Of course, these were not all
commercial grades of black. Our objectives were two-fold. One, to
establish the operating parameters to insure we fully understood the
ranges of particle size, structure, and surface area which could be
produced as independent variables; and, two, to establish grades of

black that were progressive with respect to one another in regard to these three properties.

As mentioned earlier, particle size is the key to carbon black quality. It influences the cost, the grade, and the black's ability to reinforce rubber. Recognizing the importance of this property, and the fact that the only absolute method for determining this property was electron microscopy, we were working on improved tint tests which could be directly related to particle size. It was desirable to have a test which would be inexpensive, easy to conduct, and one which would give reliable results. This large number of samples of blacks with progressive properties offered an ideal proving ground. We included with this study a reevaluation of the relationship between electron microscope particle size and photoelectric reflectance test for indicating particle size developed in our laboratories.

In the course of reevaluating the electron microscope techniques, it was found that the primary differences in duplicate sample determinations occur in the large and small particle tails of the distribution curves (Fig. 12). This indicated an inconsistent ability to disperse larger particles for analysis and errors from overlooking smaller particles. By changing the technique to printing the micrograph on film, feeding the counts and diameters into a computer pickup, with automatic printout of the distribution data and curves, the problems of reproducibility were overcome.

In the final analysis 60 distinctly different carbon blacks were selected to give a good separation of variables. Included with these were several standard commercial grades, several blacks obtained from other suppliers, and many special, noncommercial grades.

These 60 blacks fell into eight general particle sizes:

1) less than 23 nm, with DBP ranging from 103 to 128 and surface area (I_2 No.) from 104 to 139.

2) 25 nm. = 9 samples
DBP = 78 to 129, I_2No. = 100 to 125

3) 30 nm.
DBP = 68 to 148, I_2No. = 80 to 92

4) 40 nm.
DBP = 70 to 116, I_2No. = 47 to 62

5) 50 nm.
DBP = 63 to 153, I_2No. = 30 to 60

6) 60 nm.
DBP = 84 to 123, I_2No. = 27 to 35

7) 70 nm.
DBP = 66 to 125, I_2No. = 26 to 32

8) 80 nm.

$$DBP = 60 \text{ to } 67, \quad I_2No. = 26 \text{ to } 27$$

All but one size classification had at least five different blacks.

The details of this work revealed the exact effect of structure on the tinting strength of carbon black in addition to the effect of particle size (Fig. 13). Although the effect of structure is somewhat subtle, it is sufficient to cause a tint test between two blacks of the same particle size to appear quite different. So, in order to relate tint readings to absolute particle size the effect of DBP on the reflectance test result must be taken into account.

This work has contributed to the adoption of an ASTM tint test -- now employed as a standard part of carbon black specifications along with iodine number and DBP.

In summary, we will review carbon black classifications and their properties in rubber. The three most important properties are particle size, structure, and surface area. We will discuss only the tread grades. All of the tread grades of carbon black fall within the range of 20 nm. to 40 nm.

To relate the physical properties of carbon black to rubber properties, we tested these tread blacks in the ASTM natural rubber recipe and in an SBR 1500 test recipe. In both elastomers, we checked standard stress/strain properties of modulus, tensile strength, and hardness. In the natural rubber recipe we also tested Firestone running temperature and rebound, and Goodyear rebound. In the SBR we checked percent swell, extrusion rate, viscosity, and laboratory abrasion.

The results of ASTM natural rubber modulus, shown in (Fig. 14), demonstrate that particle size is not significant, only surface area and structure. Modulus increases with structure and decreases with higher surface area, a well-known relationship.

For SBR modulus (Fig. 15), structure and either particle size or surface area were significant. We found no correlation when both particle size and surface area were used together. The results are similar between SBR and natural rubber. Modulus is the least predictable property in rubber from the three important properties of carbon blacks. Both modulus and cure-rate are affected by the slight surface oxidation which blacks receive during the drying of finished pellets.

Tensile strength in both natural rubber and SBR is dependent upon structure and particle size as the key variables. Tensile strength increases with lower structure and with smaller particle size (Fig. 16 & 17).

All three properties of carbon black affect hardness in natural rubber compounds (Fig. 18). Larger particle size reduces hardness; higher structure increases hardness and increased surface area reduces hardness. Quite the contrary in SBR (Fig. 19), only structure is significant in regard to hardness. As DBP increases, hardness increases.

Additional testing in natural rubber demonstrated that particle size was significant in Firestone running temperature, as was structure (Fig. 20). Smaller particle size and higher structure both increase heat build-up.

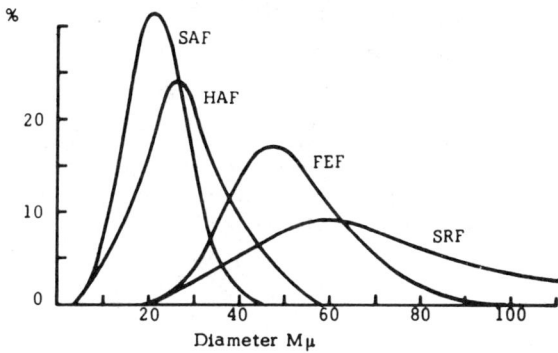

Figure 12. Particle Size Distribution

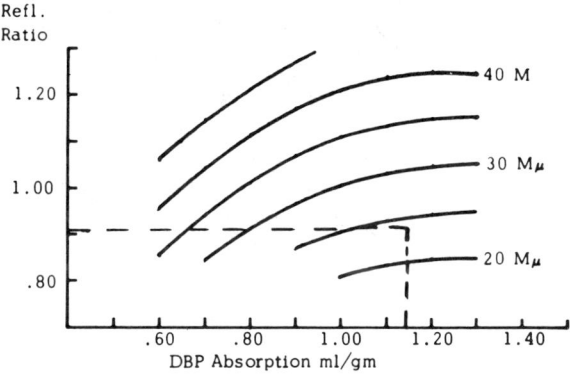

Figure 13. DBP Absorption

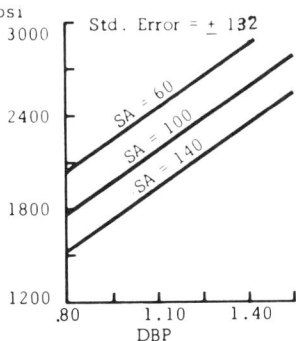

Figure 14. NR Modulus

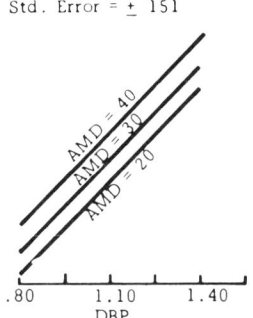

Figure 15. SBR Modulus

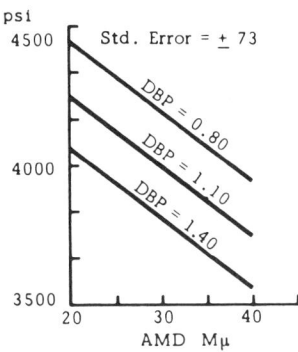

Figure 16. NR Tensile Strength

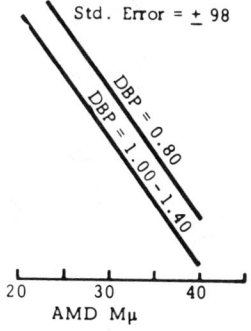

Figure 17. SBR Tensile Strength

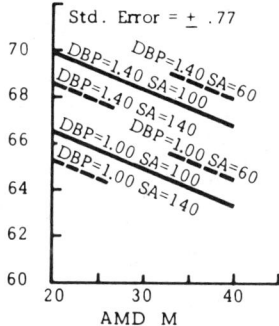

Figure 18. NR Hardness

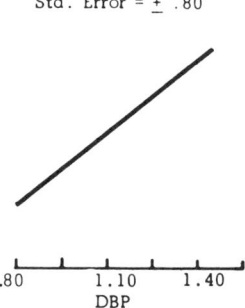

Figure 19. SBR Hardness

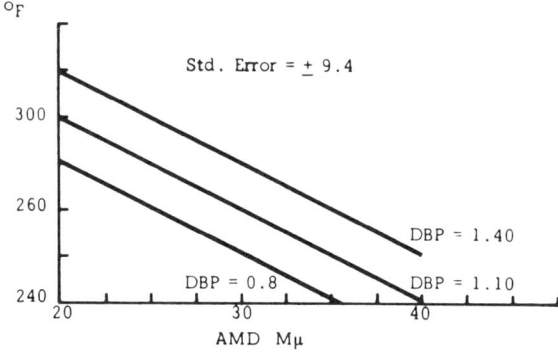

Figure 20. Firestone Running Temperature

Both increasing particle size and increasing structure cause higher Goodyear rebound, although structure has much less effect (Fig. 21).

Firestone rebound is more dependent upon surface area and particle size. Larger particles increase rebound, while a higher surface area reduces it (Fig. 22).

In SBR compounds the percent die swell increases with decreasing particle size and decreases with increasing structure (Fig. 23).

Viscosity and extrusion rate run parallel in relation to structure and reversed relative to surface area. Higher structure produces higher viscosity and higher extrusion rates. Higher surface area increases viscosity and lowers the extrusion rate (Fig. 24 & 25).

An excellent correlation was established between particle size and treadwear at equal loading and at equal hardness. In a tire test including 50 tires and 10 different blacks under both moderate and severe conditions, we established the very reliable equation.

$$\text{Treadwear Index} = \frac{A}{(\text{particle diameter})^n}$$

Correlation coefficient = 0.98
Standard Error = ± 3%

when comparing treadwear indices between blacks.

In fact, when using a low-oil SBR recipe in the laboratory and measuring abrasion loss of the same blacks used in the road test, we found excellent correlation between abrasion loss and treadwear. The same relationship indicated above holds when comparing abrasion loss of blacks versus treadwear of the same blacks and will predict the same order by either test:

$$\text{Treadwear Index} = \frac{A}{(\text{abrasion loss})^n}$$

Correlation coefficient = 0.99
Standard Error = ± 2.5%

Carbon blacks are best identified by particle size in relation to their ability to wear.

Conclusion

In order to produce a carbon black to any specification, the relationships between particle size, structure, and surface area allow the predetermination of performance parameters desired in a rubber compound, before and after vulcanization.

Once the specification has been defined, the carbon black manufacturer can produce a grade of black to that target. First, the particle size is fixed by the selection of a particular grade of black, the structure level and surface area may be targeted by the proper selection of process variables.

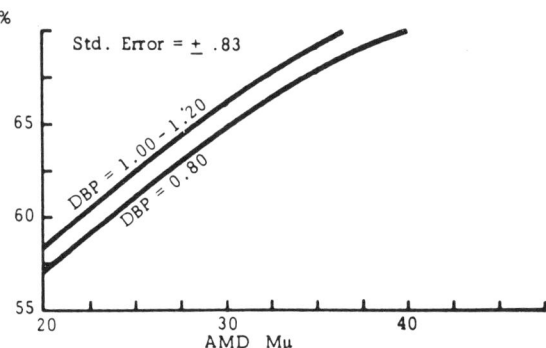

Figure 21. Goodyear Rebound

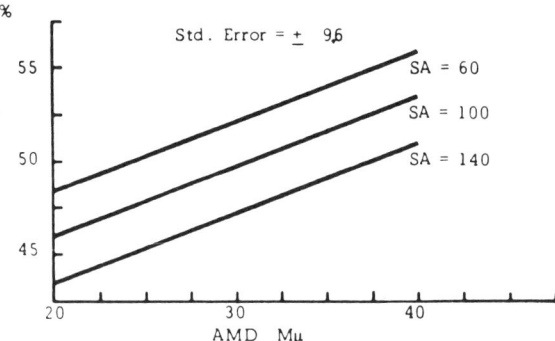

Figure 22. Firestone Rebound

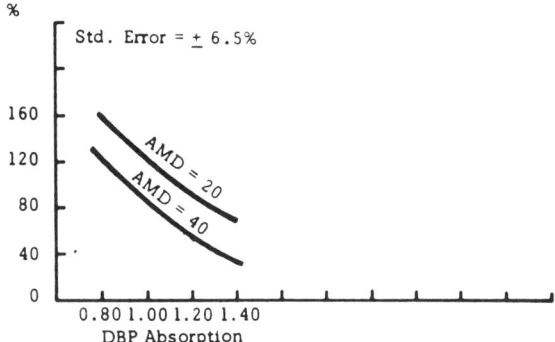

Figure 23. % Die Swell

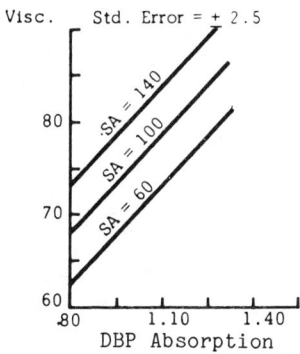

Figure 24. Viscosity

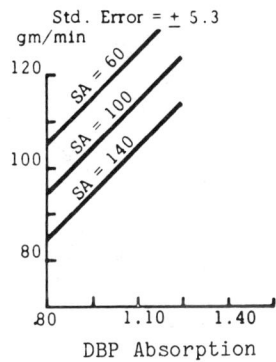

Figure 25. Extrusion Rate

To illustrate: In (Fig. 26) the effects of iodine number and structure on both running temperature and modulus are combined. If structure is held at a constant level and the iodine number is varied, running temperature and modulus change simultaneously. Both running temperature and modulus increase with increased structure. The cross-hatched area in the center of the graph is the specification for a grade of ISAF. This area thus becomes a target area for quality control for a black with the particle size dictated by N-220.

In a similar example (Fig. 27) based upon an SBR formula, extrusion swell is substituted in place of running temperature. Here, again, at fixed particle size, modulus is a function of both iodine number and structure, but extrusion swell is a function of structure only. The cross-hatched area is the quality control target area for that particular black to be in specification.

These two graphs show the effect of the three fundamental characteristics of carbon black on rubber properties. The rubber properties obtained from using a particular carbon black are, therefore, dependent upon the respective properties built into the carbon black by the manufacturer. It is obvious that the carbon black manufacturer must have a well-developed process, a full knowledge of the materials he will use, and a complete understanding of the effect of process variables upon each of the properties of each grade of carbon black to be produced.

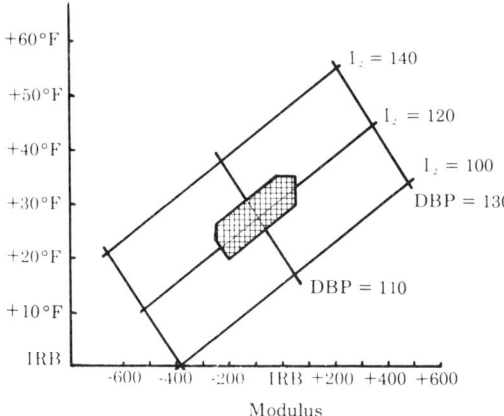

Figure 26. Relationship of Iodine Number, DBP Absorption, ASTM Modulus and Running Temperature

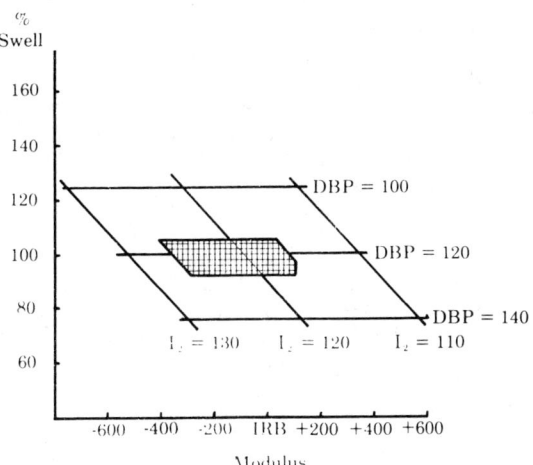

Figure 27. Relationship of Iodine Number Void Volume, SBR Modulus and Extrusion Swell

Literature Cited

1. G. A. Agoston; H. Wise; and W. A. Rosser Sixth Symposium (Intl.) on Combustion, p. 708, Reinhold Publ. Corp., New York, 1957.
2. A. G. Amelin; Colloid J. (U.S.S.R.) 29, 12 (1967).
3. R. C. Anderson Advance Chem. Series No. 20, p. 49, Am. Chem. Soc., Washington, 1958.
4. O. K. Austin "Reinforcement of Elastomers," G. Kraus, Ed., p. 287, Interscience Publ. Co. (John Wiley), New York, 1965.
5. K. A. Burgess; F. Lyon; W. S. Stoy "Encyclopedia of Polymer Science and Technology," Vol. 2, p. 820, Interscience Publ. Co. (John Wiley), New York, 1964.
6. B. D. Crittenden; R. Long Combustion and Flame 20, 359 (1973).
7. C. F. Cullis; I. A. Read; D. L. Trimm Eleventh Symposium (Intl.) on Combustion, p. 391, Combustion Institute, Pittsburgh, Pa., 1967.
8. R. A. Davies; D. B. Scully Combustion and Flame 10, 165 (1966).
9. E. M. Dannenberg J. Inst. Rubber Ind. 5, 190 (1971).
10. K. R. Dahmen; N. N. McRee "Effects of Carbon Black in Improved Wear Resistance in Tire Tread Stocks and The Effects of Compound Loading Variants on Wear Resistance," Preprint of Paper A-19 Presented to Intl. Rubber Conf., Prague, Czechoslovakia, Sept. 17-20, 1973.
11. C. P. Fenimore; G. W. Jones Combustion and Flame 13, 303 (1969).
12. A. Feugier Combustion and Flame 19, 249 (1972).
13. R. P. Fraser Sixth Symposium (Intl.) on Combustion, p. 687, Reinhold Publ. Corp., New York, 1957.
14. A. G. Gaydon; A. R. Fairbairn Fifth Symposium (Intl.) on Combustion, p. 324, Reinhold Publ. Corp., New York, 1955.
15. A. G. Gaydon; H. G. Wolfhard "Flames, Their Structure, Radiation and Temperature," Chap. VIII, Chapman & Hall Ltd., London, 1960.
16. H. J. Golding Research/Development 24 (6), 46 (1973).
17. K. H. Homann Combustion and Flame 11, 265 (1967).
18. K. H. Homann; H. G. Wagner Proc. Roy. Soc. A. 307, 141 (1968).
19. J. B. Howard Twelfth Symposium (Intl.) on Combustion, p. 877, Combustion Institute, Pittsburgh, Pa., 1969.
20. W. M. Hess; G. C. McDonald; E. Urban Rubber Chem. Technol. 46, 204 (1973).
21. W. M. Hess; V. E. Chirico "Classification of Rubber Grade Carbon Blacks from Different Suppliers," Paper Presented at Akron Rubber Group Symposium, Ohio, Jan. 22, 1971.
22. W. M. Hess; V. E. Chirico; K. A. Burgess "Morphological Characterization of Carbon Blacks in Elastomer Vulcanizates," Paper A-17 Presented to Intl. Rubber Conf., Prague, Czechoslovakia, Sept. 17-20, 1973.
23. R. D. Ingebo Sixth Symposium (Intl.) on Combustion, p. 684, Reinhold Publ. Corp., New York, 1957.
24. G. L. Johnson (with R. C. Anderson) "An Electron Microscope Study of Carbon Formation in Hydrocarbon Pyrolysis," PhD Dissertation, University of Texas, 1960, Dissertation Abs. 23, 87 (1962): University Microfilms No. 62-2586.

25. J. M. Jones; J. L. J. Rosenfeld Combustion and Flame 19, p. 427 (1972).
26. V. A. Kargin; Z. Y. Berestneva; N. Y. Safronov; V. I. Zhilkina Colloid J. (U.S.S.R.) 29, 342 (1967).
27. J. Lahaye; G. Prado; J. B. Donnet Carbon 12, 27 (1974).
28. C. L. Mantell "Carbon and Graphite Handbook," Chap. 6, Interscience Publ. Co. (John Wiley), New York, 1968.
29. A. I. Medalia; L. W. Richards J. Colloid Interface Sci. 40, 233 (1972).
30. A. I. Medalia; E. M. Dannenberg; F. A. Heckman; G. R. Cotten Rubber Chem. Technol. 46, 1239 (1973).
31. G. J. Minkoff; C. F. H. Tipper "Chemistry of Combustion Reactions," p. 237-361, Butterworths, London (1962).
32. H. B. Palmer; C. F. Cullis "Chemistry and Physics of Carbons," Chap. 5, P. L. Walker, Ed., Dekker, New York, 1965.
33. H. B. Palmer; A. Voet; J. Lahaye Carbon 6, 65 (1968).
34. B. V. Poulston; E. F. Winter Sixth Symposium (Intl.) on Combustion, p. 833, Reinhold Publ. Corp., New York, 1957.
35. W. R. Smith; D. C. Bean "Encyclopedia Chem. Technol.," Vol. 4, p. 243, Interscience Publ. Co. (John Wiley), New York, 1964.
36. C. A. Stokes "Encyclopedia Chem. Technol.," Supplement, p. 91, Interscience Publ. Co. (John Wiley), New York, 1971.
37. Arne Sjogren Fourteenth Symposium (Intl.) on Combustion, p. 919, Combustion Institute, Pittsburgh, Pa., 1973.
38. I. I. Samkhan; Y. V. Tsvetkov; V. A. Petrunichev; I. K. Glushko Colloid J. (U.S.S.R.) 33, 885 (1971).
39. F. C. Stehling; J. D. Frazee; R. C. Anderson Eighth Symposium (Intl.) on Combustion, p. 774, Williams and Wilkins Co., Baltimore, Md., 1962.
40. N. Y. Safronova; Z. Y. Berestneva; V. A. Kargin; V. K. Shmigel'skii Colloid, J. (U.S.S.R.) 29, 385 (1967).
41. R. L Schalla; G. E. McDonald Fifth Symposium (Intl.) on Combustion, p. 316, Reinhold Publ. Corp., New York, 1955.
42. P. A. Tesner Seventh Symposium (Intl.) on Combustion, p. 546, Butterworths Scientific Publications, London, 1959.
43. P. A. Tesner; H. J. Robinovitch; I. S. Rafalkes Eighth Symposium (Intl.) on Combustion, p. 801, Williams and Wilkins Co., Baltimore, Md., 1962.
44. P. A. Tesner; T. D. Snegiriova; V. G. Knorre Combustion and Flame 17, 253 (1971).
45. P. A. Tesner; E. I. Tsygankova; L. P. Guilazetdinov; V. P. Zuyev; G. V. Loshakova Combustion and Flame 17, 279 (1971).
46. T. S. Whitsel Rubber Plastics Age 45, 1209 (1964).
47. A. Williams Combustion and Flame 21, 1 (1973).
48. B. L. Wersborg; J. B. Howard; G. C. Williams Fourteenth Symposium (Intl.) on Combustion, p. 929, Combustion Institute, Pittsburgh, Pa., 1973.

RECEIVED November 4, 1985

20

High-Surface-Area Active Carbon

T. M. O'Grady[1] and A. N. Wennerberg[2]

[1] Amoco Research Center, Amoco Oil Company, Naperville, IL 60566

> This paper describes the preparation and properties of a unique active carbon having exceptionally high surface areas, over 2500 m^2/gm, and extraordinary adsorptive capacities. The carbon is made by a direct chemical activation route in which petroleum coke or other carbonaceous sources are reacted with excess potassium hydroxide at 400° to 500°C to an intermediate product that is subsequently pyrolyzed at 800° to 900°C to active carbon containing potassium salts. These are removed by water washing and the carbon is dried to produce a powdered product. A granular carbon can also be made by further processing the powdered carbon by using specialized granulation techniques. Typical properties of the carbon include Iodine Numbers of 3000 to 3600, methylene blue adsorption of 650 to 750 mg/gm, pore volumes of 2.0 to 2.6 cc/gm and less than 3.0% ash. This carbon's high adsorption capacities make it uniquely suited for numerous demanding applications in the medical area, purifications, removal of toxic substances, as catalyst carriers, etc. It will be commercially available from Anderson Development Company in mid-1985.

A unique active carbon having very high surface areas over 2500 m^2/gm, and extraordinary adsorptive capacities was developed in our laboratories. (1) This paper will describe its development, manufacture, properties, and uses. Until recently, samples of this carbon, which were provided worldwide for research and evaluation, were identified as Amoco Grades PX-21, 22, 23, and 24 in the powdered form and Amoco GX-31 and 32 in granular form. The carbon is made (Figure 1) by a direct chemical activation route in which petroleum coke or other carbonaceous sources are reacted with excess potassium hydroxide, KOH, at 400° to 500°C to an intermediate product that is subsequently pyrolyzed at 800°-900°C to active carbon and potassium salts. The salts are removed by water washing.

[2] Retired.

Materials

Although various carbonaceous sources can be used, petroleum cokes were preferred because of their low ash contents and generally higher yields of active carbon. The cokes were obtained from various sources and with widely varying properties. Sulfur contents ranged from 2.0 to 6.0 wt.%; metals, primarily nickel and vanadium, from 500 ppm to 5,000 ppm; volatile matter, from 11% to 20%. The coke quality within these ranges did not appear to affect active carbon properties. However, somewhat lower active carbon yields were noted with the higher volatile matter cokes (58-62 wt% vs. 62-65 wt%). In the pilot plant operation coke was ground to 99% smaller than 100 mesh. Commercial grade, fine crystalline KOH (90%), was used in most of the studies.

Pilot Plant Development and Process Description

Pilot plant development was carried out to provide data for process and engineering design as well as economic evaluations. The pilot plant also served to provide active carbon for research and evaluation studies. The major pieces of equipment used to carry out the pilot plant process development are depicted in their flow sequence in Figure 2. Since it was decided at the onset to concentrate on studying each process step rather than being distracted by materials flow problems, there was no actual continuous flow of material from one piece of equipment to another. Each operation was carried out and studied independently. Any material produced was stored in sealed barrels or drums and was moved to the next operation when needed. Moreover, the outputs of the various pieces of equipment were different. In terms of pounds of active carbon per hour, outputs ranged from 25 to 110.

Because of the hygroscopic nature of the KOH and the intermediate products, it was necessary to provide facilities throughout for inert gas (nitrogen) blanketing. Water not only affected the handling properties of the various intermediate process products, but also became involved chemically in the shift reaction which reduced yields and detrimentally affected active carbon properties. Because oxygen (air) had similar effects and for obvious safety reasons, oxygen was excluded from high temperature operations where reducing (H_2) atmospheres were naturally produced and maintained.

Typical pilot plant operation began by preparing an intimate mixture of ground coke and KOH in a ribbon blender in a range of KOH/coke weight ratios of 2 to 4. The blend was stored in N_2-blanketed and sealed drums.

In the pilot plant the coke-KOH mixture was charged with a screwfeeder to a precalciner which was an indirectly-fired rotary kiln. The reaction proceeded in the precalciner at internal temperatures of 400° to 500°C. The product, called precalcinate, was cooled and temporarily stored in sealed 30-gallon drums. Overall yield of solid product in this step was about 95%. The precalcination reactions produced a gas that contained mostly hydrogen with small amounts of methane and condensables such as water and tars. For process and safety reasons, a slight 1"-2" H_2O pressure was maintained in the kiln with a countercurrent gas flow.

The precalcinate was next pyrolyzed or calcined in another

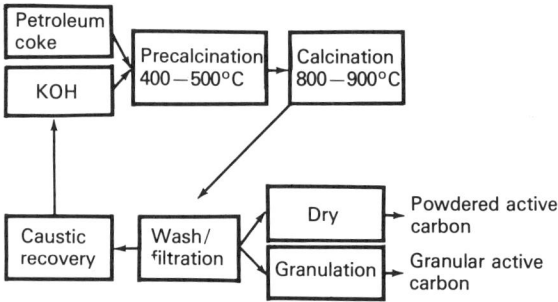

Figure 1. High surface area active carbon.

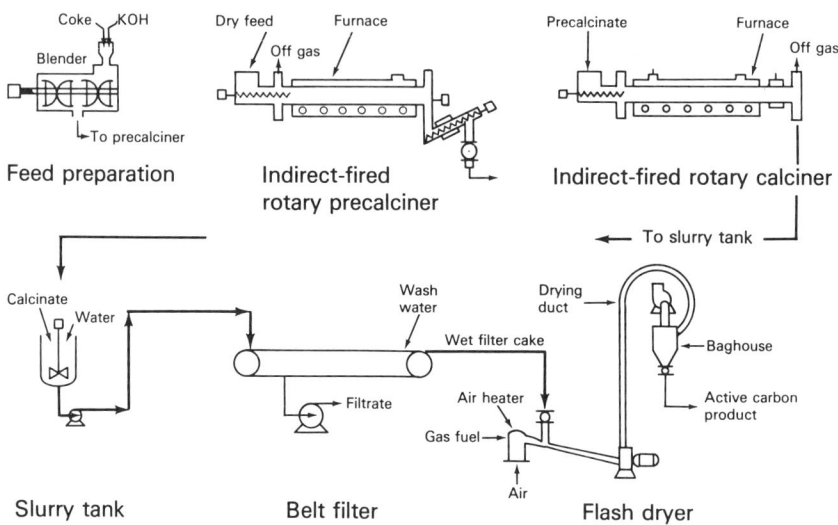

Figure 2. Powdered active carbon pilot plant.

indirectly-fired kiln at about 850°C. Somewhat less gas was produced also containing mostly hydrogen. The active carbon product contained occluded potassium salts, primarily carbonate with some sulfides and sulfate and unconverted KOH.

The calcinate was quenched in water to produce a pumpable slurry. Carbon washing was carried out on a horizontal belt vacuum filter which reduced the salt content to 4 to 5%. To further reduce the salt content, if needed, additional slurrying and washing on the belt filter was required. In commercial operation the filtrate containing potassium salts could be processed to reconstitute the KOH for recycle.

In the last process step, the wet filter cake containing about 70% water was flash dried to a final water content of 2-5%. The powdered active carbon was collected in polyethylene bags which were sealed and stored in fibre drums.

By using suitable binding agents and techniques, the powdered carbon could be produced in granular form.

From process studies carried out in this pilot plant over several years, parameters were defined to permit process and engineering designs of full commercial operations. Process economics indicated the cost of this active carbon would be greater than currently available commercial carbons. However, hundreds of evaluations in a full spectrum of applications showed that this active carbon would be competitive in many of those applications on a cost-performance basis. It will be commercially available mid-85 in various forms, trademarked as SUPER-A, from Anderson Development Company in Adrian, Michigan.

Properties

Tables I and II list major typical physical and adsorptive properties of the powdered active carbon. Effective surface area, measured by the BET method using a Digisorb 2500, is consistently in the range of 3000 to 3400 m^2/gm. This exceeds the theoretical area of about 2600 m^2/gm as calculated by the area of one gram of a graphitic plane because of multilayer adsorption and pore filling in a highly microporous structure.

As shown by H. Marsh, et al [2], using phase contrast, high resolution electron microscopy (JEOL 100C), this carbon has a cage-like structure in which the individual cages are so sized that they exhibit properties of super microporosity, i.e., essentially complete filling of the individual cages by the adsorbate at low concentration to give high effective surface areas and large micropore volumes. The cages are also substantially homogeneous in size as can be seen by low magnification image photomicrographs (x142,000) in Figure 3. At higher magnification (x3,116,000, Figure 4) the individual cages are clearly evident and appear to be formed by walls of folded carbonaceous graphitic-type lamellae 1-3 carbonaceous sheets in thickness. Using X-ray diffraction techniques, Konnert et al. [3] similarly describes the structure as being composed of distorted ribbons of one or very few graphite-like layers. More detailed discussions on this subject can be found in these papers by Marsh and Konnert.

Notwithstanding the high microporosity of this carbon, it

Table I. Typical Physical Properties
Powdered Active Carbon

Surface Area, BET m^2/gm	3000-3400
Average Pore Diameter, Å	23-25
Pore Volume, >20Å Diam., cc/gm	0.7-0.9
<20Å Diam., cc/gm	1.3-1.7
Total	2.0-2.6
Bulk Density, gm/cc	0.29-0.32
Particle Size (1)	
<149 wt%	98-99
<74 wt%	75-95
<44 wt%	65-75
<37 wt%	50-65
Ash, Wt%	2-3

(1) Alpine Jet

Table II. Typical Adsorptive Properties
High Surface Area Active Carbons

Iodine No.	3000-3600
Methylene Blue Adsorption, mg/gm	650-750
Phenol No. (1)	11-15
Total Organic Carbon Index (TOCI) (2)	300-800

(1) ANWAB-600-75

(2) Measured from isotherms using municipal and industrial wastewater, TOCI is the TOC adsorption capacity of this active carbon relative to Aqua Nuchar A or Filtrasorb 300 given a value of 100 as measured at initial equilibrium concentration.

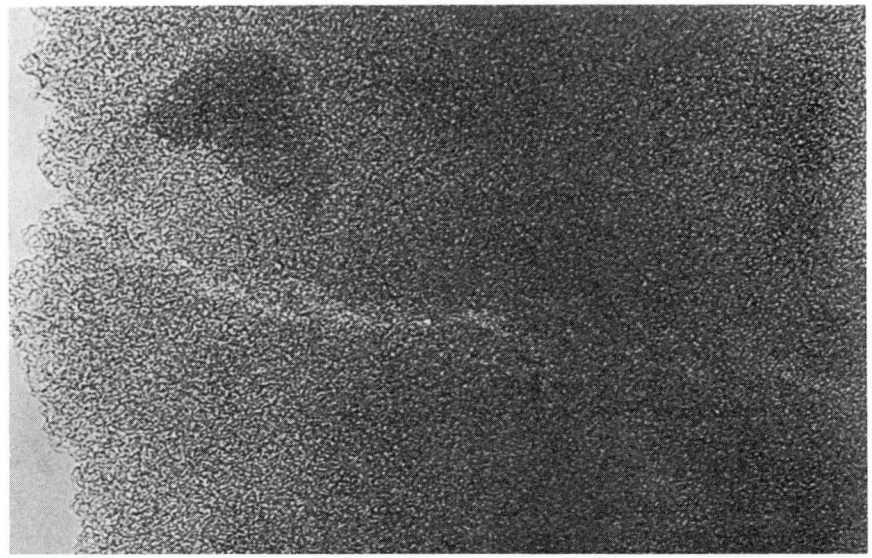

Figure 3. High surface area active carbon. Total magnification: x 145,480.

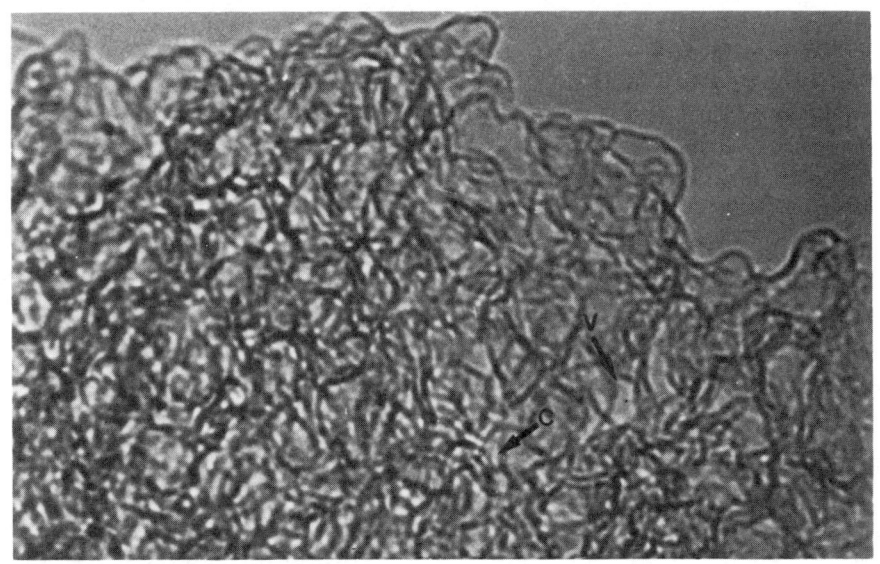

Figure 4. High surface area active carbon. Total magnification: x 2,929,040.

appears that swelling of the carbon may occur and at some conditions it can adsorb molecules larger than the pore sizes would seem to allow. This has been a passing observation with adsorbates such as humic acids, high molecular weight industrial organic wastes, and large dye molecules. Pore volume and adsorptive properties are typically two to four times higher than any available commercial carbon. For example, the Total Organic Carbon Index (TOCI), which is a measure of the carbon's ability to remove dissolved organic contaminants from municipal and industrial wastewater, indicates an adsorptive capacity over three to six times greater than standard commercial carbons. The high adsorption capacity is further demonstrated by the Iodine Number, methylene blue adsorption, phenol number, and by numerous other adsorbates tested in our laboratories and elsewhere.

Because the carbon is made by a controlled chemical activation, its quality is consistently reproduced with little variation. Adsorptive and physical properties are essentially the same from lot to lot.

Uses

This carbon has been evaluated over several years by nearly a hundred industrial, government and academic organizations worldwide in several hundred tests for numerous applications. In these tests, average performance ratios (factor of improved performance) of this carbon with other carbons (from 14 manufacturers) ranged from 1.5 to as high as 30, most in the 2 to 4 range. Table III is an overall list of applications tested.

Of considerable recent interest has been the use of this carbon for poison and drug overdose control for both humans and animals. This has appeared in publications by D. O. Cooney [4, 5, 6], W. L. Thompson [7], A. Picchioni [8], J. N. Huckin [9], D. C. Chung [10], and others. Recent work has shown this carbon 100% effective in relatively small dosages for many common organic poisons. Other medical applications range from toxin adsorption to the concentration of drugs from biological fluids.

Other general applications have included catalyst carriers and decolorizing for gold recovery from cyanide leach solutions. Of particular interest is its effective use for military and industrial clothing where protection from noxious vapors is necessary. Recent studies have shown its ability to separate chemical isomers, such as the separation of dioxin isomers. It has been used worldwide for analytical methods for determining trace (PPB) toxic substances in water.

Summary

An exceptional active carbon has been developed with a high effective surface area and high adsorptive capacity. It is prepared by a controlled chemical activation route using potassium hydroxide and a carbonaceous source, usually petroleum coke, to give a consistent quality product. It has been tested in a gamut of conventional and new uses with performance ratios averaging 2 to 4 times better than other grades of active carbon. Because of its unique structure and properties, it is likely that many new uses will be developed as it now becomes commercially available.

Table III. Applications Tested

Catalyst Support	Vapor Applications
General	Air Filters
	Cigarette Filters
Decolorizing	Fibers and Cloth
	Gas Masks and Respirators
Dry Cleaning Fluids	General
Food and Beverage	Industrial Emissions
Pharmaceutical	Odor Control
Syrups and Sugar	Automotive Applications
Medical Applications	Water Treatment (Potable)
Hemodialysis	Municipal
General	Toxics
Oral Antidotes	
Hemoperfusion	Wastewater
Metallurgy	Chemical
	General
Removal and Separation	Municipal
	Pulp and Paper
Special Applications	Refinery
	Textile
Taste and Odor	
General	

Literature Cited

1. A. N. Wennerberg and T. M. O'Grady, U.S. Patent 4,082,694.
2. H. Marsh, D. Crawford, T. M. O'Grady, and A. N. Wennerberg, Carbon 15, 419 (1982).
3. J. H. Konnert and P. D'Antonio, Carbon 21, 193 (1983).
4. D. O. Cooney, Clin. Toxicol. 11, 387 (1977).
5. D. O. Cooney and Kane, Clin. Toxicol. (1979).
6. D. O. Cooney, "Activated Carbon Antidotal and other Medical Uses," Marcel Dekker, Inc., New York (1980).
7. W. R. Van de Graaff, W. L. Thompson, I. Sunshine, D. Fretthold, F. Leickly, and H. Dayton, J. Pharmacol. Exp. Ther. 221, 656 (1982).
8. A. Picchioni and L. Chin, "The Antidotal Properties of a Super-Active Charcoal," presented at the Annual Meeting of American Association of Poison Centers-American Academy of Clinical Toxicology, Salt Lake City, August 5, 1981. (Being submitted for publication to the Journal of Clinical Toxicology.)
9. J. N. Huckins, et al., J. of the A.O.C.A. 61, No. 1 (1978).
10. D. C. Chung, J. E. Murphy and T. W. Taylor, J. Toxicol.: Clin. Toxicol. 19 (2), 219 (1982).

RECEIVED September 10, 1985

21

Dispersion of Metallic Derivates on Carbon Supports

Pierre Ehrburger and Jacques Lahaye

Centre de Recherches sur la Physico-Chimie des Surfaces Solides, CNRS,
24 avenue du Président Kennedy, 68200 Mulhouse, France

> The dispersion of metallic derivates on a carbon surface
> is dependent upon the surface properties of the carbon
> support. Edge carbon atoms which constitute the majority
> of the active surface area of graphitized carbon blacks
> may interact with the catalyst precursor or the catalyst
> itself thus providing a higher state of dispersion.
> Moreover specific interactions of the metallic derivates
> with the edge carbon atoms tend to decrease the sinte-
> ring rate of the supported solid.

During the last two decades the use of carbon as a catalyst support
has gained growing interest. Depending on the precursor and on the
processing, the carbonaceous materials will present different types
of porosity and of crystalline texture thus enabling wide possibilities for tailoring a catalyst carrier (1). For instance activated
carbons, developing high porosities and surface areas are commonly
used as supports for noble metals in heterogeneous catalysis (2-4).
Metallic oxides like the well known chromium and copper oxides for
the removal of toxics from the air are also currently dispersed on
carbons with high surface areas. More recently, supported molybdenum
sulfide appears to be very promising for the hydrodesulfurization of
petroleum feedstocks (5). Other applications for carbon supports are
in the field of electrochemistry. Carbon electrodes with few amounts
of noble metals (Pt, Ag) or metal phthalocyanines are found in
metal-air batteries or alkaline fuel cells (6-8). Supported metals
or oxides are also of prime importance for changing the reactivity
of the carrier itself like for instance in catalytic gasification of
carbonaceous materials (9-10). The most important applications of
carbon supported catalysts are indicated in Table I. It is now well
recognized that the efficiency of a metal catalyst is enhanced by
increasing its dispersion which may be defined as the ratio of the
number of surface metal atoms to the total number of metal atoms.
A simple way to achieve high degree of dispersion consists in depo-
siting the catalyst on high surface area supports. Therefore the
primary roles of the support are to provide a structural framework
which enables the formation of a highly dispersed phase and to
increase its stability when submitted to sintering conditions.

0097-6156/86/0303-0310$06.00/0
© 1986 American Chemical Society

Table I. Dispersion of Metallic Derivates on Carbons

-Electrocatalysis
Metals (Pt, Ag, Ni) - Metal Phthalocyanines
-Removal of toxics
Copper/Chromium oxides
-Change of reactivity of the carbon support
Catalytic gasification
Inhibition of oxidation

Several methods have been proposed for preparing highly dispersed catalyst on carbon. The three most common procedures are impregnation, ionic exchange and vapor phase condensation. The former consists in wetting the carbon by a solution of an appropriate salt (11, 12) while the second is based on the exchange of carbon surface groups with cationic complexes (13, 14). In the vapor phase condensation the support is contacted with the volatile metal or metallic derivate (15, 16). When necessary, the supported precursor is then reduced into the metallic state.

It has also been established that the support may interact with the catalyst thus changing its catalytic activity (10, 17). Systematic investigations of the influence of the support on the catalyst properties (dispersion, sintering, catalyst-support interactions) have been carried out with inorganic materials like silica, alumina and zeolithes (18). In contrast, similar studies have not yet been reported with carbon supports. This paper deals with the factors affecting the dispersion of a catalyst on carbon blacks and its sintering behavior. The selected system is iron-phthalocyanine (PcFe) supported on graphitized carbon blacks which is an efficient catalyst for the electro-chemical reduction of oxygen (7).

Carbon support.

The surface of a graphitized carbon consists of two types of crystalline planes, the basal planes and the prismatic or edge planes. The reactivity of the carbon atoms is usually higher in the edge planes than in the basal planes. In particular, edge carbon atoms chemisorb oxygen and give rise to oxygenated functional groups. It may be expected that the dispersion of PcFe will depend on the extent of edge carbon atoms planes. In the present study two types of carbon surfaces have been selected :
- a homogeneous surface which is build-up with basal planes
  (graphitized furnace black Vulcan 3, V3G)
- a heterogeneous surface consisting in both types of planes and
  obtained by activation in air of the homogeneous substrate
  (activated V3G).

The activation procedure has been described elsewhere (16). During the gasification, the total surface area (S_{BET}) as well as the active surface area (S_a), i.e. the area of the edge carbon atoms planes increase. The surface area data of both samples are shown in Table II.

Table II. Total Surface Area and Active Surface Area of the Carbon Supports

Carbon	S_{BET} (m²/g)	S_a (m²/g)
V3G	73	–
Activated V3G (50 % burn-off)	100	4.3

It is seen that the surface of V3G consists mainly of basal planes whereas about 4 % of the area of the activated V3G sample is build-up with edge carbon atoms.

Deposition of the catalyst.

The chemical structure of iron-phthalocyanine is shown on Figure 1. PcFe is a planar molecule with a central cavity in which the iron atom is located. Metal phthalocyanines may be sublimated and the deposition of PcFe on the carbon support has been done by vapor phase condensation (16). During the condensation, the carbon supports are kept at 235°C. Condensation rates were in the range 2-4 mg PcFe per hour. Samples with PcFe content ranging from 0.5 to 30 % in weight have been prepared with both carbon substrates.

Dispersion of PcFe.

Surface area of supported PcFe. The surface area of PcFe deposits has been determined by measuring the amount of water molecules irreversibly adsorbed on their surface. We have shown that each surface molecule of the PcFe crystal retains on the average one molecule of water according to the following reaction (16, 19).

$$PcFe_{surface} + H_2O \rightarrow PcFe_{surface} - H_2O$$

The PcFe-H_2O bond is stable under vacuum up to 340 K. At higher temperatures, the adsorbed water desorbs quantitatively. The mean surface area of an adsorption site for water has been estimated to 0.83 nm² from the lattice parameters of PcFe β. This value is in fact close to the geometric area of a single PcFe molecule.

The surface area developed by PcFe is measured for different amounts of PcFe deposited on both carbon supports, as shown on Figure 2. As the two carbon substrates have different total surface areas, the amount of supported PcFe is expressed in mg per m² of carbon surface in order to facilitate the comparison between the two supports. On the homogeneous carbon support, the surface area of PcFe increases only slightly and reaches a maximum for a content of about 7 % by weight. For higher amounts of PcFe, there is a small decrease in its surface area. The results for the heterogeneous carbon are strikingly different. The surface areas of PcFe are higher on the activated V3G and reach a maximum value of 0.25 m²/m² for a load of 20 % in weight. There is a more pronounced growth of the surface of the PcFe aggregates on the heterogeneous carbon substrate. Considering a spherical particle shape, the mean number of PcFe

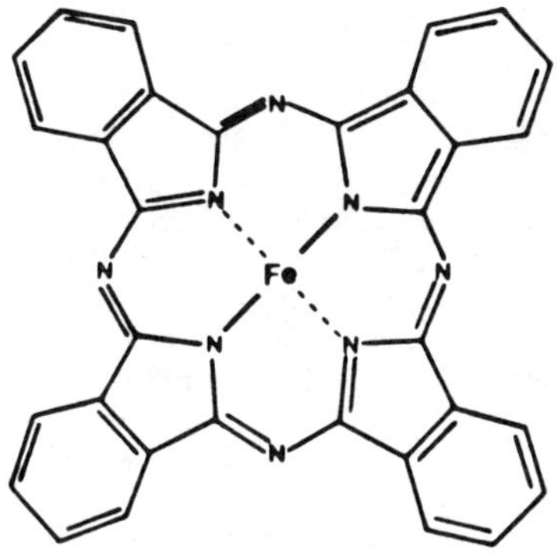

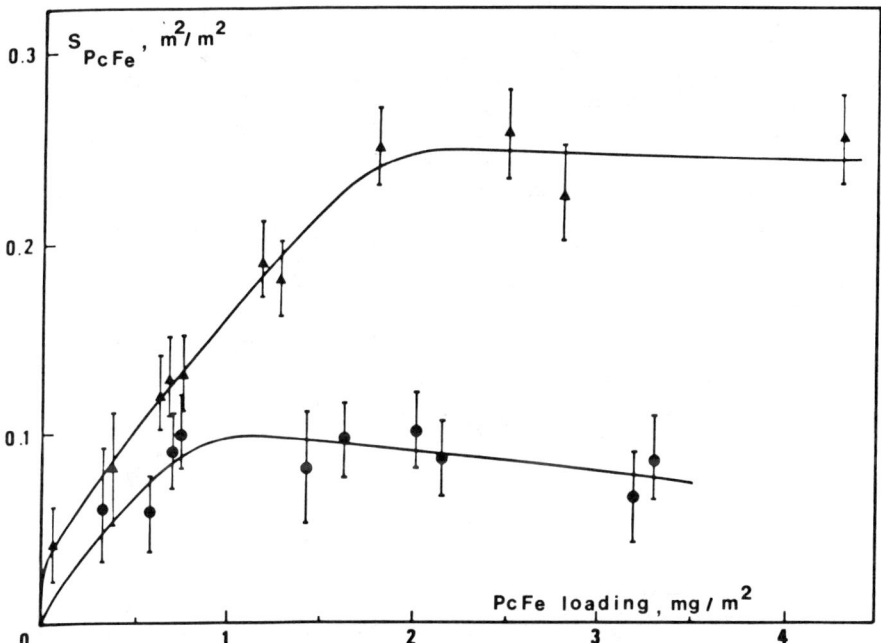

Figure 1. Chemical structure of iron-phthalocyanine.

Figure 2. Surface area of PcFe as a function of loading on V3G (●) and activated V3G (▲). "Reproduced with permission from Ref. 16. Copyright 1983, 'Academic Press'".

particles may be determined from the relation:

$$n = \frac{S^3 \rho^2}{36\pi m^2} \quad (1)$$

S : surface area of PcFe per unit area of carbon support
m : weight of deposited PcFe per unit area of carbon support
n : mean particle number of PcFe per unit area of carbon support
ρ : density of PcFe (1.52)

Figure 3 shows the change of n as a function of the amount of deposited PcFe. The number of PcFe particles does not vary significantly on the activated V3G support for loading less than 20 % in weight whereas it decreases by a factor 10 on the graphitized V3G carbon. Moreover, by extrapolating both curves at low content of PcFe, it appears that the number of PcFe nuclei is about the same on both substrates. Thus the surface properties of the carbon affect more extensively the growth process of the particles than the initial number of nuclei.

Crystallite size of PcFe particles. The mean crystallite diameters of the PcFe particles calculated from the broadening of the (100) and (10$\bar{2}$) reflexions, are given in Table III. It is seen that the crystallite sizes in both directions are similar which suggest that there is no significant anisotropy in the shape of the PcFe crystallites.

Table III. Mean Crystallite Diameter of PcFe, $\bar{d}_B$ determined by X-Ray Diffraction

Carbon support	PcFe content (mg/m²)	$\bar{d}_B$ (nm)	
		(100)	10$\bar{2}$)
V3G	0.71	9	10
	0.77	10	10
	1.36	19	15
	2.03	20	28
	2.14	16	23
Activated V3G	0.80	20	20
	1.17	25	24
	1.27	17	23
	1.80	30	36
	2.50	40	35
	2.80	34	40
	4.30	37	37

The data of Table II also indicate that for a given PcFe content the crystallite size is greater on the activated carbon. This result however does not mean that the actual particle size of PcFe is greater on the activated carbon. As will be shown below, the particles of PcFe may contain several crystallites. The mean crystallite diameter $\bar{d}_B$ calculated from the (100) reflexion is compared to the mean particle size $\bar{d}_S$ determined from the water uptake measurements in Figure 4 for both carbon substrates. It appears that $\bar{d}_B$ and $\bar{d}_S$ are in fairly good agreement in the case of the heterogeneous carbon

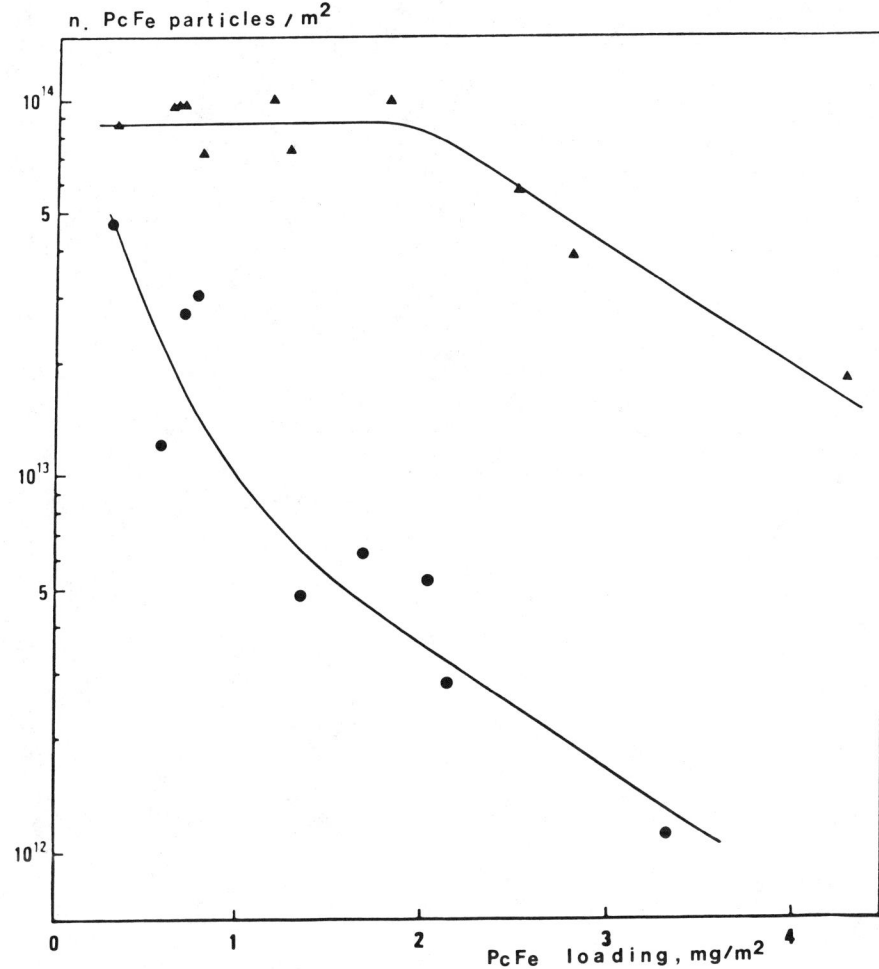

Figure 3. Number of PcFe particles per unit area of carbon support as a function of PcFe loading : V3G (●), activated V3G (▲). "Reproduced with permission from Ref. 16. Copyright 1983, 'Academic Press'".

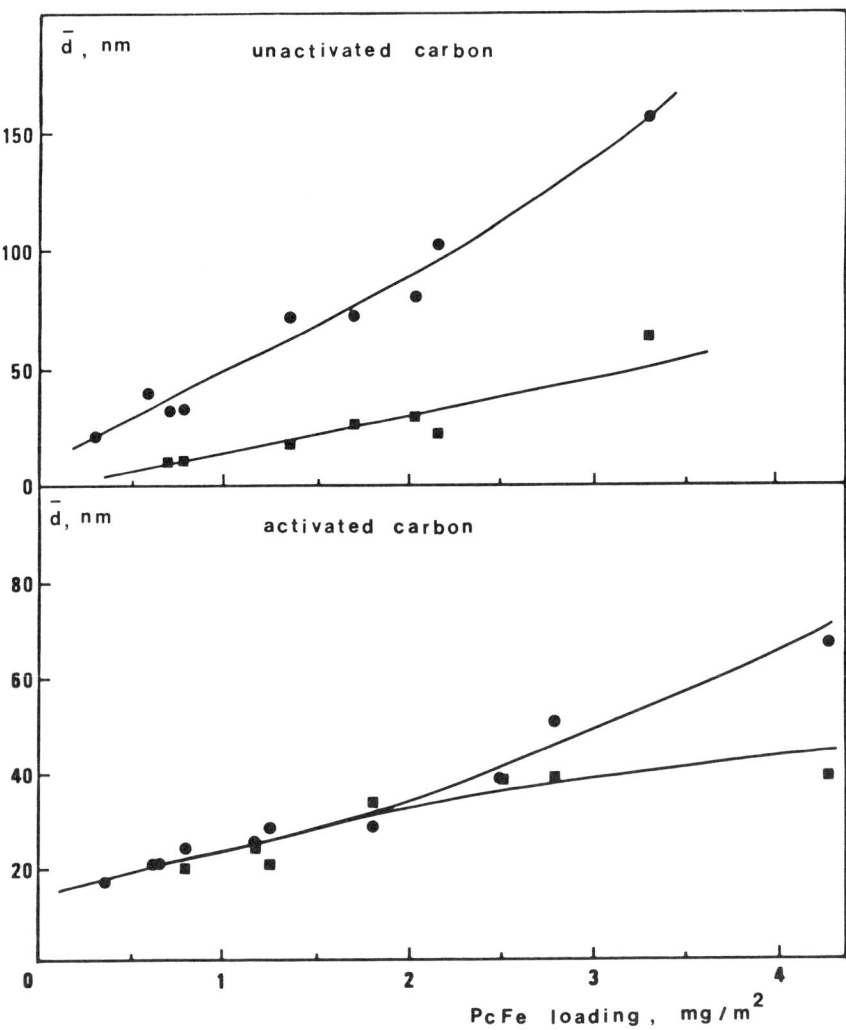

Figure 4. Comparison between the mean particle size $\bar{d}_S$ (●) and the mean crystallite size $\bar{d}_B$ (■) of supported PcFe. "Reproduced with permission from Ref. 16. Copyright 1983, 'Academic Press'".

support for PcFe content less than 20 % in weight. In contrast, $\bar{d}_B$ is always smaller than $\bar{d}_S$ with the homogeneous carbon, the discrepancy increasing with the PcFe content. These results may be explained by the fact that the PcFe particles are essentially formed by a single crystallite on the activated carbon surface whereas the PcFe aggregates are polycrystalline on the graphitized V3G surface. Hence, the surface properties of the carbon support also affect the crystalline texture of the PcFe deposits.

Growth of PcFe Particles. During the deposition process, the PcFe particles behave differently on the two carbon supports. They are polycrystalline on the homogeneous carbon surface and their number decreases with the PcFe loading, i.e. with the deposition time. These facts point to a partial coalescence of the PcFe particles on the graphitized V3G surface. Therefore the surface area developed by PcFe will depend on two mechanisms acting in an opposite way : growth of the nuclei by capture of PcFe monomers and coalescence of the particles. Coalescence of supported clusters may occur in two ways, statically due to the overlapping of two growing particles and/or dynamically as a result of particle migration. Experimental evidences of the motion of clusters on a solid surface are now well established (20-22). Hence the smaller surface area developed by PcFe on the homogeneous carbon would result from a higher migration rate of the particles as compared to their rate on a heterogeneous carbon.

Sintering of Supported PcFe

Sintering of supported solids can occur by two distinct mechanisms : particle migration and coalescence as already mentioned above and interparticle transport of atoms or molecules (Ostwald ripening) (23). Interparticle transport may be possible either by surface diffusion across the support or by vapor phase transport. Depending on the supported systems and on the sintering conditions the particles may grow predominantly via one of these possible routes. Supported PcFe deposits have been sintered in experimental conditions close to that of the condensation process (T = 235°C, residual pressure 10^{-5} torr). The surface areas of PcFe after 2 and 3 hours of sintering are shown in Table IV. It is seen that the sintering effect is more pronounced with a homogeneous carbon support.

Table IV. Sintering of Supported PcFe Particles

Support	Sintering time (h)	Surface area of PcFe (m^2/m^2)
V3G	as prepared	0.095
	2	0.032
	3	0.017
Activated V3G	as prepared	0.130
	2	0.091
	3	0.085

The rate of sintering dS/dt is generally related to the time by a rate-power law

$$\frac{dS}{dt} = -KS^n \qquad (2)$$

or by integrating

$$S^{1-n} - S_o^{1-n} = K't \qquad (3)$$

S and S_o are respectively the surface areas of the dispersed solid at time t and t = o.

K' is a constant and n is an integrator ranging generally between 2 and 12 (18). The values of K' and n have been determined for both systems using a least square fitting method for equation 3. The sintering rates of PcFe on the homogeneous and heterogeneous carbon respectively are:

$$\left(\frac{dS}{dt}\right)_{hom.} = -15.3\ S^2 \qquad S\ in\ m^2/m^2,\ t\ in\ hour$$

$$\left(\frac{dS}{dt}\right)_{het.} = -1.3\ 10^4\ S^6$$

The initial sintering rate of PcFe may be estimated by taking $S = 0.1\ m^2/m^2$. Hence, the corresponding relative sintering rate on the two supports is:

$$\left(\frac{dS}{dt}\right)_{hom.} / \left(\frac{dS}{dt}\right)_{het.} = 12$$

The sintering rate is more than one order of magnitude higher on V3G than on the activated carbon V3G at the maximum coverage of V3G with PcFe. This result confirms qualitatively the hypothesis of a more pronounced coalescence rate of PcFe on the homogeneous support during the deposition process. Sintering mechanism on graphitized carbon blacks have also been investigated with other catalysts like platinum. Bett et al. (24) showed that the sintering of supported Pt may be due to surface diffusion of metal particles. The Pt crystallites migrate from trap sites on the support and coalesce on the carbon surface. More recently, Ehrburger and Walker (25) studied the particle size distribution of Pt particles on activated and non-activated V3G supports. On the homogeneous carbon substrate, the particle size follows a log-normal distribution which is consistent with a sintering mechanism based on particle migration followed by coalescence. On the activated V3G support, the sintering behavior of Pt is different. Particle growth is detectable at much higher temperature than on the homogeneous carbon and it was concluded that the metal clusters are trapped on surface heterogeneities. Consequently, sintering would only start when the Pt particles have enough energy to escape from the trapping sites. These few examples clearly show that the surface properties of the carbons influence the dispersion and the sintering of supported catalysts.

Catalyst-Carbon Support Interactions.

Interactions between the catalyst and the carbon supports obviously depend on the type of carbonaceous material used. In the case of activated carbon, the highly porous texture acts as a framework for dispersion of the catalyst and the pores may be considered as trapping sites which hinder sintering. For less porous carbon like amorphous carbon film or carbon blacks the concept of trapping sites is still applicable. Phillips,et.al. (26) developed this concept for explaining the sintering mechanism of gold clusters on carbon. Later on, Bett, et. al. (24) suggested that the boundaries of basal planes may be the trapping sites required for the immobilization of platinum particles. Ehrburger, et. al. (27) also found that the degree of dispersion of platinum increases with increasing surface heterogeneity of graphitized supports. Considering Pt particles size distribution and sintering behaviour, they attributed a higher degree of dispersion of the metal to a stronger interaction of the chloroplatinic precursor with the surface heterogeneities (28). More recently Benedetti, et. al. (29) also attributed changes in the dispersion of palladium on charcoals to surface heterogeneity effects.

In the case of iron-phthalocyanine, it was possible to identify the nature of its interaction with the support. PcFe exchanges coordinating bonds with a variety of molecules like pyridine and quinoline (30). A complex $PcFe-H_2O$ with a bond between the metal atom and the water molecule has also been evidenced in CCl_4 (31). Since the active surface of carbon is usually covered with oxygen complexes (mainly carbonyl groups) a higher stability of the PcFe particles on activated V3G may originate from interfacial bonds with oxygen groups. A possible interaction between oxygen surface groups and PcFe molecules is drawn schematically on Figure 5. In order to verify this hypothesis, the oxygen groups of the carbon have been removed by H_2 treatment at 950°C. Following this treatment, the carbon was cooled in H_2 and under these conditions, the active surface area will be covered with chemisorbed hydrogen. Two samples having respectively 14.5 and 26 % PcFe in weight have been prepared with the hydrogen treated heterogeneous carbon. The corresponding surface areas of PcFe are compared to the previous one on Figure 6. It is seen that after removal of the oxygen complexes the dispersion of PcFe decreases significantly and that the carbon behaves like the initial V3G support. Therefore, the edge carbon atoms via the chemisorbed oxygen play an important role in the dispersion and sintering of PcFe. Melendres (32) investigated carbon-supported PcFe by Mössbauer and Raman spectroscopy and found changes in molecular electronic structure in going from pure PcFe to carbon-supported PcFe. These observations provide further physical evidence of catalyst-support interaction and confirm our own results.

Conclusion.

The surface properties of carbon play an important role in the dispersion and the sintering behavior of supported catalyst. The study of iron-phthalocyanine is of particularly interest for the investigation of catalyst-carbon support interactions. The coalescence of PcFe particles on graphitized carbon is considerably lowered by the

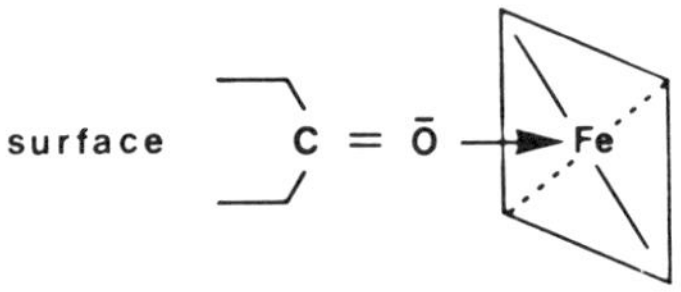

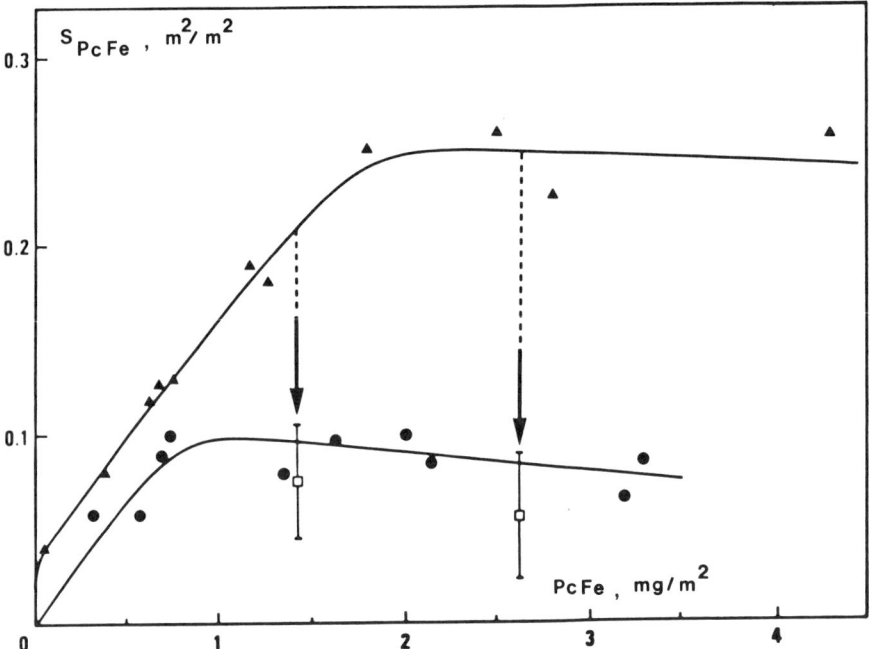

Figure 5. Interaction between an oxygen complex and a PcFe molecule. "Reproduced with permission from Ref. 16. Copyright 1983, 'Academic Press'".

Figure 6. Surface area of PcFe supported on the activated and hydrogen treated V3G (□) compared to the corresponding data on V3G (●) and activated V3G (▲). "Reproduced with permission from Ref. 16. Copyright 1983, 'Academic Press'".

presence of oxygenated groups located on edge carbon atoms which act as anchors of the supported particles. It is relevant to note that only a few percent of active surface area ensures a higher dispersion of PcFe on carbonaceous supports.

Literature Cited

1. Walker, P.L., Jr. Proc. Fifth London Int. Carbon and Graphite Conference, Society of Chemical Industry 1978, vol.1, p. 427.
2. Pope, D.; Smith, W.L.; Eastlake, M.J.; Moss, R.L. J. Catal. 1971, 22, 72.
3. Parkarsh, S.; Gadallah, F.F.; Chakrabartty, S.K. Carbon 1979, 17, 403.
4. Koopman, P.G.J.; Kieboom, A.P.G.; Van Bekkum, H. Colloids and Surfaces 1981, 3, 1.
5. Vissers, J.P.R.; Lensing, T.J.; de Beer, V.H.J.; Prins, R. Proc. 16th Biennial Conference on Carbon, American Carbon Society 1983, p. 607.
6. Tseung, A.C.C.; Wong, L.L. J. Applied Electrochem. 1972, 2, 211.
7. Alt, H.; Binder, H.; Sandstede, G. J. Catal. 1973, 28, 8.
8. Appleby, A.J.; Savy, M. Electrochim. Acta 1976, 21, 567.
9. Mc Kee, D.W. In "Chemistry and Physics of Carbon"; Walker P.L. Jr.; Thrower, P.A., Eds; Marcel Dekker: New York, 1981; Vol. I, p. 1.
10. Baker, R.T.K.; Sherwood, R.D.; Simoens, A.J.; Derouane, E.G. In "Metal-Support and Metal-Additives Effects in Catalysis"; Imelik, B., Ed.; Studies in Surface Science and Catalysis n° 11, Elsevier: Amsterdam, 1982; p. 149.
11. Bartholomew, C.H.; Boudart, M. J. Catal. 1972, 25, 173.
12. Cavalier, J.C.; Chornet, E.; Beaureguard, B.; Coquard, G. Carbon 1978, 16, 21.
13. Ehrburger, P.; Dentzer, J.; Lahaye, J. 15th Biennial Conference on Carbon, American Carbon Society 1981, p. 254.
14. Ehrburger, P.; Dentzer, J.; Lahaye, J. Proc. 16th Biennial Conference on Carbon, American Carbon Society 1983, p. 347.
15. Baker, R.T.K.; Sherwood, R.D.; Dumesic, J.A. J. Catal. 1980, 66, 56.
16. Ehrburger, P.; Mongilardi, A.; Lahaye, J. J. Colloid Interface Sci. 1983, 91, 151.
17. Baker, R.T.K.; Sherwood, R.D. J. Catal. 1981, 70, 198.
18. Anderson, J.R. In "Structure of Metallic Catalysts"; Academic Press; London, 1975, p. 244.
19. Ehrburger, P.; Mongilardi, A.; Lahaye, J. C.R. Acad. Sci. Paris 1979, C, 289.
20. Baker, R.T.K.; Skiba, P., Jr. Carbon 1977, 15, 233.
21. Masson, A.; Metois, J.J.; Kern, R. Surface Science 1971, 27, 463
22. Robins, J.L.; Donhoe, A.J. Thin Solid Films 1972, 12, 255.
23. Wynblatt, P. In "Growth and Properties of Metal Clusters"; Bourdon, J. Ed; Elsevier: Amsterdam, 1980, p. 15.
24. Bett, J.A.; Kinoshita, K.; Stonehart, P. J. Catal. 1974, 35, 307
25. Ehrburger, P.; Walker, P.L., Jr. J. Catal. 1978, 55, 63.
26. Phillips, W.B.; Desloge, E.A.; Skofronick, J.G. J. Appl. Phys. 1968, 39, 3210.

27. Ehrburger, P.; Mahajan, O.; Walker, P.L., Jr J. Catal. 1977, 43, 61.
28. Ehrburger, P.; Walker, P.L., Jr. In "Growth and Properties of Metal Clusters"; Bourdon, J. Ed; Elsevier: Amsterdam, 1980, p. 175.
29. Benedetti, A.; Cocco, G.; Enzo, S.; Piccalaga, G.; Schiffini, L. J. Chim. Phys. 1981, 78, 961.
30. Giesemann, H. J. Prakt. Chem. 1956, 4, 169.
31. Stymne, B.; Sauvage, F.X.; Wettermark, G. Spectrochim. Acta 1979, 35A, 1195.
32. Melendres, C.A. J. Phys. Chem. 1980, 84, 1936.

RECEIVED February 12, 1985

# Progress of Pitch-Based Carbon Fiber in Japan

Sugio Ōtani and Asao Ōya

Faculty of Technology, Gunma University, Kiryu, Gunma 376, Japan

Carbon fiber technology has developed rapidly in Japan during the last decade. The origins of pitch-based fiber trace back to observations of lignin deformation during pyrolysis. Although only the low-modulus general performance carbon fiber (GPCF) made by spinning isotropic pitch has been commercialized thus far in Japan, extensive development efforts are in progress on the high-modulus high performance carbon fiber (HPCF) produced by spinning mesophase pitches. Current efforts recognize two chemical factors that govern the viscous behavior and reactivity of mesophase pitch: the extent of alicyclic structure, and the hydrogen transfer between mesophase molecules. Two approaches to preparation of spinnable mesophase pitch are the Gundai "dormant mesophase" method, in which the pitch is hydrogenated just at the point of the mesophase transformation, and the Kyukoshi method, which employs tetrahydroquinoline as the hydrogenating agent.

Pitch-based carbon fiber was invented in our laboratory at Gunma University in the summer of 1963 (1). At that time we were attempting to prepare active carbon from lignin powder. One day we found whisker-like carbon near the wall of a flask in which lignin powder had been heated in air to 500°C; see Figure 1. We speculated that the lignin powder had melted incrementally, starting from the wall of the flask, and that the molten lignin had been stretched to form fibrous regions by the sintering of the lignin near the center of the flask. Eventually the stretched lignin became infusible through further heating in air to higher temperature.

A spinning experiment was undertaken to test this speculation. Molten lignin, produced by rapid heating of the lignin powder, was found to be quite spinnable. After spinning, the fiber was easily stabilized by heating to 300°C in air, and then carbonized by heating to 1000°C under nitrogen. Some of this fiber is illustrated by Figure 2.

In those days, the mechanisms of carbonization of polyvinyl chloride (PVC) were also under study in our laboratory. We found that PVC transformed to a beautiful lustrous pitch upon heating to 400°C under nitrogen. This PVC pitch could be spun quite easily, by comparison with molten lignin, and thus the pitch-based carbon fiber was first prepared in essentially the same way as the lignin-based carbon fiber. We recognize now that we were fortunate in first using PVC pitch. We later tried many other pitches as raw materials for carbon fiber, but PVC pitch was the only one that could be spun without any pretreatment. This was in 1963. We immediately applied for a patent (2), and the fundamentals of pitch preparation and spinning as well as the structure of the finished fibers were published in 1965 (3).

During the ensuing five years, a number of pitches were developed as carbon fiber precursors through use of various pretreatment techniques (4-9). We explored four basic pretreatment processes, sometimes in combination: (a) polymerization and/or aromatization by heat treatment, (b) removal of volatile species by distillation under atmospheric or reduced pressure, (c) removal of infusible matter by solvent fractionation, and (d) acceleration of polymerization by adding a radical initiator. The mechanical properties of the resulting fibers were in the ranges of 0.8 - 1.8 GPa (115 - 260 kpsi) for tensile strength and 20 - 50 GPa (2.9 - 7.3 Mpsi) for Young's modulus. These methods were subsequently developed by Kureha Chemical Industries Company, and fibers were commercialized in 1970 as the General Performance Carbon Fibers (GPCF) KCF-100 and KCF-200. At present these are the only continuous-strand low-modulus carbon fibers produced commercially from a pitch base.

Meanwhile carbon fibers derived from polyacrylonitrile (PAN) were also under development, and around 1963 methods of heat treating under stress were found to enhance sharply the mechanical properties. This success resulted in rapid growth of PAN-based carbon fiber technology for high-performance applications (Young's modulus greater than 200 GPA, 30 Mpsi). However, similar methods were not suitable for mass production of pitch-based high-performance carbon fiber (HPCF), and other approaches were aggressively explored to learn how to produce high-modulus carbon fibers from inexpensive pitch precursors.

A number of organic compounds were carbonized and graphitized in our laboratory as we sought to reveal the factors that govern the graphitizability of carbon materials. Among these was tetrabenzo (a,c,h,j) phenazine (abbreviation: PZ) with the molecular structure shown in Figure 3. The melting point of a large condensed polycyclic compound such as PZ is very sensitive to trace amounts of impurity and thus depends on the preparation procedure. On heating at 530 to 590°C for one hour under nitrogen, PZ with a melting point of 465 - 485°C was converted into a lustrous pitch that melted in the range of 300 to 380°C. On cooling, this pitch exhibited the strong preferred orientation of a mesophase pitch, as shown in Figure 4. In 1961, the PZ pitch was used to prepare carbon fibers. As expected, the fiber displayed strong preferred orientation without any special treatment (10,11).

In the course of GPCF development, as stated earlier, the precursor material was changed from PVC pitch to other pitches.

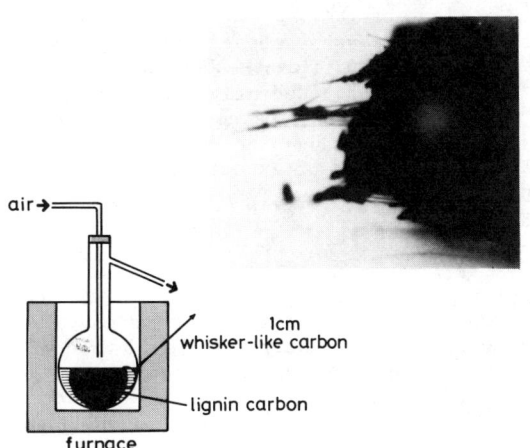

Figure 1. Whisker-like carbon from lignin.

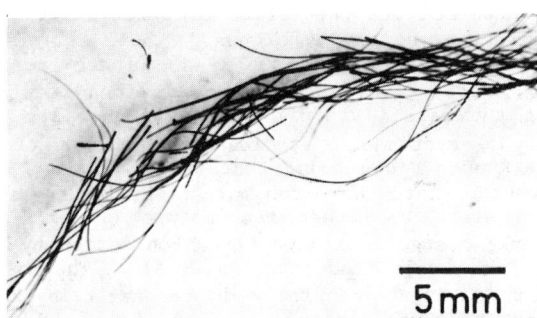

Figure 2. Carbon fiber prepared by spinning molten lignin.

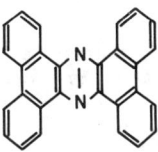

Figure 3. Molecular structure of tetrabenzo (a,c,h,j) phenazine (PZ).

Similarly, PZ pitch as precursor for HPCF was replaced by other mesophase pitches (12). At this point in time, as is well-known, Singer (13) and Lewis (14) of the Union Carbide Corporation developed similar methods. Mesophase carbon fiber progressed more rapidly in the USA than in Japan because Japanese defense and aerospace needs were less demanding. Recently, however, the drive toward higher-added-value products from the heavy fractions of coal and petroleum has intensified, and pitch-based carbon fibers, including HPCF, are now the subjects of extensive investigation in many Japanese laboratories.

Pitch Chemistry

The principal problem in pitch-based carbon fiber is the control of the properties of the precursor pitch. Studies of pitch chemistry have contributed significantly to the development of pitch-based carbon fiber, including some investigations whose practical purpose was unrelated to carbon fiber. Since about 1969, a basic understanding of pitch chemistry has been pursued aggressively in Japan, and three studies of particular significance to carbon fiber are summarized here.

Mesophase Model and the Importance of Alicyclic Structure. The research group at Kyushu University led by Takeshita and Mochida has sought to control the properties of pitch materials by catalytic and co-carbonization techniques. Their progress in these areas may be summarized as follows: (i) In earlier work by the present authors (15), $AlCl_3$ additions were found to be effective in increasing the carbon yield of pitch without loss of graphitizability. In systematic studies of the use of the $AlCl_3$ catalyst, the Kyushu group found that this catalyst can introduce alicyclic structure into the pitch molecules, leading to extension of the liquid state to higher temperature (16,17). (ii) This group also works energetically on co-carbonization by using using organic compounds, coals, and pitch materials. The most interesting conclusion is that partially hydrogenated pyrene is more reactive than non-hydrogenated pyrene (18). (iii) Through extensive analytical work, the Kyushu group also developed the so-called "spider web" model for mesophase molecules (19); see Figure 5. This model provides clear working concepts for the typical constituent molecules of the mesophase.

Carbonization in Molten Salt Media. Ōta and Ōtani (20) of Gunma University developed a novel carbonization method in which aromatic compounds, such as naphthalene, are carbonized homogeneously in molten salts with the catalytic action of $AlCl_3$. The molten salt first used was $AlCl_3$-NaCl-KCl (60:26:14 in molar ratio), which has a melting point of 95°C. More recently $AlCl_3$-$C_6H_5NC_2H_5Br$ (67:33 in molar ratio) has been used; this mixture is liquid at room temperature. When this carbonization technique is used, polymerization initiates below 100°C, and mesophase pitch with much alicyclic structure forms at temperatures as low as 230°C. Semicoke forms at just 300°C. By using various chloroalkanes as coupling reagents, polymeric compounds can be obtained at temperatures as low as 80°C. By varying the nature and amount of

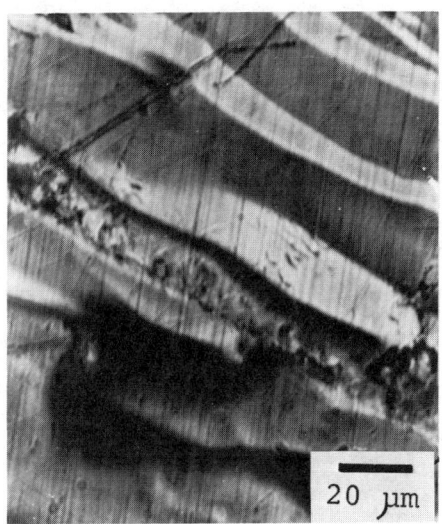

Figure 4. Polarized-light micrograph of PZ pitch.

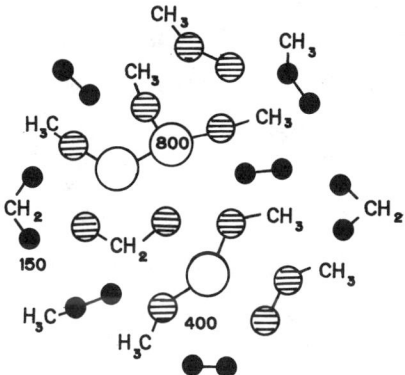

Figure 5. The "spider web" model for the constituent molecules of mesophase, after Mochida et al. (19).

the coupling reagent, the graphitizability of the resulting coke can be controlled to a large extent. Although some problems remain to be solved (such as the separation of pitch from molten salt), these novel methods are expected to contribute significantly to the development of pitch-based carbon fibers as well as other applications in the near future.

Characterization Techniques for Pitch Materials. Among a number of characterization techniques developed in Japan, the technique due to the members of the Society of Heavy Oil, led by Kunugi, stands out as particularly useful. The analytical data are treated by computer methods to construct average molecular structures for the carbonaceous materials. Sanada's group in Hokkaido University used high-temperature NMR and ESR data obtained by in situ measurements of pitch materials in molten salt (21). Much information on mesophase behavior during the heat-treatment process was obtained in this way.

Another significant technique developed by Sanada's group is the characterization of pitch for its electron donor ability, which is estimated by the amount of hydrogen transferred from pitch to anthracene after the mixture has been heated to 400°C (22). The present authors later showed that the electron acceptor ability of pitch can be estimated in a similar manner by using a mixture of pitch and dihydroanthracene (23). The details of hydrogen transfer between pitch molecules is an important topic for study to understand the initial stages of carbonization processes.

Recent Developments of Pitch-Based Carbon Fiber in Japan

A number of investigations of the preparation of pitch-based carbon fiber are in progress in industrial laboratories in Japan. However, aside from patents, only the developments by Honda and Yamada's group at the Kyushu Industrial Research Institute and by the present authors at Gunma University have been published.

Pitch-Based General Performance Carbon Fiber (GPCF). As described in the introduction, continuous-strand GPCF has been produced commercially only by the Kureha Chemical Industries Company. Public attention has recently been attracted to this type of carbon fiber by the success in using carbon-fiber-reinforced concrete in the construction of the Arsasheed Monument in Iraq (24) by the Kashima Construction (Kashima Kensetsu) Co. Future construction projects in Japan plan to utilize further this type of fiber-reinforced concrete. Such applications may lead to mass consumption of fiber if its price can be brought below $9/kg ($4/lb). The authors believe that some substantial reductions in the price of the general-performance fiber, perhaps to $6.5/kg ($3/lb), may occur in the near future.

High-Performance Carbon Fiber (HPCF) from Non-Hydrogenated Pitch. Pitch-based high-performance fibers from commercially available pitches were first prepared from mesophase pitches derived from naphtha-cracking pitches (12). In general these mesophase pitches have quite high softening points and are not suitable for smooth spinning. Prior to 1970, the mesophase content in such pitches was thought to be equal to the quinoline-insoluble (QI) content. In 1976 we found that, in pitches derived from naphthalene and anthracene with the use of $AlCl_3$ catalyst, the optically anisotropic regions were far larger than the QI content (25). By heating naphtha-tar pitches or atmospheric-reduced pitches with $AlCl_3$, Yoshimura (26) prepared mesophase pitches with good spinnability and with softening points as low as 200 to 300°C.

In those days, pitch chemistry had not advanced sufficiently to understand fundamentally the foregoing phenomena. The preparation of mesophase pitch with low softening point (the so-called "soft mesophase pitch") was based on direct experiment. Nevertheless through extensive and serious efforts, it became possible to prepare soft mesophase pitches from naphtha tars, decant oils from fluidized catalytic crackers (FCC), atmospheric-reduced crude oils, and other pitch-like materials. A typical example of these preparation procedures is the following. A purified FCC or naphtha pitch is heated at 400°C for one hour under methane to convert the pitch to a mesophase content of 23.6% (27,28). The mesophase separated by sedimentation has a softening point of 226°C; it is spun at 320°C, and the fiber is stabilized in air and finally carbonized by rapid heating at 100 to 1600°C/min (29).

High-Performance Carbon Fiber (HPCF) from Hydrogenated Pitch. Alicyclic molecular structure and the extent of hydrogen transfer between molecules have been progressively recognized as important factors to control the properties of the precursor pitch for fiber spinning. Three experimental methods of pitch preparation were explored as our understanding of pitch chemistry was developed. The first method is the so-called "dormant mesophase" or Gunma University (Gundai) method (30,31) in which the mesophase pitch is initially prepared from petroleum asphalt by ordinary pyrolysis procedures; this pitch is then hydrogenated to convert it to an isotropic pitch under conditions that avoid decomposition reactions, and finally converted again to mesophase pitch by another thermal treatment. The dormant mesophase method (also known as the Gundai method) is outlined and compared in Figure 6 with the second method, known as the Kyukoshi method (32) because the process was developed by the Kyushu Industrial Research Institute (Kyukoshi). In this second method, naphtha or coal-tar pitches are hydrogenated by using tetrahydroquinoline (THQ) solvent and then converted to mesophase pitch by rapid heating to 450 to 500°C. The third method uses precursor pitches prepared from hydrogenated coal-tar pitch or an SRC (solvent-refined coal) pitch subjected to a hydrocracking technique.

A feature of the Gundai preparation method is shown in Figure 7. The mesophase pitches indicated by DA240(A) were obtained by reheating a hydrogenated pitch derived from a pitch containing 3% mesophase; the DA240(B) pitches were similarly

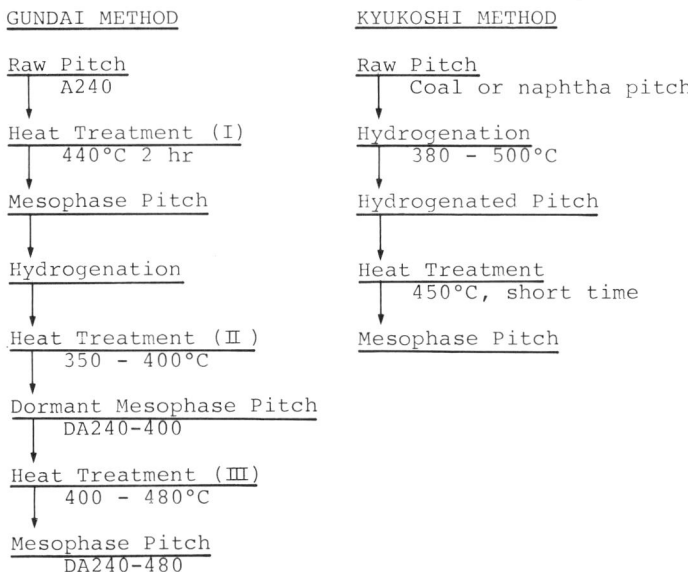

Figure 6. Flow charts for the preparation of fiber-spinning pitches by the Gundai (dormant mesophase) and Kyukoshi methods.

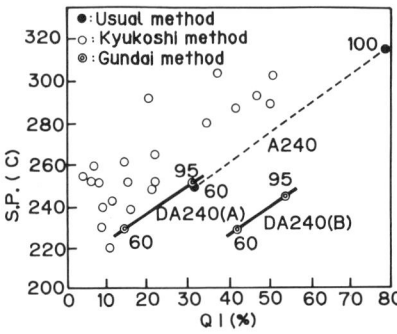

Figure 7. Relations between softening points, quinoline-insoluble contents, and mesophase contents of precursor pitches prepared by the Gundai and Kyukoshi methods. The volume percentage of mesophase is indicated by the numbers adjacent to some points.

prepared from a 5% mesophase pitch. In both cases, the initial mesophase pitches were obtained from Ashland A240 petroleum pitch by heating at 400°C. The numbers adjacent to some symbols in Figure 7 refer to the mesophase contents measured optically and expressed in volume-%. The DA240 mesophase pitches have lower QI contents and lower softening points than those of mesophase pitches prepared by the usual pyrolysis procedures, and Figure 7 shows that the softening points also tend to be lower than for pitches prepared by the Kyukoshi method. Thus the Gundai method pitches are characterized by low softening points despite their high QI contents.

To reveal the hydrogenation effects more clearly, measurements of radical concentration and the amount of transferred hydrogen were compared for pitches prepared by the Gundai method and by ordinary pyrolysis; see Table I. Mesophase appears at nearly the same temperature in both methods. For measurements made just before the mesophase appears, the pitch prepared by the Gundai method exhibits a larger amount of transferred hydrogen and a lower radical concentration. These characteristics must cause the lower softening points at high QI contents. These differences disappear upon thermal treatment to 480°C.

Table I. Comparison of Transferred Hydrogen and Radical Concentrations in Dormant and Ordinary Mesophase Pitch

	Transferred Hydrogen (mg/g)	Radical Concentration (/g)
Ordinary Mesophase Pitch (A240)		
After 400°C for 2 hr	3.88×10^{-2}	97.4×10^{18}
After 480°C for 20 min	2.30×10^{-2}	22.7×10^{18}
Dormant Mesophase Pitch (DA240)		
After 400°C for 2 hr	42.6×10^{-2}	2.9×10^{18}
After 480°C for 30 min	2.52×10^{-2}	21.9×10^{18}

Three points are noteworthy for the Kyukoshi method. (i) Rapid heating of the hydrogenated pitch to above 450°C produces a pitch suitable for smooth spinning. (ii) This pitch appears to be isotropic at the spinning temperature of 370°C. (iii) The viscosity-temperature relationship, plotted in Figure 8 in terms of the Andrade equation

$$\eta = A e^{B/T}$$

shows a change in slope at a transition temperature T_s, which is dependent on the pitch. Fiber spun near T_s exhibits radial structure, but without an open wedge. Fiber spun at lower temperatures develops the open-wedge radial structure, while fiber spun at higher temperature displays either random or onion-skin structures, as sketched schematically in Figure 8.

In respect to processes of oxidation stabilization and carbonization of the spun mesophase fibers, we are aware of no further developments except for the rapid carbonization technique previously noted (29).

The mechanical properties as a function of heat treatment temperature are shown in Figure 9 for fibers prepared by the Kyukoshi method. Fiber heat-treated to 2000°C or higher have strengths above 3GPa (435 kpsi) and tensile moduli of the order of 500 GPa (72 Mpsi). The present authors now believe that, in the near future, it will be possible to produce carbon fiber of equivalent properties by selection of suitable raw pitch materials and by development of specialized pretreatment procedures for the pitch to replace the extensive hydrogenation technique described here.

The Future of Pitch-Based Carbon Fiber

In comparison with the USA, the aerospace and defense industries of Japan are quite small. This is the principal reason for the relatively slow commercialization of pitch-based high-performance carbon fiber (HPCF) in Japan. As incentive for the HPCF industry, other fields of applications must be sought. In general, the automotive industry is thought to be the most promising field, but several well-known conditions must be satisfied. The cost of fiber must be decreased, mass production processing of fiber must be established, and improved molding techniques for the composites should be developed. Furthermore, new types of applications should be considered; for example, a GPCF cloth reinforcement for phenolic resin has been used for several years as a wear ring in the suspension of a dump truck. Such applications suggest that carbon fiber can be extended beyond primary structures to such areas as filler for engineering plastics, electromagnetic shields, and so on.

In the case of general performance carbon fiber (GPCF), carbon-fiber-reinforced concrete is a very promising application. As shown in Figure 10, Akihama et al. (33) accomplished remarkable improvements in the mechanical properties of concrete by adding chopped GPCF. By skillful unidirectional alignment of GPCF in the cement mortar, Furukawa et al. (34) obtained increases in mechanical strength by factors of 2 to 3 for fiber additions of one percent or less. As noted earlier, new applications of fibers in the construction industry are increasing. These should bring about mass consumption of the general performance fiber, but lower costs for all types of fiber must be achieved. In this respect, the pitch-based fiber is in a more favorable situation than PAN-based fiber. Through establishment of mass production facilities, GPCF should play a role as natural leader to HPCF, with favorable effects in cost reductions of not only GPCF but also HPCF.

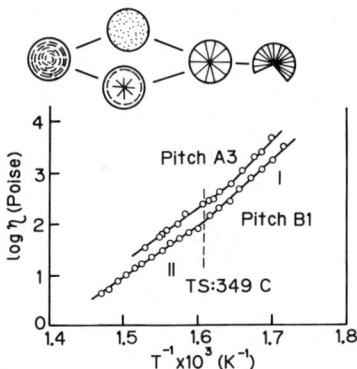

Figure 8. Viscosity of mesophase pitches prepared by the Kyukoshi method, with schematic microstructures of fibers spun at temperatures in the range of 300 to 400°C.

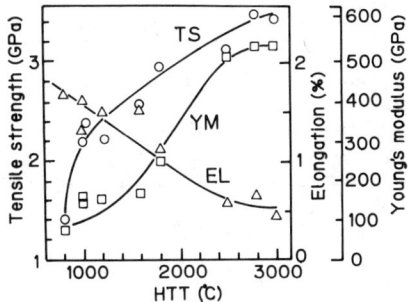

Figure 9. Mechanical properties, as a function of heat treatment temperature, of carbon fibers spun from mesophase pitch prepared by the Kyukoshi method.

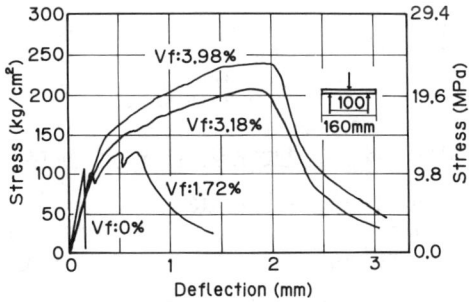

Figure 10. Flexural performance of carbon-fiber-reinforced concrete. V_f is the volume fraction of carbon fiber.

Literature Cited

1. Ōtani, S. Mol. Cryst. Liq. Cryst. 1981, 63, 249.
2. Ōtani, S. Japanese Patent 1966-15728.
3. Ōtani, S. Carbon 1965, 3, 31.
4. Ōtani, S. Carbon 1965, 3, 213.
5. Ōtani, S.; Yamada, K. J. Chem. Soc. Japan, Ind. Section 1966, 69, 626.
6. Ōtani, S.; Yamada, K.; Koitabashi, T.; Yokoyama, A. Carbon 1966, 4, 425.
7. Ōtani, S. Carbon 1967, 5, 219.
8. Ōtani, S.; Yokoyama, A. Bull. Chem. Soc. Japan 1969, 42, 1417.
9. Ōtani, S. Japanese Patent 1969-2511.
10. Ōtani, S.; Watanabe, S.; Ogino, H. Bull. Chem. Soc. Japan 1972, 45, 3715.
11. Ōtani, S. Japanese Patent 1979-8634.
12. Fujimaki, H.; Kodama, F.; Sakaguchi, T.; Okuda, K. Tanso 1975, No. 80, 3.
13. Singer, L. S. Carbon 1978, 16, 408.
14. Lewis, I. C.; McHenry, E. R.; Singer, L. S. U.S. Patent 3 976 729, 1976.
15. Ōtani, S.; Ōya, A. J. Chem. Soc. Japan, Ind. Section 1970, 73, 493.
16. Mochida, I.; Nakamura, E.; Maeda, K.; Takeshita, K. Carbon 1975, 13, 489.
17. Mochida, I.; Nakamura, E.; Maeda, K.; Takeshita, K. Carbon 1976, 14, 123.
18. Mochida, I.; Tamura, K.; Korai, Y.; Fujitsu, H.; Takeshita, K. Carbon 1982, 20, 231.
19. Mochida, I.; Maeda, K.; Takeshita, L. Carbon 1978, 16, 459.
20. Ota, E.; Ōtani, S. Ext. Abstr. 4th Int. Symp. on Molten Salts No. 755. 1983.
21. Miyazawa, K.; Yokono, T.; Sanada, Y. Carbon 1979, 17, 223.
22. Obara, T.; Yokono, T; Miyazawa, K.; Sanada, Y. Carbon 1981, 19, 263.
23. Park, D. Y.; Ōya, A., Ōtani, S. Fuel 1983, 62, 1499.
24. Tatsuhana, M.; Hirata, J.; Matsui, J. J. Chem. Phys. 1984, 81, 711.
25. Ōtani, S.; Endo, T.; Ōta, E.; Ōya, A. Tanso 1976, No. 87.
26. Yoshimura, S.; et al. Japanese Patent 1978-7533.
27. Watanabe, S. Japan Patent 1983-156020.
28. Watanabe, S. Japan Patent 1983-154701.
29. Watanabe, S. Japan Patent 1983-156022.
30. Ōtani, S.; Kikuchi, A. Japan Soc. for Promotion of Science, 117-163-A-2.
31. Ōtani, S. Japan Patent 1982-100186.
32. Imamura, T.; Shibata, M.; Yamada, Y.; Arita, S.; Honda, H. Proc. 10th Mtg. Carbon Soc. Japan, A-8 ~ A-11, 1-3 Dec. 1983, Tokyo.
33. Akihama, S.; Suenaga, T.; Sakano, T. Concrete Journal 1982, 20, 75.
34. Furukawa, S.; Ōtani, S.; Kojima, A. Proc. 10th Mtg. Carbon Soc. Japan, D-9, 1-3 Dec. 1983, Tokyo.

RECEIVED November 19, 1985

Growth of Carbon Fibers in Stainless Steel Tubes by Natural Gas Pyrolysis

G. G. Tibbetts

Physics Department, General Motors Research Laboratories, Warren, MI 48090-9055

> Formation of uniform, macroscopic carbon fibers by pyrolysis of hydrocarbons occurs in two principal stages: 1) growth of long submicron filaments by the interaction of nanometer-sized transition metal particles with a decomposing hydrocarbon gas, and 2) thickening of the filaments to diameters on the order of micrometers by subsequent deposition of pyrolysis products. Stainless steel tubes provide a suitable environment for both of these processes. As the steel begins to carburize, its surface fragments to produce metal particles. These particles can catalyze filament growth because gas phase hydrocarbon concentrations are suitably low due to substantial absorption by the walls. Thickening of the filaments to macroscopic fibers takes place after the walls are saturated with carbon. The concentration in the gas phase increases markedly, depositing pyrolytic carbon on previously-grown filaments to thicken them to the required diameter. Uniform 10 μm fibers as long as 20 cm and of average modulus 1.8×10^{11} Pa have been grown by this method.

In a 1953 study of material deposited on blast furnace brickwork, Davis et al. ([1]) observed the presence of twisted carbon filaments about 0.01 μm thick. Since then, similar microscopic carbon filaments formed during the decomposition of CO and hydrocarbons have been observed by many other investigators. Typically, these filaments are formed by the decomposition of CO or other gaseous hydrocarbons on iron subgroup metal catalyst particles (Ni, Co, Fe, and Cr) ([2]).

Considerably fewer reports of macroscopic carbon fibers (i.e., exceeding 1 μm in diameter) have been published. Iley and Riley ([3]) (1948) grew visible fibers by decomposing methane, propane, and ethylene at 1200°C on quartz substrates. Hillert and Lang ([4]) grew a wide variety of graphitic filaments, including helical coils, twisted triads, and branched straight threads, by decomposing n-heptane in a

0097-6156/86/0303-0335$06.00/0
© 1986 American Chemical Society

silica tube at 1000°C. Occasional straight fibers with lengths up to 5 cm and diameters up to 200 μm were also observed. Koyama (5) and later Koyama and Endo (6) grew fibers by the thermal decomposition of benzene at about 1200°C. Most recently, Katsuki et al. (7) reported fiber growth from the decomposition of napthalene-hydrogen mixtures.

Methane, or rather natural gas (which may contain carbon oxides, higher hydrocarbons, and inert gases), is of great interest as a source of pyrolytically grown fibers because of its relatively low cost.

Fiber Growth

General Requirements. At General Motors carbon fibers were accidentally grown from the decomposition of natural gas in an apparatus designed to measure the diffusivity of carbon through steel tubes (8). In the apparatus of Figure 1, (a modification of one originally designed for diffusion studies by R. P. Smith (9)) the interior surface of a steel tube was saturated with carbon from pyrolyzing gaseous hydrocarbons. Simultaneously, the exterior surface was continuously decarburized by wet hydrogen flowing through a jacket surrounding the tube. For a rather broad set of conditions of natural gas flow rate and temperature, it was found that, after many hours, masses of fibers (Figure 2) grew within a 304 stainless steel (18% Cr, 8% Ni) tube of 0.8 mm wall thickness. Most of the experiments utilized experimental conditions consisting of a temperature of 970°C, room temperature flow rate of 20 cc/min of natural gas (containing 1.8% ethane and 1.7% oxides of carbon), and a room temperature flow rate of 200 cc/min of wet H_2 through the surrounding jacket.

Figure 3 shows video images of fiber growth made through a window at the top of the growth tube; each dark circle defines the inner wall of the tube at a different time. Figure 3a was made after the growth tube had been exposed to the experimental conditions cited above for 9.5 h. Very thin fibers first became visible within 30 minutes (Figure 3b). These fibers continued to thicken with time and thus became more visible as the experiment was concluded.

Under some conditions one may observe a brightness increase within the volume of the growth tube during fiber growth. There are many references in the literature to a "fog" of droplets produced in organic vapors under highly pyrolyzing conditions (10). This increase in luminosity is due to thermal radiation emitted or scattered by these droplets. Figure 4 is a plot of light intensity in the growth tubes from which the appearance of this fog can be determined. Because the direction of gas flow in the experiment of Figure 4 was toward the camera, the fog is more apparent than in Figure 3, where it is not discernible in the photographs. The tube interior remained dark for 10.5 h and then substantially brightened just as the fibers appeared. The brightness rapidly increased as the fibers continued to thicken until the experiment was terminated after 16 h.

Also shown in Figure 4 is the brightness measured within a 1010 steel tube which did not grow fibers under otherwise identical conditions. No brightness increase was observed in such low-Cr mild steel tubes, and no fibers were grown. In contrast to mild steel, fused quartz tubes showed a high level of hydrocarbon fog during the entire

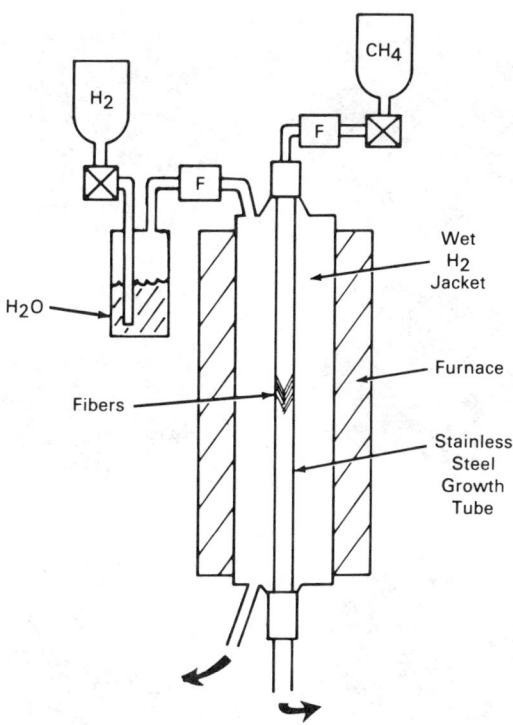

Figure 1. Schematic diagram of fiber growth apparatus including 304 stainless steel growth tube. Flowmeters are labeled "F".

Figure 2. Carbon fibers produced from natural gas.

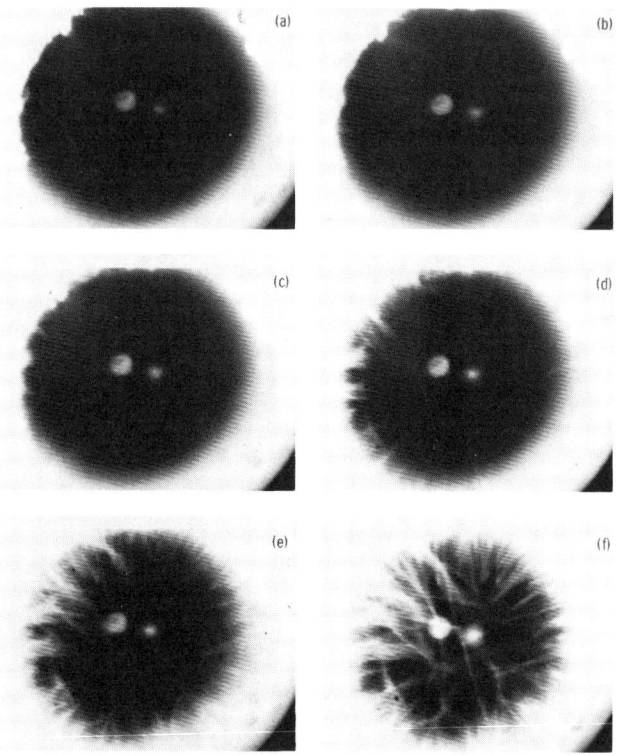

Figure 3. Micrographs of the interior of an 11 mm inside diameter growth tube at 970 C after a) 9-1/2 h, b) 10 h, c) 10-1/2 h, d) 11 h, e) 11-1/2 h, f) 12 h. Reproduced with permission from reference 8. Copyright 1983 American Institute of Physics.

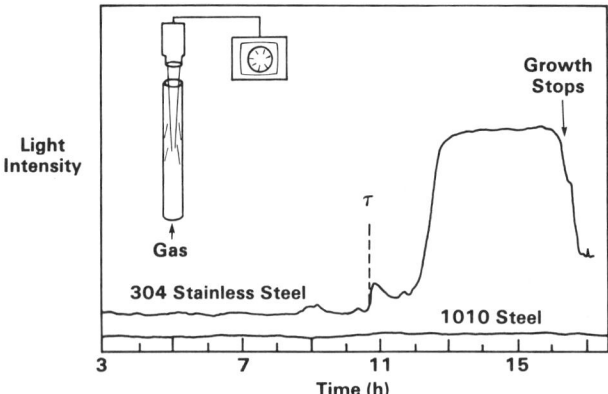

Figure 4. Light intensity in the center of the growth tube (obtained by photo-diode measurement from the videotaped image) as a function of time for 304 stainless and 1010 steel.

course of the experiment, yet no fibers could be grown within fused quartz tubes under the same conditions of flow rate and temperature that were successfully used with stainless steel tubes. Even when these tubes were provided with suitable nuclei for filament growth, no fibers could be grown.

These observations suggested that the onset of the fog was related to the saturation of the walls of the growth tube by carbon. In order to elucidate this effect in 304 stainless steel, a set of experiments was performed at 970°C and several different natural gas flow rates F to measure τ, the time elapsed before the onset of fogging. The results, plotted in Figure 5, show that τ is as small as 3 h for rapid gas flow through the system and as long as 17 h for slow flow. The observed linear increase of τ with 1/F corresponds to the behavior expected if a fixed concentration of carbon atoms [C] were required in order to saturate the steel surface at time τ,

$$[C] \propto \tau \cdot F \ . \tag{1}$$

Measurements of the temperature dependence of τ yielded an activation energy consistent with this picture. Thus, the time τ required for the appearance of fogging corresponds to the time necessary to saturate the growth tube's surface with carbon.

Pyrolysis Regimes. Results of a further experiment show how pyrolysis conditions change after time τ when the tube interior saturates. A number of thick 304 stainless steel wires (0.31 mm in diameter) were supported in the furnace during a fiber growth experiment and allowed to a drop out of the hot zone after different periods. Thus, the mass increase of these wires could be determined as a function of the length of time they remained in contact with pyrolyzing natural gas in the growth tube.

The measurements were performed under standard conditions except that the flow rate of natural gas was 50 cm^3/min. The mass of the wires (top panel of Figure 6) shows an initial sharp rise as they carburize, followed by a much slower rise during the time required to carburize the growth tube of wall thickness 0.8 mm. In a separate experiment, it was shown (Figure 6b) that, after 4.6 hours, fogging begins. At exactly that time, the mass of the stainless steel wires began to increase rapidly again, corresponding to deposition of an ever thickening layer of pyrolytic carbon. It is this pyrocarbon deposition which thickens any fibers present within the growth tube after fogging begins.

Measurements of the effluent gas from the growth reactor also show significant changes when fogging begins. The effluent from a fused quartz tube showed little change with time (Figure 7), in sharp contrast with the effluent from a 304 stainless steel tube. The tubes utilized in these experiments were from a different lot of stainless steel which apparently had a somewhat more reactive surface and thus required only 2.7 h to carburize at a flow rate of 50 cm^3/min. During most of this period, the concentration of ethylene and all higher hydrocarbons was below 0.1%. However, when fibers first became visible at 2.7 h, the ethylene concentration had risen to half its ultimate value. Higher hydrocarbons were beginning to be

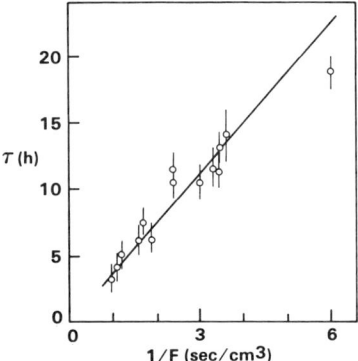

Figure 5. Time required to grow visible carbon fibers using a fresh 0.9 mm wall 304 stainless steel growth tube at 970°C, plotted as a function of reciprocal natural gas flow rate.

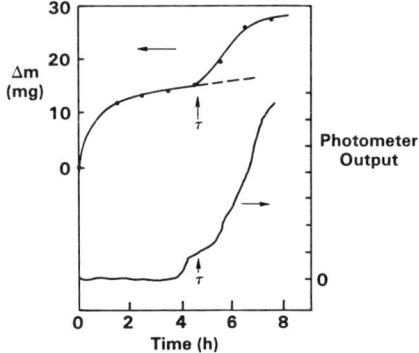

Figure 6. Top: Mass increase of stainless steel wires as a function of pyrolysis time. Bottom: Luminosity as a function of time. Arbitrary units are used.

produced in the tube in abundance; from them the complex molecules which produce pyrocarbon were beginning to appear.

The conditions appropriate for growing fibers from natural gas may not be successful with other hydrocarbon feedstocks. Using pure ethane during a standard growth run produces merely a tarry sludge. At the other extreme, pure methane will neither grow fibers nor deposit a pyrocarbon layer under the standard conditions at 970°C, but it can grow fibers near 1100°C.

Formation of Catalyst Particles by Surface Fragmentation. The phenomenon of "metal dusting corrosion" by which stainless steel surfaces are fragmented to produce submicron catalytic particles for fiber growth has been described by Bradley (11). In this process, the strongly reducing and carburizing atmosphere in the growth tube severely corrodes the natural oxide surface of the stainless steel to produce a fine dust of metal, carbide, and oxide particles. Some of the smaller particles can then act as nuclei for fiber growth.

The role of the hydrogen jacket in fiber growth will now be discussed. Gas chromatographic studies have shown that sufficient hydrogen from the jacket diffuses through the stainless steel to increase the concentration of hydrogen in the effluent gas by 2% under standard conditions. In the vicinity of the steel walls, where the filaments grown, the increase is even greater. Figure 8 shows the influence of hydrogen jacket pressure on fiber growth. A series of experiments were performed, each under standard conditions but with a different H_2 jacket pressure. The ordinate is Nc, the number of distinct fibers in focus slightly below the midplane of the furnace 3 hours after fibers first became visible -- a quantity that is proportional to total fiber yield. It is clear from these data that the hydrogen in the jacket aids in producing fibers. The top curve refers to growth experiments in which the inside of the tube is seeded with $Fe(NO_3)_3$ to provide nuclei suitable for filament growth. In this case only a very weak dependence on hydrogen pressure is noted. These data imply that wet hydrogen from the jacket aids in promoting the production of adequate metallic nuclei from the stainless steel surface. Furthermore, if adequate nuclei are already present, the hydrogen jacket is not useful in the remainder of the growth process.

Fiber Morphology and Properties

As Figure 9a shows, the fibers grow nearly parallel to each other; they appear to grow approximately along the gas streamlines. The fibers are attached by one end to a laminar pyrocarbon deposit which adheres to the tube wall. The "cobbled" appearance of this layer may originate from shorter fibers which have been buried by pyrocarbon. The pyrocarbon deposit is about 0.3% hydrogen by weight, and contains particles of all the metallic constituents of the tube.

Figure 9b shows that the fibers are quite uniform in diameter. Fibers as long as 20 cm and having diameters up to nearly 1 mm have been produced. Thickening of the fibers by chemical vapor deposition gives them an annular structure (Figure 10a), in contrast to the radial structure of a Thornel P fiber (Figure 10b).

The average modulus of the fibers is 180 GPa (26.1 Mpsi), although peak values of up to 450 GPa (65.3 Mpsi) have been measured

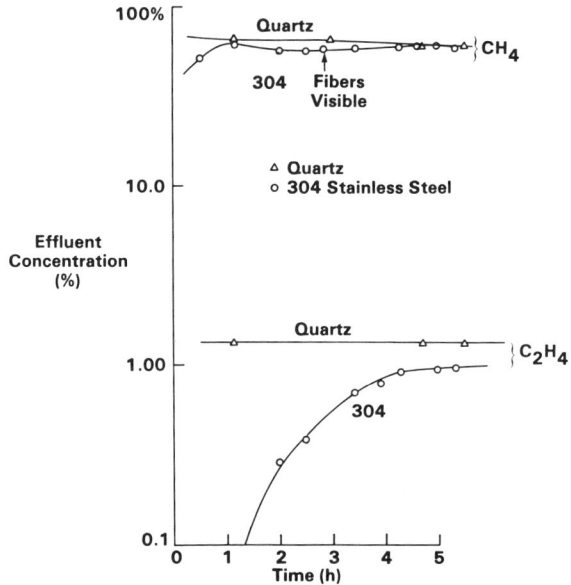

Figure 7. CH_4 and C_2H_4 effluent obtained after pyrolyzing natural gas at 970°C and 5 sec residence time in a 304 strainless steel or quartz reactor. Fibers appeared in the stainless steel reactor after 2.8 h.

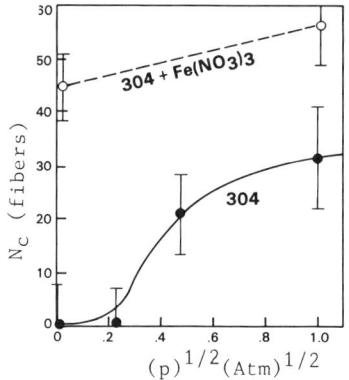

Figure 8. Number of fibers counted in the camera's focal plane 2.5 cm below the furnace centerline after a number of standard growth experiments where the hydrogen jacket pressure was held constant at the values shown. Bottom curve refers to a series of 304 tubes. In the top curve 304 tubes which had previously grown fibers and whose surfaces were treated with $Fe(NO_3)_3$ were used.

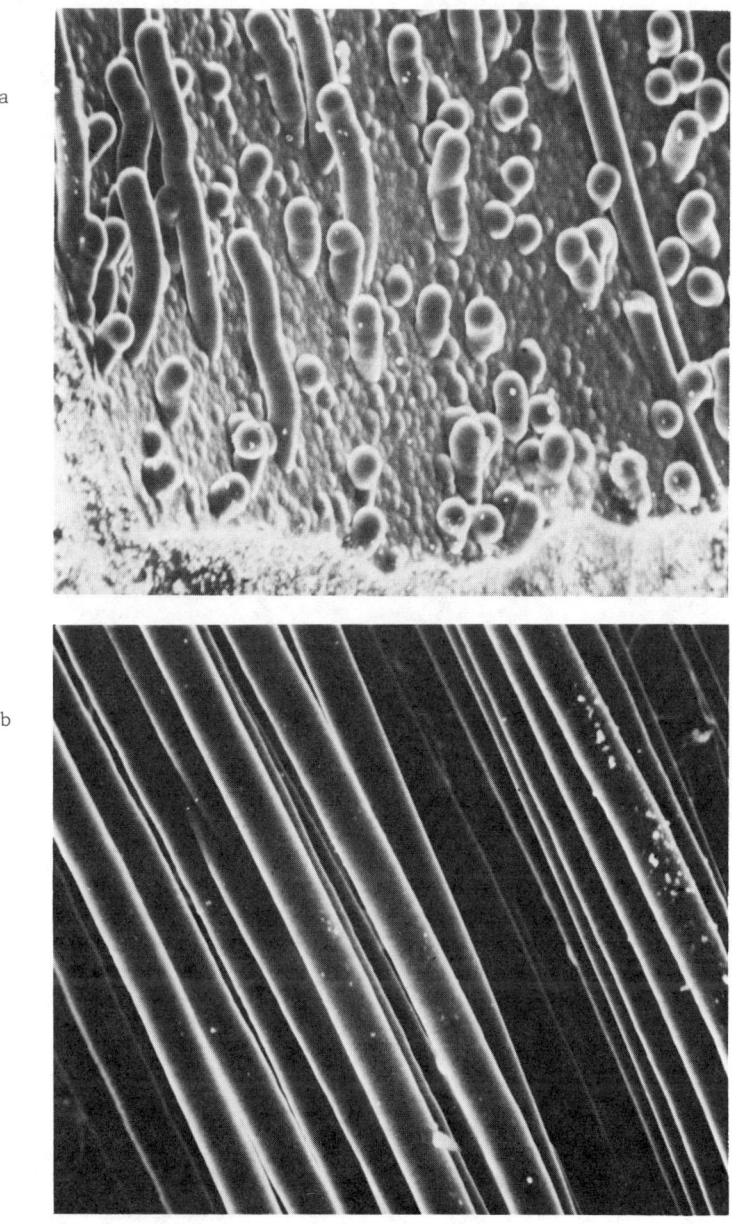

Figure 9. Scanning electron micrograph of a) fibers growing from matrix layer, b) a bundle of fibers.
Copyright 1983 American Institute of Physics.

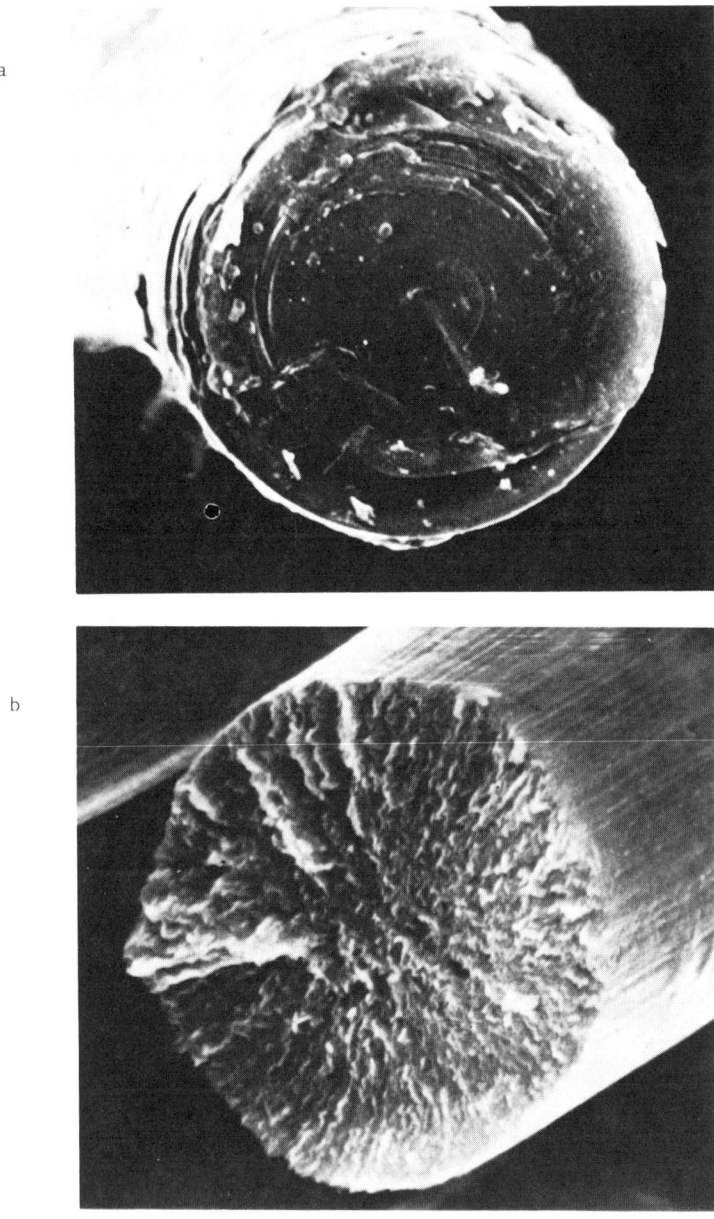

Figure 10. Scanning electron micrographs of broken ends of a) natural gas and b) Thornel P fibers.

for fibers slightly smaller than 10 μm in diameter. Fiber tensile strengths average 1.0 GPa (145 kpsi). X-ray diffraction studies show that the d_{002} spacing for fibers grown at 970°C is 0.345 nm, compared to 0.335 nm for crystalline graphite.

Conclusions

Several factors contribute to the growth of carbon fibers from natural gas in 304 stainless steel tubes surrounded by a jacket containing circulating wet hydrogen. First, the carburizing stainless steel surface fragments to produce metal particles suitable for catalyzing filament growth, particularly when hydrogen diffuses through in large quantities. Second, during carburization the stainless steel absorbs enough hydrocarbons from the gas phase to provide an atmosphere sufficiently depleted in hydrocarbons to be suitable for growth of microscopic filaments. Third, after the stainless steel is saturated, the hydrocarbon concentration climbs so that the fibers may be thickened by deposition of pyrocarbon.

Acknowledgments

I would like to acknowledge helpful discussions with J. R. Bradley, G. W. Smith, and W. E. Yetter. C. P. Beetz, Jr. and G. W. Budd made the modulus and strength measurements. I would like to thank M. G. Devour for skillful experimental help.

Literature Cited

1. Davis, W. R.; Slawson, R. J.; Rigby, G. R. Nature 1953, 171, 756.
2. A good recent review is: Baker, R. T. K.; Harris, P. S. Chemistry and Physics of Carbon, Walker, P. L.; Thrower, P. A., Eds.; Dekker: New York, 1978; vol. 14, p. 83.
3. Iley, R.; Riley, H. L. Jour. Chem. Soc. London, II 1948, 1362.
4. Hillert, M.; Lang, N Zeit Krist. 1958, 111, 24.
5. Koyama, T. Carbon 1972, 10, 757.
6. Koyama, T.; Endo, M. Oyo Buturi 1973, 42, 690 (in Japanese).
7. Katsuki, H.; Matsunaga, K.; Egashira, M.; Kawasumi, S. Carbon 1981, 19, 148.
8. Tibbetts, G. G. Appl. Phys. Lett. 1983, 42, 666.
9. Smith, R. P. Acta. Meta. 1953, 1, 578.
10. Lahaye, J.; Prado, G; Donnet, J. B. Carbon 1974, 12, 27.
11. Bradley, J. R. Extended Abstracts, 16th Conf. on Carbon 1983, 533.

RECEIVED April 24, 1985

24

Carbon-Fiber-Reinforced Carbon Composites Fabricated by Liquid Impregnation

Erich Fitzer and Antonios Gkogkidis

Institut für Chemische Technik, Universität Karlsruhe, D-7500 Karlsruhe, Federal Republic of Germany

> Although developed initially for aerospace applications, carbon/carbon composites are now finding wider uses, e.g., in nuclear reactors, automobiles, metalforming, and biomedical implants. This paper is concerned with problems and choices in fabrication, including chemical vapor deposition, but focussing principally on liquid impregnation methods. High final heat treatment temperatures for the carbon fiber are desirable to realize good translation of fiber strength into the composite. "Soft" matrix precursors, e.g., coal-tar pitch, contribute to the composite modulus by alignment of the graphitic layers parallel to the fibers. Stress cycling can destroy the matrix by internal fracturing to fine dust, but this loss of matrix can be limited by a final resin impregnation to produce a hybrid matrix. The hybrid matrix composites have mechanical properties similar to those of polymer-matrix composites, but with reduced flammability.

Carbon/carbon composites fabricated by multiple cycles of liquid impregnation and recarbonization are a typical example of modern petroleum derived carbons. In the 1975 ACS Symposium on Petroleum Derived Carbons (1), papers were presented on carbon/carbon composite materials formed by pyrolytic infiltration processes (2) or by liquid impregnation with petroleum pitch (3,4), on fabrication processes for high-modulus carbon fibers based on polyacrylonitrile (PAN) or pitch precursors (5), and on the use of carbon materials for thermostructural (6) as well as biomedical applications (7).

The present paper addresses the problems posed in these earlier contributions from a new viewpoint, namely the development of a superior carbon material that realizes more fully the strength, stiffness, and thermal properties inherent in the strong chemical bonds of carbon in the graphitic layer (8). Petroleum products, as the carbon precursors for pitch-based carbon fibers and for the carbon matrix of the composite, have proved advantageous for such superior carbon materials.

0097-6156/86/0303-0346$09.50/0
© 1986 American Chemical Society

Applications of Carbon/Carbon Composites

The performance of carbon/carbon composites became well known to the general public by the repeated successful landings of the American space shuttles, in which 2D carbon/carbon composites, reinforced two-dimensionally by a fiber web or fabric, are used for structural parts that are critical during reentry, such as the nose cap and the leading edges of the wings; see Figure 1 (9).

Because of its high strength and infusibility even at very high temperatures, as well as its low density, carbon-fiber-reinforced carbon is most suitable as disc brake material for supersonic aircraft such as the civilian CONCORDE and nearly all military jets; see Figure 2. For this application, 2D carbon/carbon composites are preferred. The performance characteristics of disc materials as tested by SEP France are compared in Figure 3 (SEPCARB is the trade name for carbon/carbon composites fabricated by Societé Europeene de Propulsion). The high tolerable concentration of consumed energy in the carbon/carbon brake discs should be noted (10).

In the near future, it is planned to replace the conventional asbestos brake linings for automobiles by modified carbon/carbon composites (11,12).

Carbon/carbon materials are currently used in the most critical parts of solid-fuel rocket engines. A test nozzle for the French ARIANE rocket is shown in cross section after firing in Figure 4 to demonstrate the undamaged surface of the 3D composite within the throat of the nozzle (10).

Carbon/carbon composites are also used as refractory components in gas-cooled high-temperature nuclear reactors, e.g., in the heat exchanger between the primary and secondary helium cooling circuits (13).

Gas turbines have been built with blades made from ceramic materials and using 1D-reinforced carbon-carbon circumferential rings developed by DFVLR, Stuttgart (14) (DFVLR = Deutsche Forschungs- und Versuchsanstalt fur Luft- und Raumfahrt); see Figure 5. In this application the carbon-carbon ring provides compressive prestress on the ceramic turbine blades, which are sensitive to tensile stress.

A recent successful application of carbon/carbon composites is the tool for superplastic forging of titanium illustrated by Figure 6; tubes up to 1.5 m in length can be forged at temperatures up to 1000°C, thus offering a rapid alternative fabrication technique to present production methods, e.g., riveted tubes (15). Contact brushes for electrical commutators, made with carbon fibers and carbon/carbon composites (16), are opening another new field of application. Furthermore, pistons in diesel engines have been proposed to be made from carbon/carbon composites (17).

Finally the use of carbon-fiber-reinforced carbon materials for implantation purposes in human medicine should be mentioned. One of the most impressive applications is that for hip joints. In Figure 7 a state-of-the-art hip joint made from a cobalt alloy with a polyethylene socket is compared with a design concept employing a carbon-fiber-reinforced carbon stem in a polygranular carbon socket. Such a stem design offers the possibility to approach closely the femur structure by appropriate combination of 1D-, 2D,

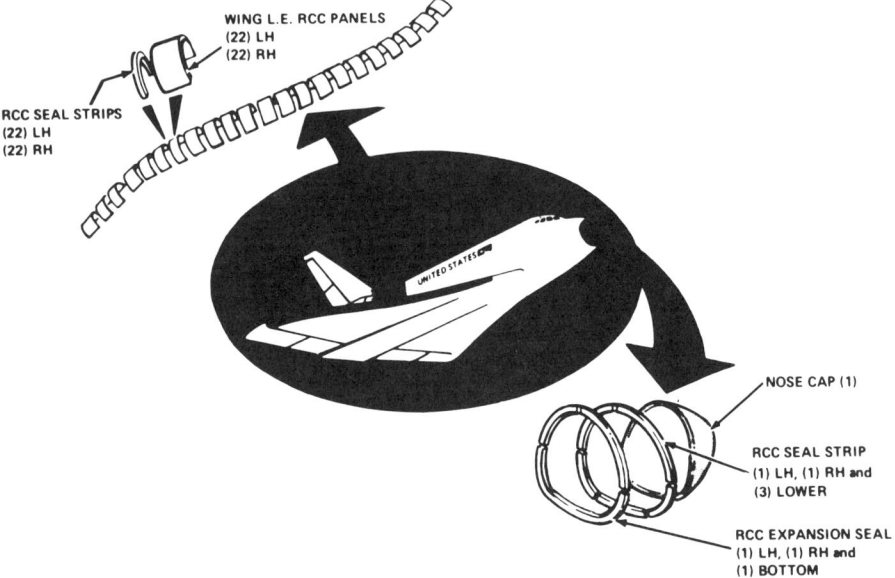

Figure 1. The leading edge (L.E.) of the wing and the nose cap of the COLUMBIA space shuttle (9) are made of reinforced carbon composites (RCC).

Figure 2. An aircraft disc brake fabricated from carbon/carbon composites (10).

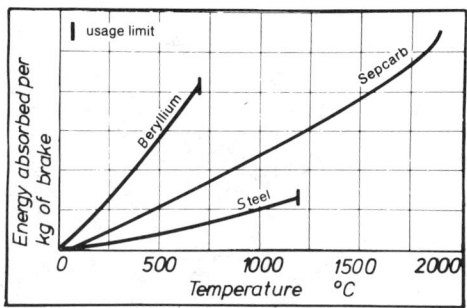

Figure 3. Comparative performance of three disc materials for aircraft brakes (10).

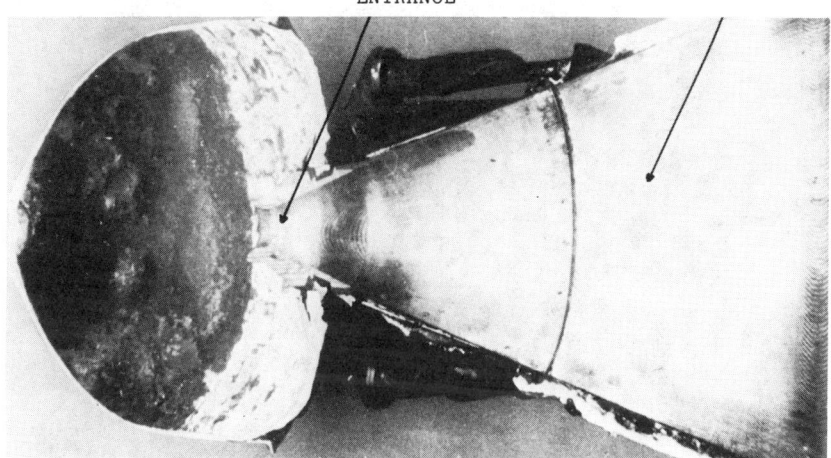

Figure 4. Rocket nozzle of the ARIANE rocket, observed in cross-section after firing (10).

Figure 5. A turbine rotor made with ceramic blades and a carbon-carbon composite ring to withstand tensile stresses (14).

Figure 6. Above: a tool for the superplastic forging of titanium, made entirely from carbon/carbon composites (15). Below: aircraft exhaust manifold formed from titanium by superplastic forging with the carbon/carbon tool.

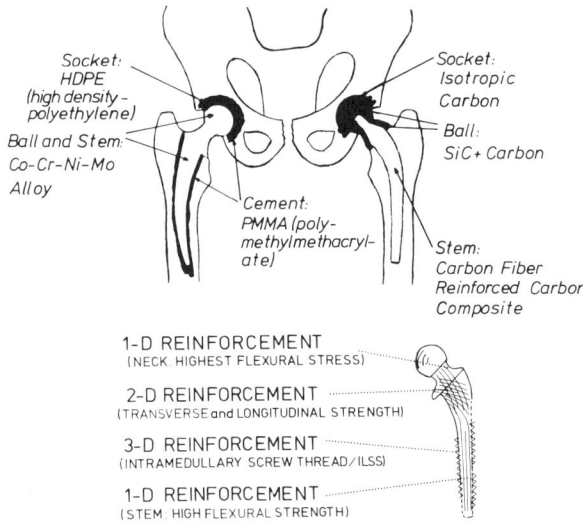

Figure 7. Comparison of a conventional hip joint fabricated from cobalt alloys with a joint design using carbon-fiber-reinforced composites tailored to meet specific local stress requirements (18).

and 3D-reinforced composites (18,19). Some values of flexural strength and Young's modulus achievable by various types of carbon/carbon composites are shown in Figure 8 (20). A wide range of strength and stiffness in various directions can be designed into such tailored materials. The properties of the Co-Cr-Mo alloys that are used for medical implants are compared with mechanical data for human bones to demonstrate that carbon/carbon composites are promising candidates for implantation in the future.

Fabrication Processes for Carbon/Carbon Composites

Processes for fabrication of carbon/carbon composites by "gas-phase impregnation" and by "liquid-phase impregnation" were described in the 1975 Symposium (1). The vapor impregnation process was first used for fabrication of the special aircraft brakes required for the commercial jet CONCORDE, and is still in use, although most new carbon/carbon parts and even aircraft brakes are now produced by the more economical pitch-impregnation processes.

The gas-phase impregnation process (in-pore deposition of pyrolytic carbon) is difficult to perform because of the tendency for pore closing instead of pore filling. Figure 9 shows the results of chemical vapor impregnation (CVI) experiments with SiC deposition in tube-like model pores in a polygranular graphite body. SiC deposition was used for better recognizability of the CV deposits. The tendency to deposit in the pore entrances can readily be understood if the overall deposition rate of the heterogeneous reaction is controlled by the chemical deposition rate and not by transport phenomena (21). Transport control occurs mainly at the higher deposition temperatures.

In Figure 10 (left-hand side) experimental results on penetration depth in polycrystalline graphite are shown by dashed lines as a function of deposition temperature. The decrease in penetration depth can be precalculated for various pore diameters if the diffusion behavior of the gaseous species and the rate of the chemical deposition reaction are known. For this calculation the dimensionless Damköhler II number (identical with the square of the Thiele modulus) can be used as indicated in the right-hand side of Figure 10. The results of precalculation are shown in the left-hand side and indicate that the experimental data fit the precalculated data quite well if the decreases in pore diameter during CVI are taken into consideration (22-25).

Pore filling without pore blocking can more easily be achieved by liquid impregnation; see Figure 11 (26). Furthermore, pore filling can be improved by impregnation and carbonization at high pressure. In recent years, we have performed several basic studies (27-29) of the liquid impregnation process.

The liquid impregnation process for carbon/carbon composites is similar to the industrial fabrication process for the high-density graphite electrodes required for ultra-high-power steel furnaces. Figures 12 and 13 show the analogies as well as the differences between the two processes. In both cases, the heterogeneous structure consists of two constituents: (a) primary carbon, which is introduced as elemental carbon during the process step of "green fabrication," and (b) secondary carbon, formed by thermal degradation of the carbonaceous binder during the process

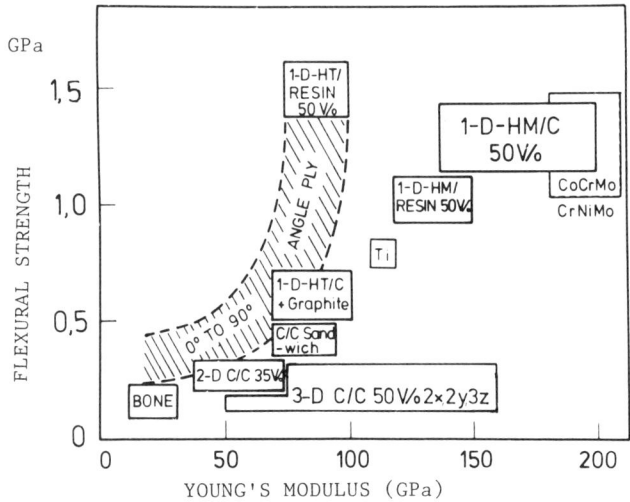

Figure 8. Mechanical properties of various carbon-fiber-reinforced composites compared to bone and some biomedical alloys (20). HM and HT refer, respectively, to fibers of high modulus and high tensile strength.

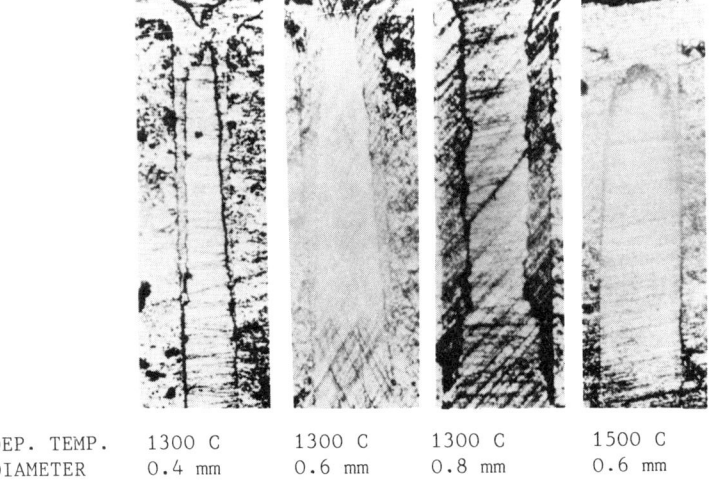

DEP. TEMP.	1300 C	1300 C	1300 C	1500 C
DIAMETER	0.4 mm	0.6 mm	0.8 mm	0.6 mm

Figure 9. Pore filling (second and third micrographs) and pore blocking (first and fourth micrographs) of model pores in a graphite body, as functions of pore diameter and reaction temperature. SiC deposition was used to distinguish deposits clearly (21).

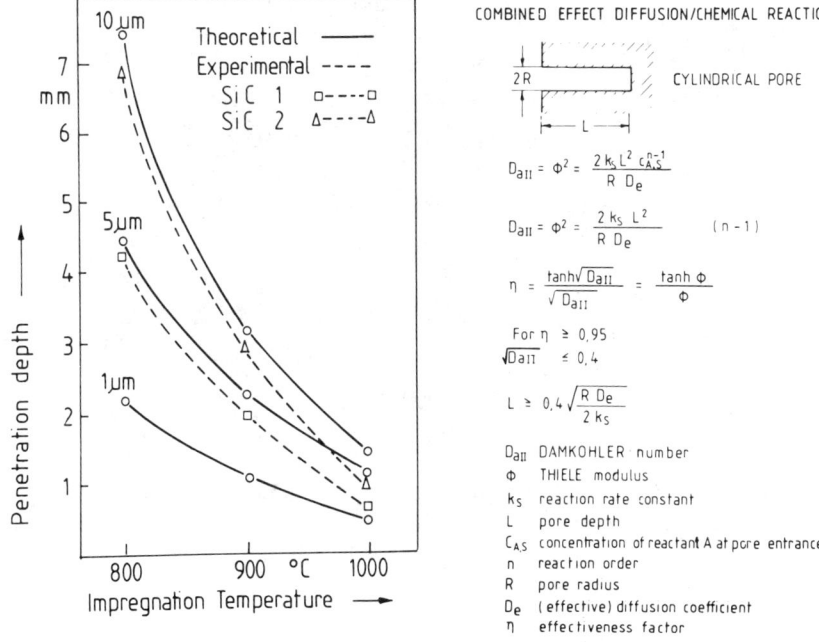

Figure 10. Penetration depth for chemical vapor deposition into cylindrical pores with radii of 1, 5, and 10 μm. Left-hand side: comparison of experimental data with theoretical calculations. Right-hand side: calculation of combined effects of diffusion and chemical reaction during the CVD process, using the Damköhler number (22-25).

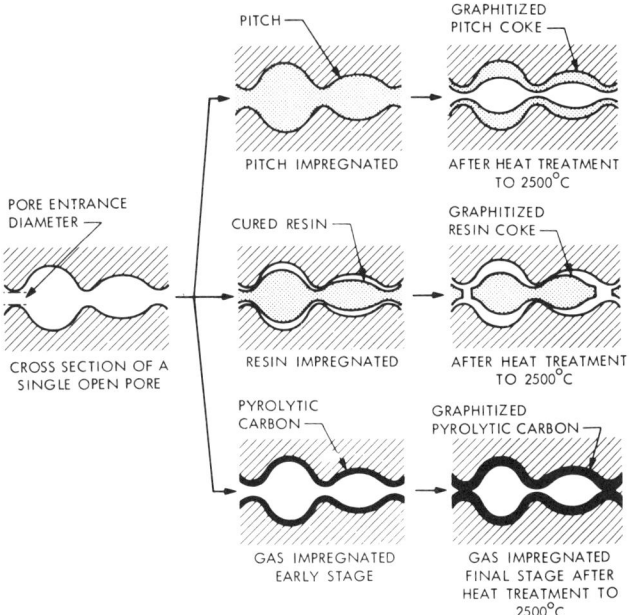

Figure 11. Schematic mechanisms of pore filling and pore blocking by liquid impregnation and by chemical vapor deposition (26).

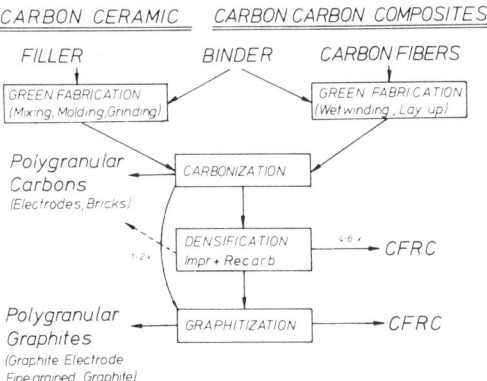

Figure 12. Comparison of production processes for synthetic polygranular graphites and carbon/carbon composites.

step of "baking". In polygranular carbons, the primary carbon is represented by the filler grains, the secondary carbon by the binder bridges between the grain particles. Carbon/carbon composites consist of carbon fibers as primary carbon and the carbon matrix as secondary carbon. The major difference between the two products is seen in the distribution of the structural elements. In the case of the composites both primary and secondary carbons are continuous phases, whereas in granular carbons only the binder coke is a continuous phase. The advantages of the continuous filler phase in the carbon/carbon composites appear in high bulk strength, better thermal and electrical conductivity, and good resistance to fracture propagation.

In principle, the liquid impregnation process consists of the formation of a fiber skeleton by pyrolysis of a temporary binder (as in the fabrication of conventional carbon ceramics), then liquid impregnation with pitch or other liquid carbon precursors and recarbonization, followed by multiple repetition of the latter process steps. Owing to shrinkage of the resin or pitch during curing and carbonization, new slit-like pores are formed in each cycle, and these can be refilled by liquid matrix precursor for subsequent recarbonization. In order to achieve composites with sufficiently high density, strength, modulus, impact resistance, and thermal conductivity, this multi-step impregnation-recarbonization process must be optimized with respect to the process conditions as well as the fiber and precursor materials.

Our studies (8,20,30-34) were performed with unidirectionally reinforced model composites because only with this arrangement can the translation of fiber strength in the composite be evaluated by comparison of precalculated and experimental results. The flexure tests were made by a three-point bending machine with a ratio of span-length to specimen thickness of 40. Data for the mechanical properties of the fiber were obtained by single-filament-testing with a gage-length of 30 mm.

Systematic Studies of the Liquid Impregnation Process

The Influence of the Matrix Precursor. The first requirement for a suitable matrix precursor is high carbon yield, which must be achievable under simple pyrolysis conditions. Figure 14 (left-hand side) shows weight loss as a function of pyrolysis temperature for several matrix precursors; practical precursors that are commercially available include coal-tar and petroleum pitches, phenolic resins, polyimides, and the para-polyphenyleneacetylene resin Hercules HA 43 (35,36). The structural formulas of some polymer binders are shown in Figure 15.

The carbon yield of pitch can be greatly increased by pyrolysis under high gas pressure (31,37); see Figure 14 (right-hand side). High carbon yields can also be obtained at room pressure if the initial dehydrogenation of the pitch is achieved by addition of elemental sulfur before thermal decomposition (32,33,38-40).

There are two further requirements for a suitable matrix precursor. Firstly, the carbonization shrinkage of the matrix should not damage the carbon fiber of the skeleton. Secondly, the porosity formed by carbonization of the precursor should be open

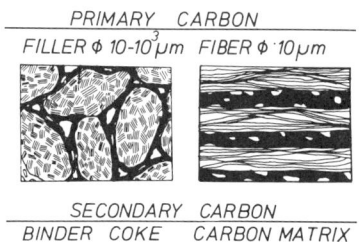

Figure 13. The "two-phase" structures of synthetic polygranular graphites and carbon/carbon composites.

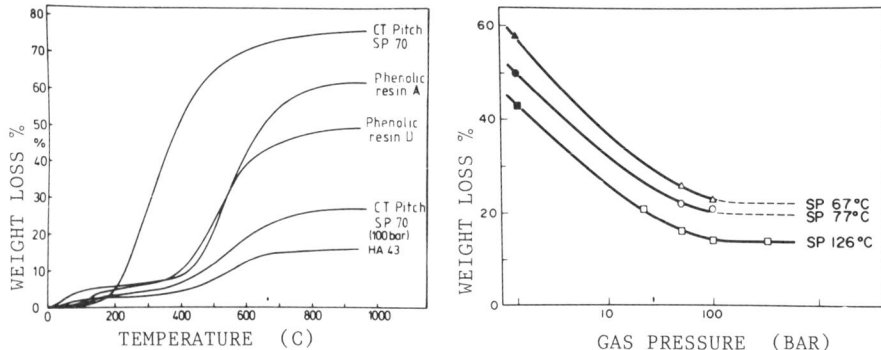

Figure 14. Weight losses in the carbonization of various precursors. Left-hand side: weight losses in room-pressure carbonization at 2°C/min (20). Right-hand side: Influence of gas pressure on three coal-tar pitches with various softening points (SP), pyrolyzed to 600°C at 10°C/min. (35).

and accessible to further impregnation. Only under these preconditions can further improvement of density and mechanical properties be achieved by subsequent processing cycles.

The suitability of carbon matrices for further impregnation is shown in Figure 16 for two groups of precursors with quite different pyrolysis behavior. Crosslinked resins with high molecular weight and large monomer units, such as polyimides and polyphenyleneacetylene, achieve good strength after only one cycle of impregnation and carbonization. This results from high carbon yield and isotropic shrinkage without damage to the fiber. However, the mechanical properties achieved after the first carbonization will not be increased by further impregnation and carbonization cycles (8,32,36).

The second group of matrix precursors is represented by the phenolic resins as well as by petroleum and coal-tar pitches; when carbonized under normal pressure, these form porous carbon matrices with only low mechanical strength of the composites after the first carbonization. However, strong improvements in the mechanical properties can be achieved by repeated impregnation and carbonization cycles.

An intermediate position can be seen for pitches that are carbonized under pressure or by addition of elemental sulfur. Because of the high cost of precursor materials of the first group, the selection of precursor type and processing must be optimized carefully (20,32).

Carbon Fibers and Their Influence on Carbon/Carbon Composites.
Carbon fibers produced by spinning polyacrylonitrile (PAN) consist of polymer carbon and are completely non-graphitizable, as can be recognized in X-ray diffraction patterns from the weak two-dimensional 10 and 11 reflections. In contrast, graphitizable carbon is formed from mesophase pitch. Although mesophase-pitch-based fibers are oxidized during the stabilization process, they do not completely lose the tendency to graphitize; this can be recognized by the modulation of the two-dimensional 10 reflection into 100 and 101 diffraction lines and the 11 reflection into 110 and 112 lines (41,43).

Carbon fibers can have extremely anisotropic mechanical properties, depending on the intensity of the preferred orientation of the graphitic layers in the direction parallel with the fiber axis. This orientation can be seen in the TEM micrographs of Figure 17.

As described by the elastic constants (44,45),

$$C_{11} = 1060 \text{ GPa}$$
$$C_{33} = 36.5 \text{ GPa}$$
$$C_{44} = 5.4 \text{ GPa}$$

the graphite crystal is very shear sensitive, as illustrated in Figure 18 for a natural graphite flake. Graphite flakes can be bent easily in spite of the high tensile strength of the graphite layer, and, upon milling, the layers are bent and twisted in manifold ways because of the low shear resistance (46).

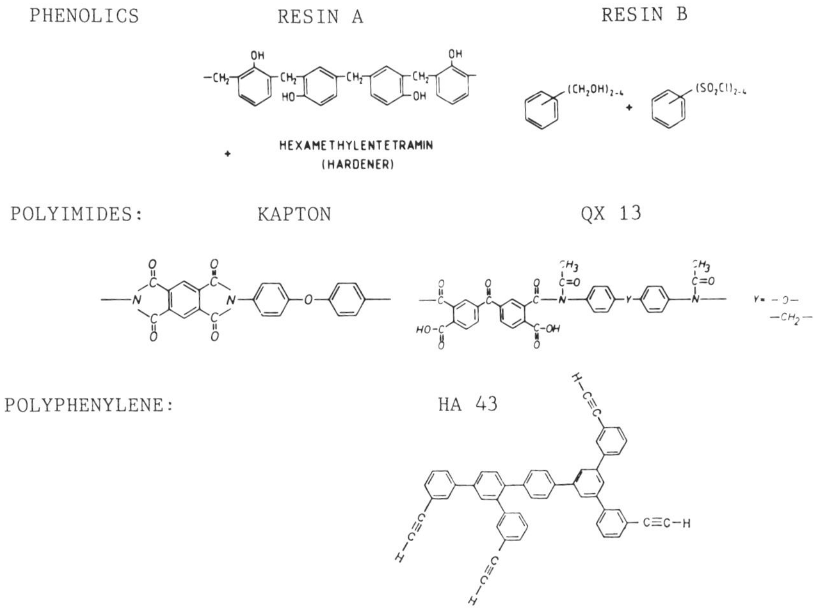

PHENOLICS   RESIN A   RESIN B

POLYIMIDES: KAPTON   QX 13

POLYPHENYLENE: HA 43

Figure 15. Some polymers used as matrix precursors for carbon/carbon composites (37).

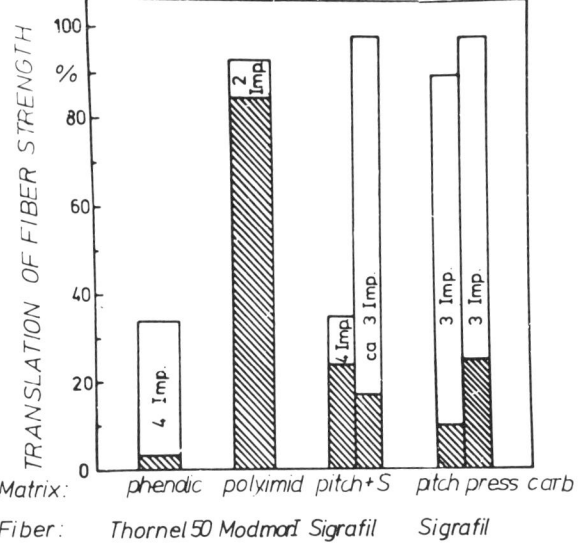

Figure 16. Translation of fiber strength, after the first impregnation and carbonization (shaded bar), and after repeated densification cycles (32). Thornel 50, Modmor I, and Sigrafil HM fibers are high-modulus PAN-based fibers; Sigrafil HF is a high-strength PAN-based fiber.

POLYACRYLO-
NITRILE
(PAN)
based

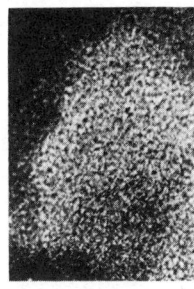 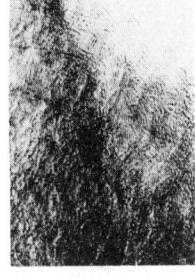 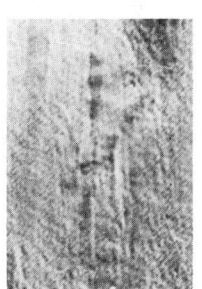

      HT as delevered   post heat treatment   HM as delivered
         (1400 °C)   (1800 °C)   (greater than 2800 °C)

MESOPHASE
PITCH (MMP)
based

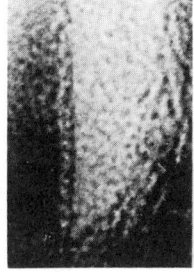 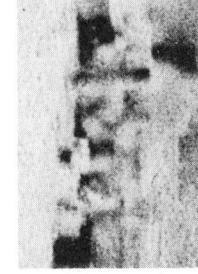 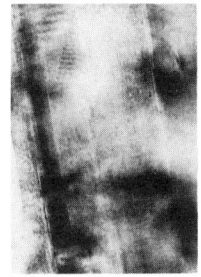

     P55 as delivered   post heat treatment   P100 as delivered
                       (2100 °C)   (greater than 2800 °C)

Figure 17. Transmission electron micrographs of PAN-based and mesophase-pitch-based carbon fibers, as delivered and after heat treatment (42,43). HT and HM refer to high strength and high modulus PAN-based fibers; P55 and P100 refer to mesophase-pitch-based fibers, with tensile moduli of 55 and 100 Mpsi, respectively.

In carbon fibers, the less-than-perfect preferred orientation and minor imperfections in the graphitic layers diminish the stiffness to less than 300 GPa, but improve the resistance to shear and increase the tensile strength to values as high as 4 GPa (47). The preferred orientation is introduced during fabrication of the carbon fibers by orientation of the polymer molecules and is intensified by heat treatment following the carbonization step. Heat treatment at graphitizing temperatures produces the high-modulus (Type HM) fibers. The high-tensile-strength (Type HT) fibers are produced by heat treatments near 1400°C.

Figure 19 summarizes the mechanical properties of the various types of carbon fibers (48). Future work on composites must consider not only the basic fiber types HT and HM but also the new generation of high tensile strength fibers (sometimes described as medium modulus fibers) and the ultra-high-modulus fibers based on mesophase pitch. There is some confusion in the nomenclature; to be precise, we note that all fibers, except the high-modulus mesophase-pitch-based fibers, are carbon and not graphite fibers. Unfortunately in American practice even the HT-type fibers are sometimes called graphite fibers. In Europe only the term carbon fibers is used. An earlier classification developed by British workers for PAN-based fibers and still in use distinguishes Type I and Type II fibers: Type I refers to high-modulus fiber and Type II to high-strength fiber. Another point of nomenclature concerns the Type S fibers; for most polymer-matrix composites, surface treatment of the fiber by electrochemical, wet chemical, or thermal oxidation is required to achieve good mechanical properties in the composite. These surface-treated fibers are often indicated by an S included in their commercial designation.

In recent work (41,49), we have shown that the surface properties of carbon fiber are changed by high-temperature heat treatment. The high tensile strength fibers have chemically more active surfaces, whereas the high-modulus fibers, and especially those based on mesophase pitch, have very smooth graphite-like non-reactive surfaces. Despite the high-temperature heat treatment applied in the manufacture of high-modulus PAN-based carbon fibers, they are structurally non-graphitic because they are built up from non-graphitizing polymer carbon. Structurally, the high-modulus mesophase-pitch-based fibers show graphitization effects.

As shown in Figure 20, the bulk strength of 1D-composites is superior if high-modulus fibers are used instead of the high-strength types. M40 and T300 are, respectively, high-modulus and high-strength PAN-based fibers manufactured by Toray Industries. This superior composite strength can be expressed in terms of the translation (sometimes termed utilization) of the fiber strength into the composite; the translation reaches values above 90% in the case of high-modulus fibers. With high-strength fibers, only 20% translation of fiber strength into the composites is achieved. Surface treatment of the fiber decreases the poor translation of fiber strength to even lower values, as shown in Figure 20. This latter effect can be reduced if the surface-treated fibers are cleaned by heat treatment at 1000°C in nitrogen atmosphere (20).

Recent work (50) with model 1D-composites made with phenolic resin as matrix precursor shows that heat treatment of commercial HT fibers (from Toho Beslon) improves the translation of fiber

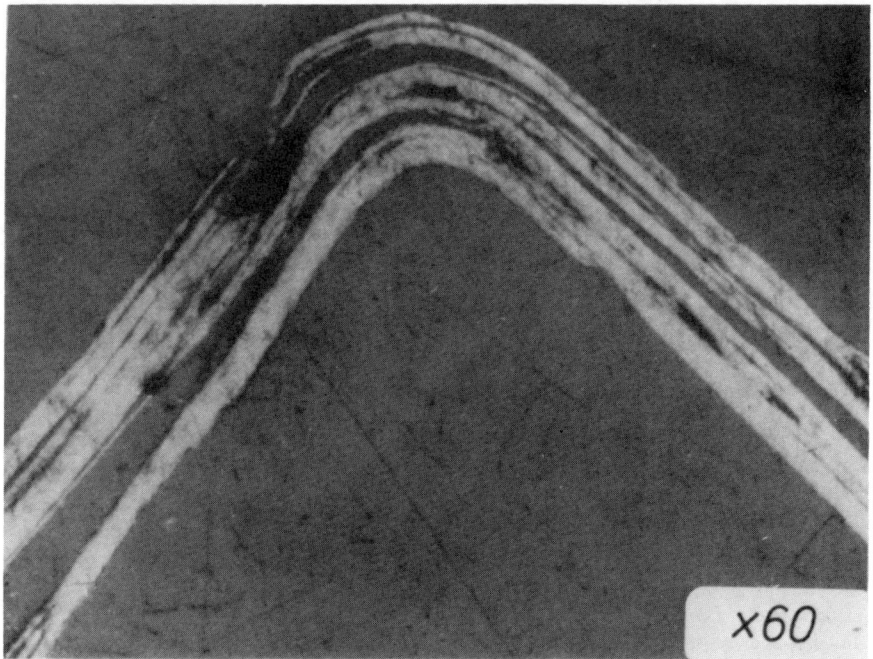

Figure 18. Shear deformation in a bent flake of natural graphite (46).

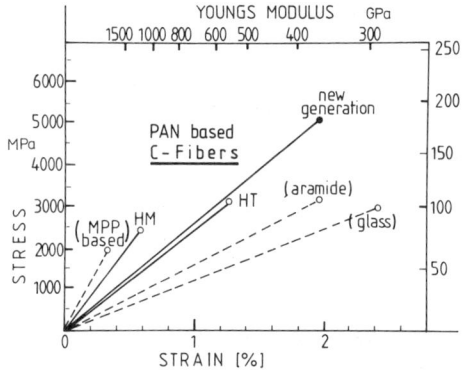

Figure 19. Tensile behavior of carbon fibers, based on polyacrylonitrile (PAN) or mesophase pitch (MPP), and compared with polyaramide and glass fibers (48).

strength into the composite. Maximum values of 75% were obtained with fibers heat treated up to 1800°C. Further increases in heat treatment temperature, to 2300°C, gave no further improvement in translation.

Figure 21 compares the translations of fiber strength into phenolic matrix unidirectional composites for high-modulus (Type I) and high-strength (Type II) PAN-based fibers in both surface-treated and untreated conditions. The left-hand plot refers to the rigidized skeleton after the first impregnation and carbonization to 1000°C; the right-hand plot refers to composites after four densification cycles. The weight loss on heat treatment of the fiber to 600°C provides a measure of the surface groups on the fiber. The superior translation of the high modulus Type I fibers is only achieved after densification cycling, not in the singly impregnated skeleton (20,51).

The cross-sectional shrinkage of unidirectional (UD) model composites in the first carbonization offers a good indication of the suitability of a selected combination of fiber and matrix precursor. Table I gives shrinkage observations for composites fabricated with polyphenyleneacetylene (HA 43) as matrix precursor and with three types of PAN-based fibers from the same manufacturer (Toray). The results indicate that low cross-sectional shrinkage because of poor adhesion between the matrix and the high-modulus untreated fiber is advantageous. This result can also be described as a tendency of the matrix to shrink away from the fibers during carbonization. The slit-shaped pores formed by this shrinkage are filled in subsequent impregnation. After recarbonization, two different layers of matrix can be observed, as shown by the scanning electron micrograph of Figure 22.

The amount of cross-sectional shrinkage decreases with increasing yield of the matrix precursor, as shown in Figure 23. In this figure, the low carbon yields of 50 to 60% are due to phenolic resin matrices, and the yields of 60 to 65% refer to matrices of coal-tar pitch with sulfur additions. Higher carbon yields (~ 80%) are obtained by pressure carbonization (100 bars) of coal-tar pitch, while the highest yields (~ 85%) refer to the HA 43 resin precursor. It should be noted that the effect of carbon yield cannot compensate for the dominant factors determining shrinkage behavior, namely the fiber type and surface treatment.

The importance of the fiber surface can also be seen in Figures 24 and 25 in the translations of fiber strength into the initial skeleton as well as the final composite after four densification cycles. The temperature attained in carbonization after the initial rigidization with phenolic resin has some influence, but the absolute level of strength translation is dominated by the type of fiber. The mesophase-pitch-based P55 fibers have proved to be superior, relative to the high-modulus PAN-based fibers M40 and HM3, because of the very smooth and nonreactive surface. The interlaminar shear strength (ILSS) was measured in a test similar to the three-point flexure test but with short span dimensions (span to thickness ratio = 5). The composites prepared with mesophase-pitch-based fibers also showed superior ILSS values, but the absolute values of bulk strength are lower because of the lower intrinsic fiber strength (34).

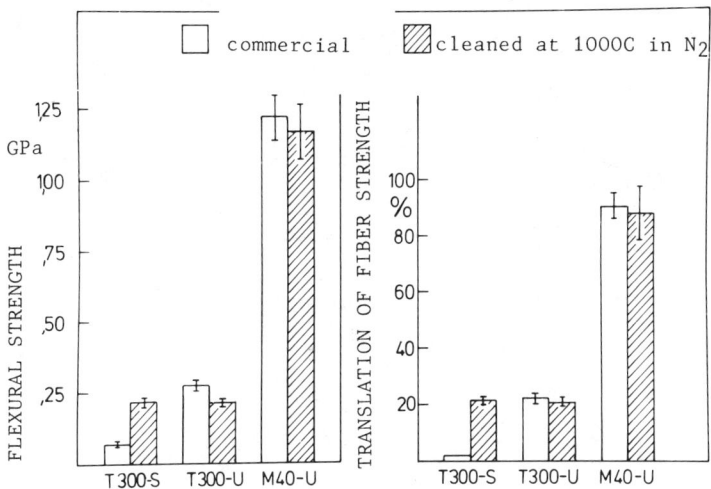

Figure 20. Flexural strength and translation of fiber strength in unidirectional composites (50 vol-% fiber) prepared by four densification cycles with HA 43 (polyphenyleneacetylene) matrix precursor (20). T300 and M40 refer to high-strength and high-modulus PAN-based fibers; U and S refer to untreated and surface-treated fibers, respectively.

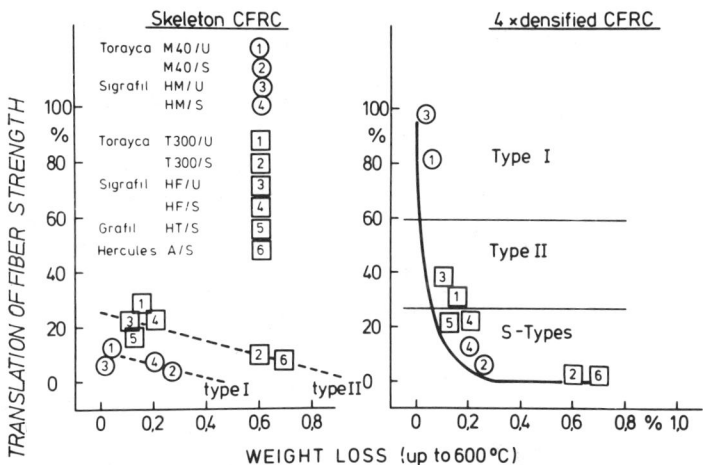

Figure 21. Translation (utilization) of fiber strength as a function of fiber type and surface treatment (51). Flexural tests on unidirectionally reinforced composites made with phenolic-based matrices carbonized to 1000°C.

Table I. Carbonization Shrinkage and Translation of Fiber Strength
In Unidirectional C/C Composites Fabricated with
Polyphenyleneacetylene (HA 43) as Matrix Precursor ([20])

Fiber Description		Shrinkage	Translation
Designation	Type	($\Delta Q/Q_o$, %)	n_F, %
T300 90A	Type II, S	11.3	10
T300 99A	Type II, U	9.0	23
M40 99A	Type I, U	1.5	90

$$n_F \text{ (translation)} = \frac{\sigma_{c/c}}{\sigma_F \cdot \emptyset_F}$$

Q_o = transverse cross section of composite after first impregnation, before carbonization

ΔQ = change in transverse cross section caused by first carbonization

$\sigma_{c/c}$ = flexural strength of composite after four densification cycles, measured by the three-point long-beam test.

σ_F = tensile strength of the carbon filaments, single filament tests on 30 mm gage length.

$\emptyset_F$ = carbon fiber content of composite = 55%.

S = surface treated fiber

U = untreated fiber

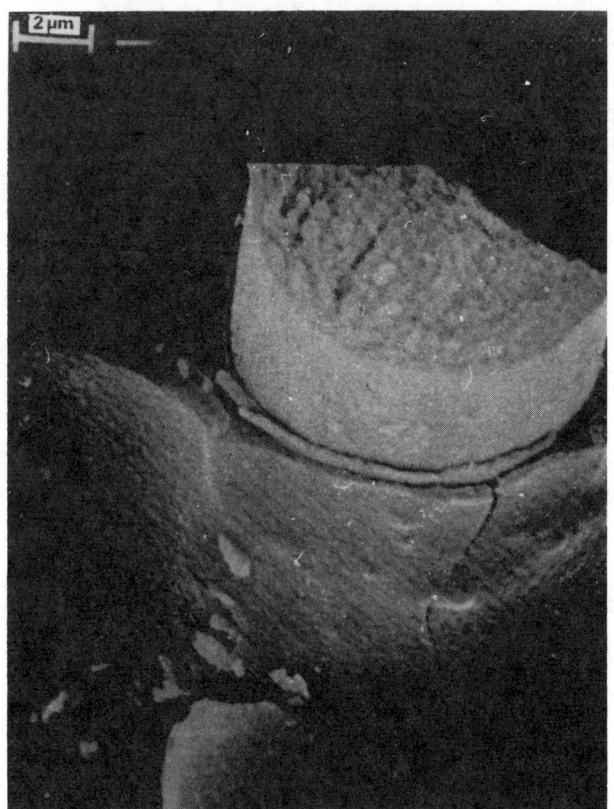

Figure 22. Two phases of carbon matrix caused by filling of slit-shaped pores during reimpregnation (20,48).

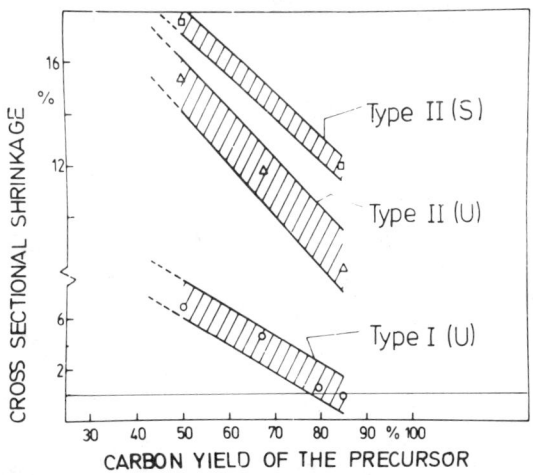

Figure 23. Cross-sectional shrinkage as a function of carbon yield in the first carbonization. Unidirectional composites fabricated with PAN-based fibers and various matrix precursors.

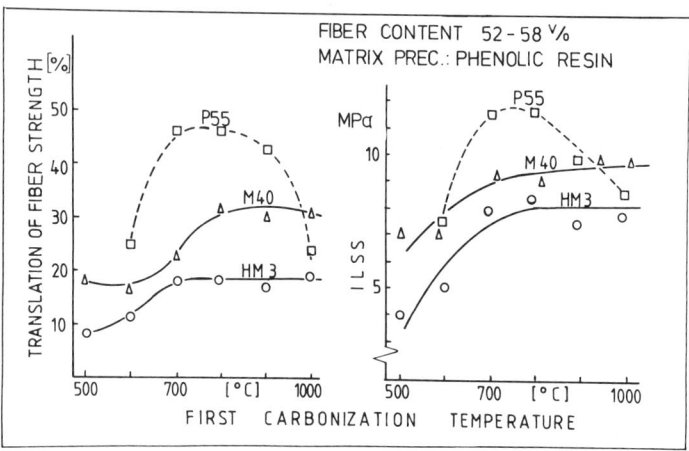

Figure 24. Translation of fiber strength and interlaminar shear stress (ILSS) of carbon/carbon skeletons rigidized with a phenolic resin, as a function of the carbonization temperature (34).

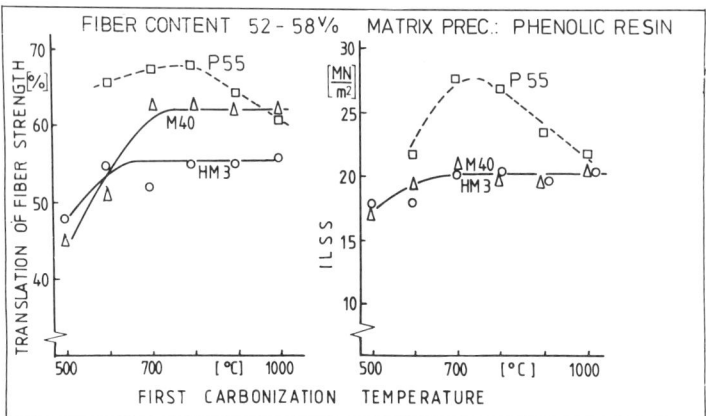

Figure 25. Translation of fiber strength and interlaminar shear stress (ILSS) of the carbon/carbon composites of Figure 24 after four impregnation and carbonization (1000°C) cycles with coal-tar pitch and 12.5% sulfur (34).

Process Parameters. As in conventional carbon and graphite technology, the heating rates and final heat treatment temperatures are most important process parameters. Figure 26 shows the effect of heating rates up to 300°C/h on the cross-sectional shrinkage of unidirectional composites with high-modulus fibers and coal-tar pitch as matrix precursor carbonized under pressure and/or with sulfur additions. As expected the shrinkage increases with increasing heating rate. Carbonization under pressure inhibits gas evolution from the composite and reduces the shrinkage (20).

Graphitizing heat treatments cause shrinkage of the carbonized matrix and strength decreases of the composites, as shown in Figure 27. The loss in final strength can be reduced by intermediate graphitizing heat treatment between the impregnation/carbonization cycles. This intermediate heat treatment opens slit-shaped pores and improves the ease of impregnation (52).

Most interesting is the influence of the graphitizing heat treatment on the bulk Young's modulus of the composite. Up to 180% translation of the fiber modulus is found if the contribution of the matrix is neglected for the calculation. It can be concluded that the matrix itself can contribute to the composite modulus in an amount of the same order as the fiber (32). The micrographs of Figure 28 show that the preferred orientation of the matrix carbon is qualitatively the same as that of the fibers. The X-ray diffraction results support the optical evaluations. These observations apply only if pitch or pitch/sulfur is used as matrix precursor; the preferred orientation is caused by the intermediate mesophase transformation (8). Process parameters such as low pyrolysis pressure (<100 bar), high heating rates (>100 °C/h), or final heat treatment at temperatures less than 1800°C can lead to lower modulus values of the composite.

As the mesophase has a much lower volatile content than the original binder pitch, and because the mesophase is still deformable by molding, a carbonization process was developed by Brückmann (19) in which mechanical pressure is applied after the green composite has been pyrolyzed at 450°C for 5 h. A tensile strength of about 800 MPa and a Young's modulus of about 150 GPa were achieved without subsequent impregnation processes. These values were measured after final baking to 950°C.

Final Properties of Carbon/Carbon Composites

Recent tests have revealed surprisingly good fatigue and creep resistance for carbon/carbon composites. Figure 29 presents some results of torsion and flexure tests in which the fatigue properties of carbon-fiber-reinforced carbon (CFRC) 3D composites are compared with those of carbon-fiber-reinforced polymer (CFRP) 3D composites (53).

The strength of carbon/carbon composites is retained at test temperatures as high as 2000°C, as expected if a protective atmosphere is provided. Our results (31,54) on unidirectional composites are compared in Figure 30 with data from the literature (55) on 3D-composites, pyrolytic graphite, and the fine-grained graphite ATJ.

Just as for other carbon and graphite materials, carbon/carbon composites oxidize in air at 500°C and above, as shown in

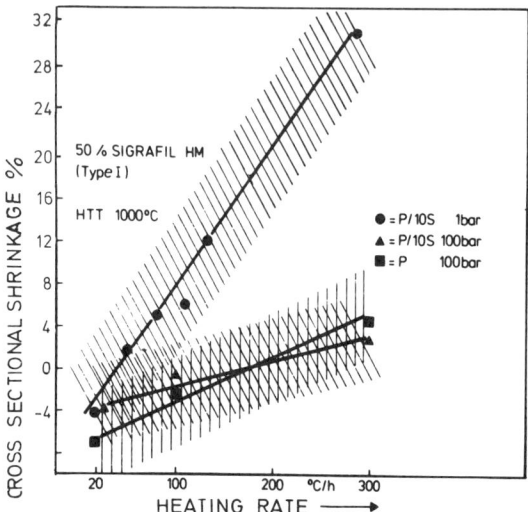

Figure 26. Cross-sectional shrinkage as a function of heating rate during carbonization of unidirectional composites fabricated with high-modulus PAN-based fibers and pitch matrix under three conditions of pressure and sulfur content (20).

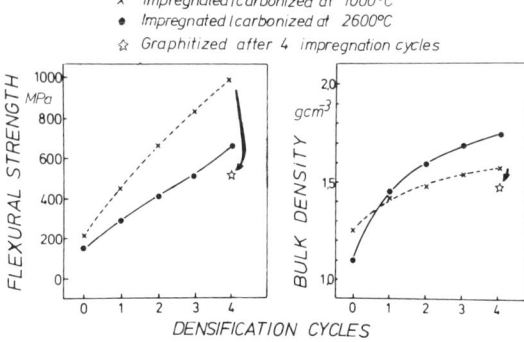

Figure 27. Flexural strength and bulk density of unidirectional composites subjected to graphitizing heat treatments either for each densification cycle or after the densification cycles were completed (20,52). The matrix was a coal-tar binder pitch with 10 wt.-% sulfur; the high-modulus PAN-based fiber was Sigrafil HM.

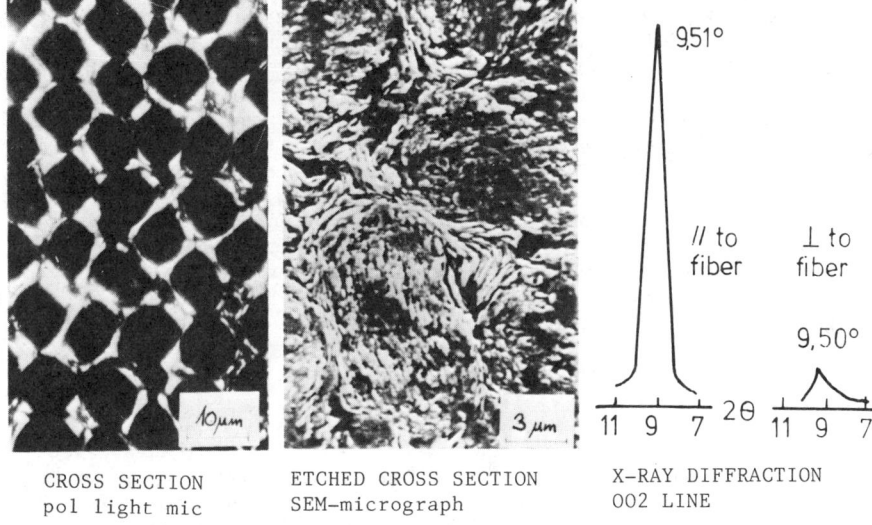

CROSS SECTION
pol light mic

ETCHED CROSS SECTION
SEM-micrograph

X-RAY DIFFRACTION
002 LINE

Figure 28. A pitch-based matrix that pyrolyzes via the mesophase transformation develops a strong preferred orientation parallel to the fiber axis (8).

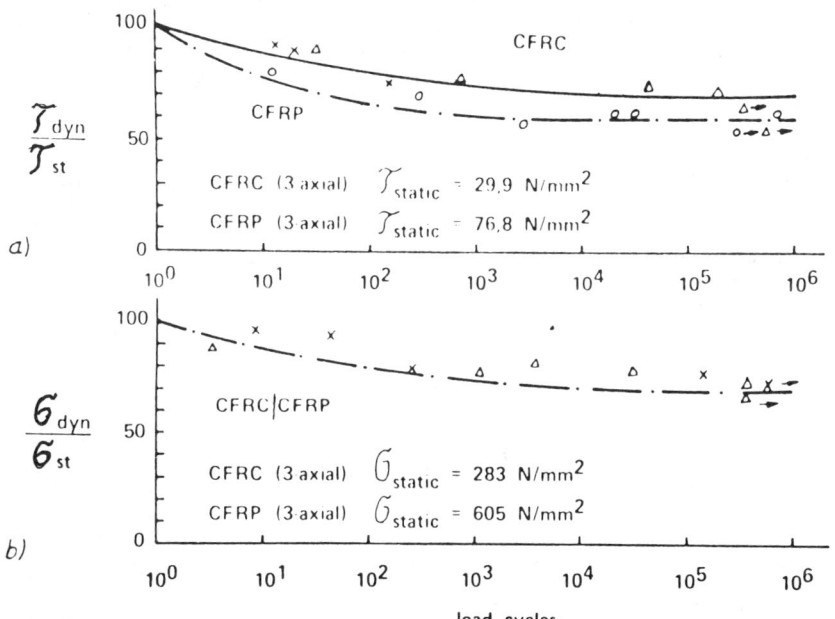

Figure 29. Comparison of the fatigue behavior of 3D carbon-matrix (CFRC) and epoxy-matrix (CFRP) composites (53). (a) Dynamic torsion strength (relative to static torsion). (b) Dynamic flexure strength (relative to static flexure).

Figure 31. However, the oxidation resistance can be improved by impregnation with inhibitors, especially zinc phosphate, which prevents weight loss to about 600°C (32,49). At higher temperatures up to 1400°C, SiC coatings are proving successful, if repeated but only short-time heating is considered (56), as in the case of a reentry space vehicle. An SiC-impregnation process using tetraethylorthosilicate (TEOS) was used to impregnate the "all-carbon" parts of the space shuttle COLUMBIA (9).

Since both components of carbon/carbon composites, i.e., the carbon fibers and the carbon matrix, are brittle materials, it is surprising that the two- or three-dimensionally reinforced composites display good fracture toughness (57,58), as shown in Figure 32. This fracture toughness is due to microcracks within the matrix and to fiber-matrix debonding, which acts to stop crack propagation through the interface and to increase the energy of fracture; see Figure 33 (59).

In the case of good adhesion between fiber and matrix, brittle fracture of the fibers is observed. A further disadvantageous behavior was found in composites used in ultracentrifuges running at high temperature (60). Internal fracturing destroyed the integrity of the matrix, and loss of matrix as carbon dust was observed. These observations led to systematic studies of carbon/carbon composites with hybrid matrices (34).

Carbon/Carbon Composites with Hybrid Matrices

Composites with hybrid carbon-polymer matrices are prepared by means of the conventional carbon-precursor impregnation process but with a final impregnation by a cross-linking resin which is not subsequently carbonized. The values of strength (see Figure 34), modulus, and interlaminar shear stress (ILSS) are comparable with those of carbon fiber reinforced polymers (CFRP). An additional advantage, relative to CFRP composites, is the higher thermal stability because of the greater fraction of carbon in the matrix. The bulk strength of the hybrid-matrix composite is not strongly influenced by the first carbonization temperature, especially if the low-cost mesophase-pitch-based fibers (P55) are used; however the bulk strength is lower than composites fabricated with high-modulus PAN-based fibers (M40 or HM3) because of the lower intrinsic strength of the mesophase-pitch-based fibers.

Some results of systematic studies of the process parameters (61,62) are given in Figure 35. The composition of the hybrid matrix is indicated by the bulk density of the carbon/carbon skeleton before final impregnation with the cross-linking resins. The flexural strength and interlaminar shear stress (ILSS) decrease with increasing carbon content in the matrix; the Young's modulus, however, increases. For comparison, data are included for the carbon skeleton before the final resin impregnation; these data show the effect of the impregnation/carbonization cycles with coal-tar pitch plus sulfur.

Relative to carbon/carbon composites, the hybrid composites have a reduced thermal stability in air due to the resin content of the matrix. The flammabilities of the hybrid composites, as measured by the limiting oxygen index (LOI) test (63), are given in Figure 36. Pure carbon/carbon composites are not inflammable in

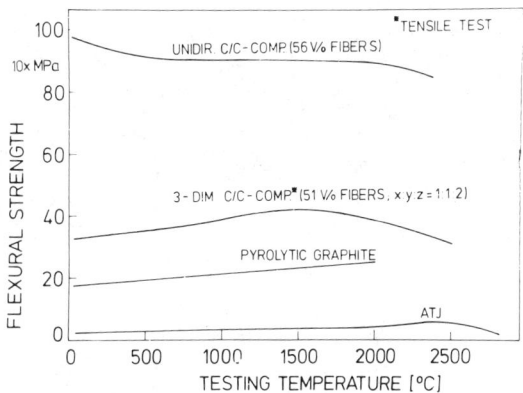

Figure 30. High-temperature strength of unidirectional carbon/carbon composites fabricated with rayon-based Thornel 75 fibers and coal-tar pitch as matrix precursor to a density of 1.51 g/cm^3 (31,54), in comparison with a 3D composite (55), pyrolytic graphite, and a commercial graphite.

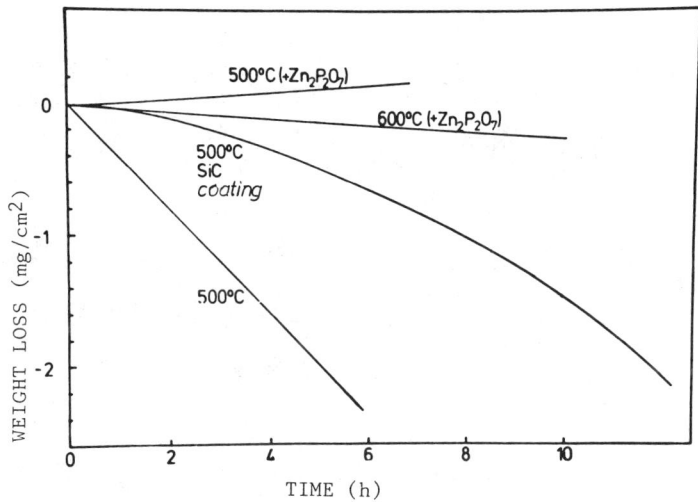

Figure 31. Weight-loss of unidirectional carbon/carbon composites by isothermal oxidation in air, as affected by $Zn_2P_2O_7$ inhibitor or by SiC coating (32,49). The composites were fabricated with 50 vol.-% high-modulus Modmor I fibers, coal-tar pitch as matrix precursor, four densification cycles, and final heat treatment to 1400°C.

Figure 32. Demonstration of the toughness of carbon/carbon composites (57).

Figure 33. Scanning electron micrographs of the fracture surface of a 3D carbon/carbon composite (59). Fiber pullout causes increased energy consumption in fracture.

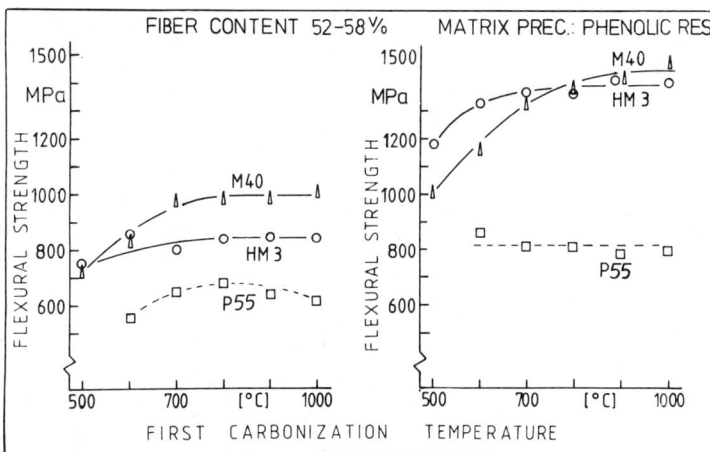

Figure 34. Comparison of the flexural strengths of unidirectional carbon/carbon composites (left-hand side) with those of hybrid composites in which the final impregnation is made with an epoxy resin (34). The composites were fabricated with high-modulus fibers rigidized with phenolic resin, and subjected to four densification cycles with coal-tar pitch plus sulfur.

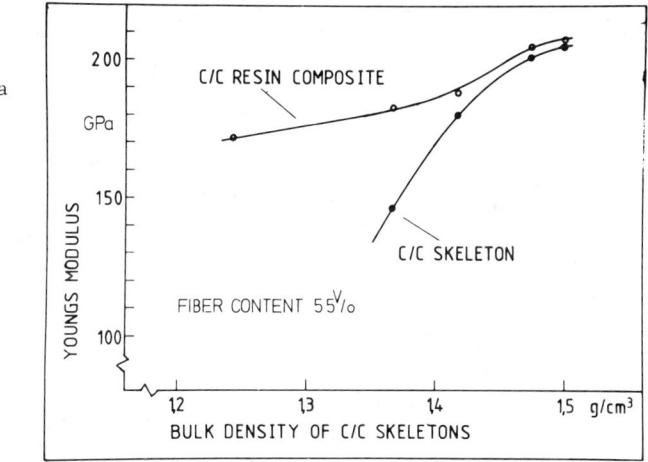

Figure 35. Mechanical properties of carbon/carbon epoxy-resin hybrid composites, compared with the properties of the composite skeletons before resin impregnation (61,62). The composite skeletons were prepared from Sigrafil HM 3 PAN-based fiber, rigidized with a phenolic resin, and densified by four cycles with coal-tar pitch plus sulfur; the carbonization temperature was 1000°C. (a) Young's modulus.

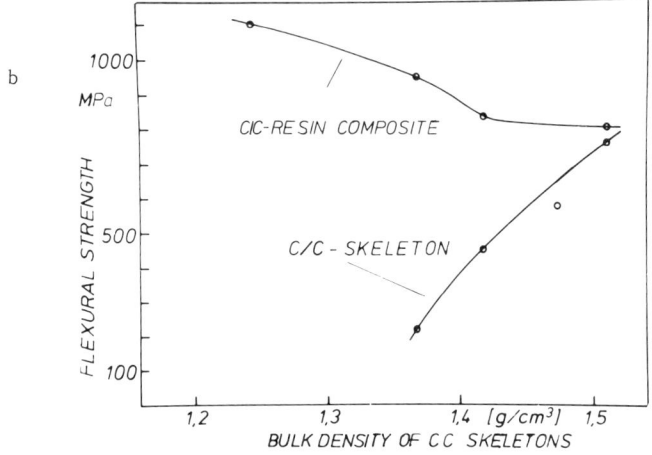

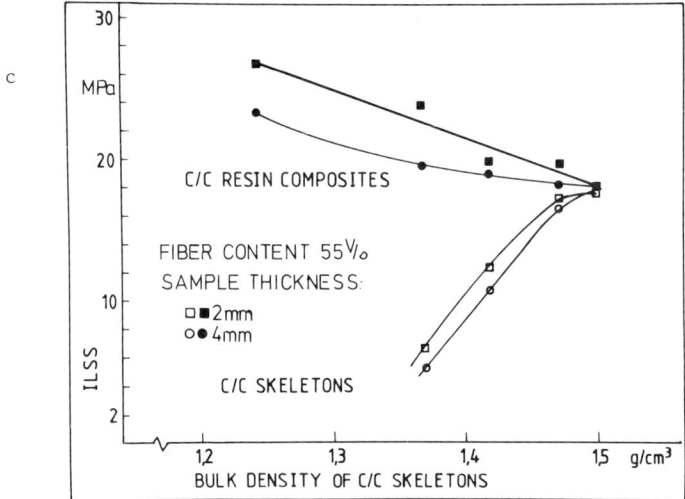

Figure 35. Mechanical properties of carbon-carbon epoxy-resin hybrid composites, compared with the properties of the composite skeletons before resin impregnation (61,62). The composite skeletons were prepared from Sigrafil HM 3 PAN based fiber, rigidized with a phenolic resin, and densified by four cycles with coal-tar pitch plus sulfur; the carbonization temperature was 1000 C. (b) Flexural strength. (c) Interlaminar shear stress, measured with two sample thicknesses.

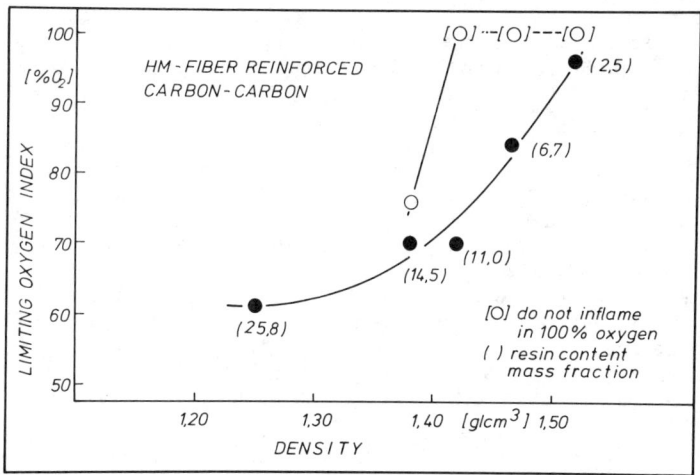

Figure 36. Limiting oxygen index (LOI) of flammability resistance for carbon/carbon composites and related hybrid composites with epoxy resin impregnant, as functions of bulk density and resin content (64).

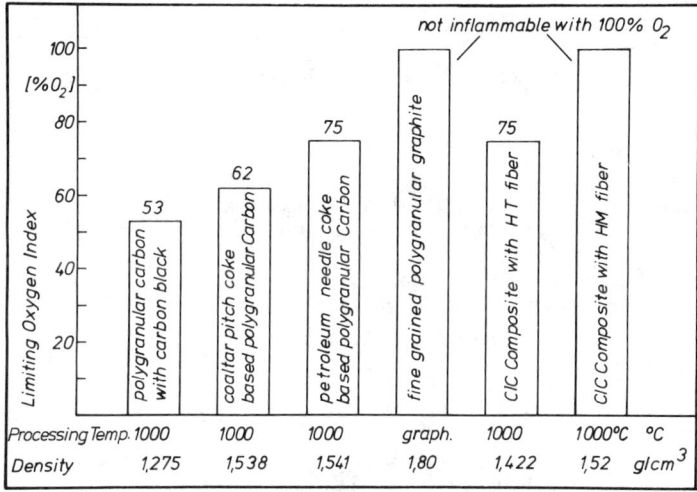

Figure 37. Limiting oxygen index (LOI) of flammability resistance of various carbon and graphitic materials in comparison with carbon/carbon composites (64).

100% oxygen (see Figure 37), while the LOI-values of the hybrid-matrix composites lie between 65 and 100% oxygen, depending on their resin content.

It is hoped that the concept of hybrid matrices will open new fields of application for carbon/carbon composites.

Literature Cited

1. Deviney, M. L.; O'Grady, T. M.; Eds. "Petroleum Derived Carbons"; ACS SYMPOSIUM SERIES No. 21, American Chemical Society: Washington, DC, 1976; 463 pages.
2. Gebhardt, J. J.; Stover, E. R.; Mueller, W.; Yodsnukis, J. In ref. 1, pp. 212-27.
3. Burns, R. L.; Cook, J. L. In ref. 1, pp. 139-54.
4. Chard, W.; Conaway, M.; Niesz, D. In ref. 1, pp. 155-71.
5. Diefendorf, R. J. In ref. 1, pp. 315-23.
6. Pratt, C. A., Jr., In ref. 1, pp. 203-11.
7. Bokros, J. C.; Akins, R. Y.; Shim, H. S. In ref. 1, pp. 237-65.
8. Fitzer, E.; Hüttner, W. J. Phys. D: Appl. Phys. 1981, 14, 347.
9. Curry, D. M.; Scott, H. C.; Webster, C. N. 24th National SAMPE Symposium 1979, p. 1524.
10. By courtesy of SEP (Societe Europeene de Propulsion), Puteaux, France, 1982).
11. Gkogkidis, A. Ph.D. Thesis, Univ. Karlsruhe, in prepn.
12. Hüttner, W.; Kehr, D. Ext. Abstr., 15th Conf. on Carbon 1983, p. 329.
13. Rose, P. G.; Gruber, U.; Gerr, W. Preprints, Carbon '80 1980, p. 639.
14. Kochendörfer, R.; Hüttner, U. Deutsches Bundes Patent (DBP) 25 07 512, 1977.
15. By courtesy of Dr. H. Böder, Sigri GmbH, Meitingen, Fed. Rep. of Germany.
16. Swinnerton, B. R. G. Wear 1982, 78, 81.
17. Schäper, S. Audi NSU Union Auslegeschrift 29 16 706, 1979.
18. Brückmann, H.; Hüttner, W.; Maurer, H. J.; Hüttinger, K. J. Sprechsaal 1981, 2, 71.
19. Bruckmann, H. Ph.D. Thesis, Univ. Karlsruhe, 1979.
20. Hüttner, W. Ph.D. Thesis, Univ. Karlsruhe, 1980.
21. Fitzer, E. Proc. Int. Symp. on Factors in Densification and Sintering of Oxide and Non-Oxide Ceramics, Hakone (Tokyo Inst. of Tech.) 1978; pp. 40-76.
22. Fitzer, E.; Kehr, D.; Sahebkar, M. Chem. Ing. Techn. 1973, 21, 1244.
23. Damköhler, G. In "Der Chemie-Ingenieur"; Eucken, A.; Jakob, J.; Akademische Verlagsgesellschaft: 1937, pp. 359-485.
24. Thiele, E. W. Ind. Eng. Chem. 1939, 31, 916.
25. Fitzer, E.; Fritz, W.; Gadow, R. Proc. Int. Symp. on Ceramic Components for Engines, Somiya, S., Ed.; KTK Publishing: Tokyo, 1983; p. 508.
26. Kotlensky, W. V. Chem. and Phys. of Carbon 1973, 9, 173.
27. Fitzer, E.; Heym, M. Ext. Abstr., 13th Conf. on Carbon 1977, p. 128.

28. Fitzer, E.; Geigl, K. H.; Heym, M. Ext. Abstr., 13th Conf. on Carbon 1977, p. 168.
29. Fitzer, E.; Hüttner, W. Sprechsaal 1980, 113, 452 and 919.
30. Schäfer, W. Ph.D. Thesis, Univ. Karlsruhe, 1975.
31. Terwiesch, B. Ph.D. Thesis, Univ. Karlsruhe, 1972.
32. Heym, M. Ph.D. Thesis, Univ. Karlsruhe, 1976.
33. Karlisch, K. Ph.D. Thesis, Univ. Karlsruhe, 1977.
34. Cziollek, J. Ph.D. Thesis, Univ. Karlsruhe, 1983.
35. Fitzer, E.; Terwiesch, B. Carbon 1972, 10, 383.
36. Fitzer, E.; Geigl, K. H.; Hüttner, W. Proc. 5th London Carbon and Graphite Conf. 1978, p. 493.
37. Fitzer, E.; Müller K.; Schäfer, W. Chem. and Phys. of Carbon 1971, 7, 237.
38. Tillmanns, H. Ph.D. Thesis, Univ. Karlsruhe, 1975.
39. Fitzer, E.; Hüttinger, K. J.; Tillmans, H. Preprints, Carbon '72 1972, p. 14.
40. Fitzer, E.; Hüttinger, K. J.; Tillmans, H. Ext Abstr., 11th Conf. on Carbon 1973, p. 112.
41. Fitzer, E.; Frohs, W. Ext. Abstr. 17th Conf. on Carbon 1985.
42. Frohs, W. Ph.D. Thesis, Univ. Karlsruhe, in preparation.
43. Oberlin, A. Carbon 1984, 22, 521.
44. Blakslee, O. L.; Proctor, D. G.; Seldin, G. B.; Spence, G. B.; Weng, T. J. Appl. Phys. 1970, 41, 3373.
45. Bacon, D. J.; Nicholson, A. J. J. Phys. C 1977, 10, 2295.
46. Cooper, C. Chem. and Ind. 1982, 18, 678.
47. Johnson, D. J. Chem. and Ind. 1982, 18, 692.
48. Fitzer, E. J. Chim. Phys. 1984, 18, 678.
49. Fitzer, E.; Gkogkidis, A.; Heine, M. High Temp.-High Press. 1984, 16, 363.
50. Fitzer, E.; Gkogkidis, A.; Rajamohan, A. Ext. Abstr. 17th Conf. on Carbon 1985, p. 314.
51. Fitzer, E.; Geigl, K. H.; Huttner, W. Carbon 1980, 18, 265.
52. Fitzer, E.; Hüttner, W.; Manocha, L. M. Carbon 1980, 18, 291.
53. Hüttner, W.; Keuscher, K.; Hüttinger, K. J.; Nietert, M. In "Ceramics in Surgery"; Vincenzini, P., Ed.; Elsevier: Amsterdam, 1982, p. 225.
54. Fitzer, E.; Heym, M. High Temp.-High Press. 1978, 10, 29.
55. Levine, A.; Roetling, J. A.; Stover, E. R.; Gebhardt, J. J. Ext. Abstr., 12th Conf. on Carbon 1975.
56. Fitzer, E.; Gadow, R. Proc. Int. Symp. on Ceramic Components for Engines, Somiya, S., Ed.; KTK Publishing: Tokyo 1983; p. 505.
57. Hüttner, W. Schunk Kohlenstofftechnik, Giessen, Fed. Rep. of Germany.
58. Fitzer, E. First Indian Carbon Conference, New Delhi, 1982.
59. By courtesy of Dr. Grenier, Aerospatiale, Les Mureaux, France.
60. Private communication, Dr. Leis, Messerschmitt-Bölkow-Blohm (MBB), Munich, Fed. Rep. of Germany.
61. Fitzer, E.; Cziollek, J.; Jacobsen, G., Ext. Abstr., 15th Conf. on Carbon 1981, p. 280.
62. Cziollek, J.; Fitzer, E.; Jacobsen, G. Ext. Abstr., Carbon '82, London, 1982, p. 365.
63. ASTM Test D 2863-77.
64. Brauns, I.; Fitzer, E.; Gkogkidis, A. Ext. Abstr., Carbone '84, Bordeaux, 1984, p. 160.

RECEIVED January 21, 1986

25

Carbon–Carbon Composites
Matrix Microstructure and Its Possible Influence on Physical Properties

R. A. Meyer and S. R. Gyetvay

Materials Sciences Laboratory, The Aerospace Corporation, P.O. Box 92957, Los Angeles, CA 90009

> Processing conditions required to attain desirable composite properties can be defined more easily if the factors controlling composite microstructure are understood. Such factors include type of precursors employed and the composite's processing history. The microstructure of the matrix may contribute to the performance of the fibers and influence the properties of the composite. Reviewed are experiments to determine matrix microstructural features, how microstructural variations are achieved, and ways in which thermal expansion and fracture behavior relate to microstructure.

Carbon–carbon (C–C) composites with a variety of unique properties can be fabricated by altering the combinations of the type and distribution of filaments and the bonding matrix used. Many engineering applications can be satisfied with a composite material whose density is just 70% that of aluminum and 25% that of steel, but for which the specific strength and stiffness values are four or five times those of steel. To attain the desired properties for such applications requires an understanding of the interrelationship of the fibers and the matrix that holds them.

The newest advances in C–C composite materials have resulted from fiber improvements, such as modulus increases of two to three times and diameter decreases of more than 50% in the last ten years. Weaving techniques have also been improved, so that increased fiber content and reproducible distribution of filaments and yarns (down to 0.75 mm center-to-center spacing between yarns in 3D composites) are possible. However, because the interrelationship of the filaments and yarns with the bonding matrix is not well understood, the improved fiber properties have not yet been fully translated into a corresponding magnitude of improvement in C–C composites.

This paper is intended to stimulate more intense study of the matrix's influence on the behavior of C–C composites. Experimental observations are presented to exemplify how the matrix and its microstructure may contribute to variations in some physical properties; reasons for those variations are also proposed.

Background

A structure woven of carbon fiber (a preform) can be densified in a number of ways that will result in different types of matrix microstructures. The two usual densification methods are chemical vapor deposition (CVD) and impregnation with pitch or resin in the liquid state. The choice of densifier can determine whether the matrix will be graphitizable. Pitches and CVD are considered graphitizable, whereas most resins do not graphitize except under great stress. Composites can also be made using cloth that has been "prepregged" with a resin, then laid up into the desired configuration. After a series of forming steps, the resin-cloth layup is carbonized, possibly graphitized, and then further densified by CVD or liquid-impregnation methods. The microstructure of the matrix depends to a large extent on the composite configuration and processing conditions. Identifying those factors that influence the matrix properties should aid in defining the conditions necessary to attain desired composite properties.

Studies reviewed by Fitzer and Gkogkidis (1) on composite samples carbonized to 1000°C indicate that carbonizing the matrix contributes to high strength and low strain, changes the fracture behavior, and enhances the effective fiber strength. In contrast, lower composite strengths, higher strains to failure, and lower utilization of fiber strengths have been observed after the matrix is transformed to a more graphitic state by additional high-temperature heat treatment (HT). In comparative flexure tests performed on small 2D flexure samples before and after HT (2), we have observed decreases of the load to failure of the order of 15%; however, the strain-to-failure values were increased by 100%.

Evangelides has noted that the more graphitic matrix (derived from coal tar pitch) in composites prefers to fracture along the weaker direction parallel to the a-b planes (3). This fracturing effect is reasonable because the mechanical properties of crystalline graphite (4) are highly anisotropic; those measured in the direction of the basal layer planes ("a-direction") vary by orders of magnitude from the properties measured across the layers ("c-direction"). The a/c anisotropy ratios of tensile strength and modulus are about 20 and 150, respectively (5). The basal layer planes are held together in the c-direction by weak Van der Waals forces. Consequently, the layer planes are readily displaced by shear forces parallel to the a-direction and easily separated or cracked by tensile forces in the c-direction, and the preferred direction for fracture lies in the a-b plane. Therefore, the fracture path in the matrix depends on the layer orientation on the microstructural scale.

If the properties of carbon-carbon composites are in part determined by the properties of the matrix, then altering the matrix and observing the resulting changes in the composite's thermal-mechanical properties would be important for understanding the behavior of C-C composites. A first step toward this goal is to characterize the matrix in the yarn bundles and the preform pockets with respect to its unique fracture and thermal expansion characteristics.

The matrix pockets within a C-C composite are filled with variously oriented grains whose fracture characteristics are comparable to those of bulk graphite. Previous studies have shown that the size, orientation, and distribution of crystallites in the grains within a volume of bulk graphite determine its strain and strength properties, because the mode of fracture is controlled by the microcracking of the individual grains and the redirection of cracks by the basal plane orientation in adjacent grains and by pore distributions. Verification of this concept was made for bulk graphite by a micromechanical model that uses microstructural information to predict the stress-strain characteristics of graphite (type ATJ-S) (6). The same approach should be useful to approximate the performance of the matrix in pockets in C-C composites, provided details of the microstructure, especially the crystallinity of the matrix, have been determined.

Several procedures are available to define the degree of crystallinity of carbon and the orientation of layer planes in the matrix. X-ray diffraction can be used to measure the basal layer spacing, but fiber cannot be distinguished from matrix in the composites. The orientation and alignment of basal planes in the matrix of polished samples of C-C composites can be mapped by polarized microscopy (7,8). Similar indications can be obtained by cathodically etching the polished samples with inert gas, which reveals the lamellar structure by sputtering off carbon atoms in lower-density regions. The lamellae are packets of stacks of nearly parallel basal layer planes. Lamellar structure is relevant to the direction of crack propagation (9).

The highly anisotropic characteristics of the graphitic microstructure permit HT, in addition to changing the fracture behavior, to alter the thermal expansion. In single-crystal form, the expansivity is approximately $25 \times 10^{-6} °C^{-1}$ in the c-direction and $0.5 \times 10^{-6} °C^{-1}$ in the a-direction, giving an anisotropy factor, c/a, of about 50 to 60. Therefore, with such a high degree of anisotropy, the expansion of the matrix within a pocket between yarns or between fibers would be expected to depend on the distribution and size of grains, their orientations, and the void structure surrounding the grains.

Matrix Microstructures

The matrix microstructure can be intentionally varied by manipulating such factors as fiber spacing, heat-treatment rate and temperature, matrix precursor, and processing pressure. From microscopic examinations, a number of inferences about the importance of the distance of spacing between fibers have been drawn. Optical microscopy has revealed, for example, that with pitch impregnation processing at ambient pressure, the matrix aligns in sheaths around filaments, even below graphitization temperatures. The alignment of the lamellae as revealed by ion etching indicates that large molecules of pitch preferentially orient themselves parallel to any surface (Figure 1a). It has been reported that as the spacing between fibers increases, the alignment or "sheath" effect around the fibers diminishes and the matrix can contain disclinations and various grain orientations.

The preferred orientation of the matrix parallel to the fiber bundles appears to be affected by processing pressure during pyrolysis (400-500°C), becoming more random as impregnation pressures are increased. Even transverse lamellar orientations can result (Figure 1b). Because pitch mesophase coalescence has been inhibited by the pressure, grains of matrix may be transversely oriented to the filaments.

Heat treatment above 2000°C is known to increase the degree of graphitization, e.g., as evidenced by the formation of the lamellae. At around 2000 to 2500°C, the frequency of lamellae increases as the degree of graphitization increases. As Figure 2 illustrates for isothermal HT, when the time is increased, the boundaries (lines) between lamellae become more distinct and numerous. Furthermore, the lamellae tend to straighten out, then buckle, causing the microstructure to appear segmented or polygonized (10). X-ray determinations of the spacing between basal layer planes show decreasing values as the degree of graphitization is increased. From high-magnification observations with the scanning electron microscope (SEM), these boundaries do not appear to be cracks (e.g., Mrozowski cracks). It is thought, although not conclusively proven, that such boundaries are the result of interstitial carbon atoms and defects diffusing to these areas as the basal layers between the boundaries become more ordered and graphitic. Long-range lamellar alignment in matrix grains permits the formation of larger crystallites than for the more randomly oriented or disclinated shorter lamellar structures.

In contrast, ungraphitized matrix or fiber that has only been carbonized (HT < 1900°C) does not exhibit significant lamellar microstructure. X-ray studies confirm the different degrees of crystallographic order of the carbonized and graphitized matrix. The degree to which graphitic perfection can be attained is limited by the morphology of the matrix during pyrolysis, because the alignment of the grains is set in the matrix when it is in the liquid crystal state. But this alignment is not transformed into crystalline structure unless sufficient time is provided at temperatures >2000°C.

The precursor used for infiltration can also influence the microstructure that is generated. Generally, resins do not graphitize unless subjected to stress and high-temperature heat treatment (11). Pitch precursors, which are readily graphitized, appear to have their grain size reduced by an increase in quantity of quinoline-insoluble particles (7), by heat treating the pitch prior to infiltration, or by the use of high processing pressures during the impregnation cycle.

Variations of Physical Properties by Altering the Microstructure

The exact role of matrix microstructure in determining the physical properties of C-C composites is difficult to determine precisely because the microstructural variations and fiber-matrix interactions are highly complex. Some insight into the importance of

384   PETROLEUM-DERIVED CARBONS

a

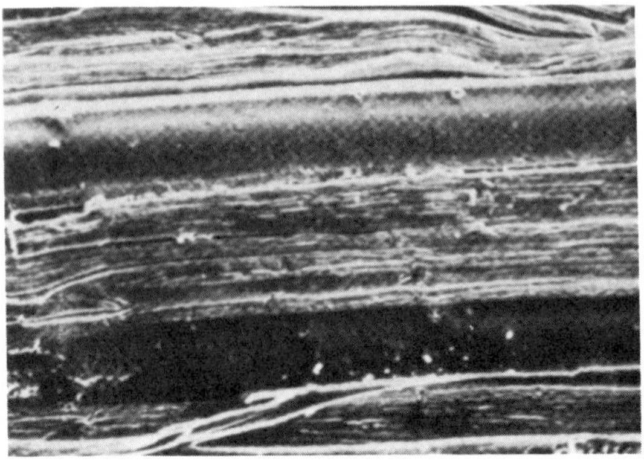

FILAMENT

ALIGNED
MATRIX

b

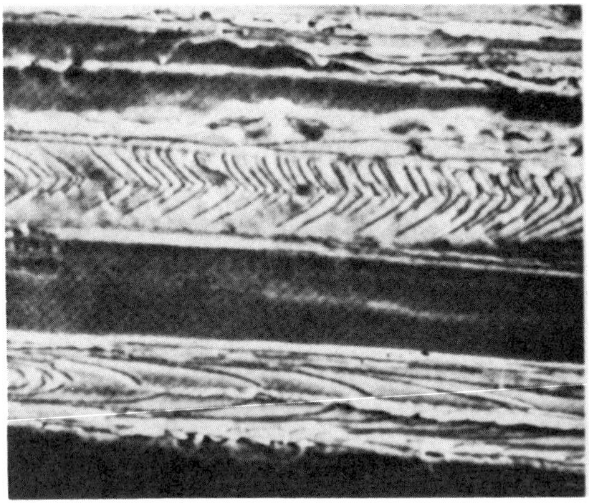

FILAMENT

TRANSVERSELY
ALIGNED
MATRIX

Figure 1. Scanning electron micrographs of (a) sheath of aligned matrix produced by low-pressure impregnation, and (b) transversely oriented matrix produced by high-pressure impregnation.

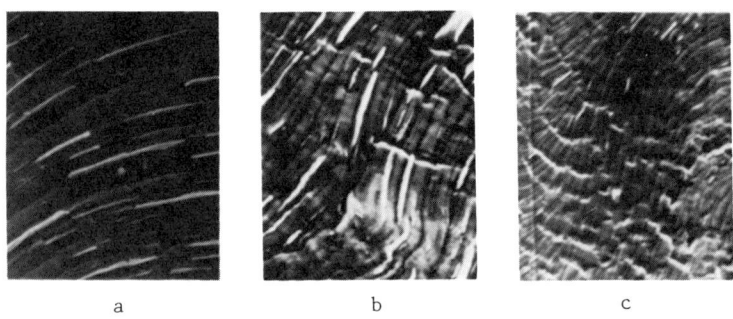

Figure 2. Scanning electron micrographs of bulk matrix graphitized at 2400°C for (a) <0.25 hr, (b) 2 hr, and (c) 4 hr.

microstructure is presented by the following property measurements and microstructural analyses.

Thermal Expansivity (ΔL/L). Experiments were performed to determine what influence microstructure has on the thermal expansion characteristics of three-dimensional (3D) C-C composites prepared with two types of coal tar pitch as matrix precursors.

Three 1-in.-cube billets were fabricated from a preform, prepared from mesophase pitch fiber (Union Carbide P55) by Fiber Materials, Inc. It was of 3D cartesian construction with a 1 to 1 to 2 ratio of yarns in the X, Y, and Z orientations (1-1-2 fiber composite), respectively. The processed billets were a high-density and a low-density Allied 15V coal tar pitch composite, and a high-density Koppers A (KPA) experimental pitch composite. The KPA impregnant had a quinoline-insoluble content of < 1%, whereas that of 15V was approximately 8%. Each billet was rigidized with pitch at atmospheric pressure and then sequenced through high-pressure densification cycles followed by graphitization until the desired density was attained. The resulting densities and thermal expansivity data are summarized in Table I.

Table I. Some Properties of 3D Cartesian C-C Billets Fabricated with Coal Tar Pitch Impregnants

	Billet Identification		
	15V High ρ	KPA High ρ	15V Low ρ
Impregnant (coal tar pitch)	Allied 15V	Koppers A	Allied 15V
Bulk Density (g/cm^3)			
By weight/volume	1.96	1.96	1.87
By porosimetry	1.96	1.97	1.89
Apparent Density (g/cm^3)			
Porosimetry	2.16	2.17	2.11
Open Pore Volume (%)	8.9	9.8	10.1
Thermal Expansion, ΔL/L × 10^{-3} (20 to 1400°C)			
X-direction	2.3	3.0	4.1
Fiber content (%)	14	11	11
Z-direction	1.8	1.7	3.9
Fiber content (%)	26	27	26

The ΔL/L values of Table I are shown in Figure 3. Apparently, variations of thermal expansion can be obtained by using different precursors, increasing the density of the composite, or varying the number of yarns in each direction. Since the three billets were cut from the same preform, the fiber type and distribution are similar prior to processing. Therefore, variations in properties are primarily attributed to differences within the matrix microstructure and distribution.

The influence of matrix is shown in Figure 3a. The KPA value of $\Delta L/L$ in the X-direction divided by that in the Z-direction is 1.76, which is highly anisotropic compared with the 1.28 for 15V. In addition, microstructural examination shows the KPA sample to have an average grain size five times larger than that of the 15V samples (Figure 4). Note also that the fiber fraction (F) ratio F_X/F_Z for the KPA sample is 0.41, in contrast to 0.56 for the 15V sample. Since both samples were rigidized and processed in the same manner, the matrix in the yarns for both KPA and 15V would be expected to expand in a similar manner, because at ambient pressures only the yarns become filled with a similarly oriented matrix during the rigidization cycle. This argument implies that the bulk matrix in the pockets, which is produced during the high-pressure cycles, must cause the $\Delta L/L$ differences here.

The reasons for such differences are complex and require further clarification. Presented here are some influences that should be considered. Large as opposed to small grains in the pockets could result in variations of $\Delta L/L$ between the X- and Z-directions, especially for small samples. Assuming each pocket between the yarns is filled with graphitic structures, there will be a higher probability of isotropic expansion within the pockets if the graphitic grain sizes are small. As a result, there will be little anisotropy for the X- and Z-directions, as shown in Figure 3a for 15V. Conversely, for larger and fewer grains in the pockets as for KPA, a high degree of anisotropy occurs between X- and Z-directions. This pitch pocket effect is modified, since only 41% of the fibers are in the X-direction; so the restraint is less than for the Z-direction.

The lower value of thermal expansion in Figure 3b for the higher-density 15V composite is, at first glance, surprising. Careful examination of the microstructure revealed that the lower-density composite had narrower gaps between yarns and matrix than did the higher-density one. As a result, the narrower gap is closed at a lower temperature and expansion initiates earlier in the lower-density composite and continues at an equal or greater rate than that of the high-density sample. The same fiber distribution restraints apply, as mentioned above. In both cases, the anisotropy values are small and equivalent.

The causes for the differences of these gap widths is not clearly understood. One possible explanation for the larger gap in the higher-density sample is the effects of creep at the high processing temperatures. During graphitization, the grains will expand and fill any available void space. As the preform is densified, it will contain fewer voids and thermal expansion of the grains will exert force against its fiber or yarn surroundings. Above 2200°C, meaningful creep of the restraining yarns has been measured by Feldman (12) and accommodation of the matrix occurs. On cooldown, the yarns do not shrink as much as the matrix and a gap is left between yarn and matrix; consequently, the higher-density composites must creep more for the same graphitization temperature, thereby producing larger gaps at room temperature.

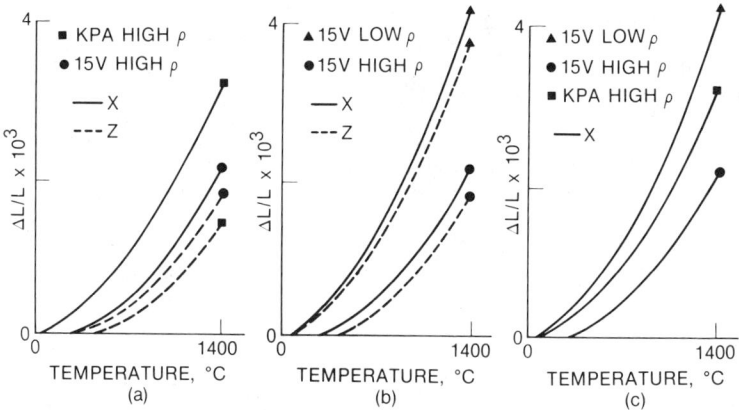

Figure 3. Relationship of thermal expansion to (a) direction in composite, for two impregnants; (b) direction in composite, for two densification levels; and (c) X-direction, for three composites.

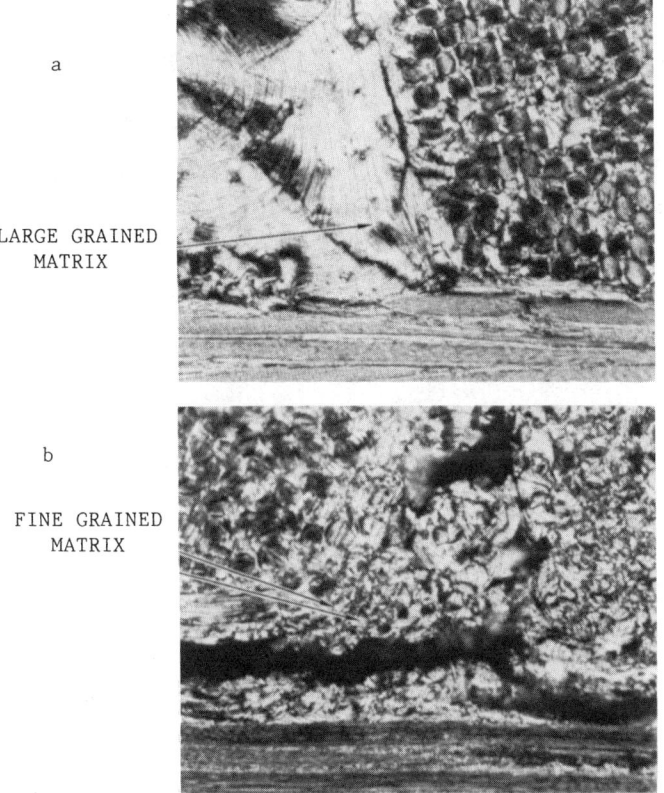

Figure 4. Optical micrographs showing (a) large-grained matrix, KPA specimen, and (b) fine-grained matrix, 15V specimen.

The role of the different finely dispersed porosity, gaps produced by graphitization, or pyrolysis steps needs to be further evaluated.

The influence of the density of the matrix is depicted in Figure 3c for the X orientation alone. These curves demonstrate the variability that may possibly be achieved by altering the matrix between yarns. More work is needed to further clarify the mechanisms for such variations. For example, the fibers may vary slightly because the high-density sample received four graphitization cycles rather than three. Other factors could contribute, such as variations of yarn distributions between samples.

Torsional Shear and Bonding Characteristics of the Matrix. The carbon or graphite matrices serve to maintain the geometric distribution and rigidity of the yarns, as well as providing a means of transferring forces between yarns. The microstructural characteristics of the matrix, such as anisotropy, grain size, and void structure, are expected to influence how efficiently the yarns are utilized. How effectively the matrix is bonded to the yarns is also important, as the following example demonstrates.

Torsional shear strength values (Figure 5) were found to be vastly different for two 3D billets (2-2-3 yarn distribution) processed identically with CVD rigidization followed by high-pressure impregnation cycles using liquid pitch. The maximum shear strength of composite A was approximately 800 psi, whereas that of composite B was 2200 psi at failure. The stress-strain curves, as measured at Southern Research Institute, are very different, as the figure evidences.

Microstructural examination of billets A and B by SEM revealed the major difference between them was in the CVD layer deposited on the fibers before additional densification steps were performed. In billet A the layer appeared to be porous (see Figure 6), whereas billet B exhibited a uniformly dense CVD coating. The remaining portions of the matrix are derived from the liquid pitch impregnations, and these looked similar for A and B. It is possible that the lower shear strength of billet A can be attributed to the more porous and probably weaker CVD layer.

Further evidence for the lack of bonding was obtained by examination of tensile samples. Numerous pulled-out fibers were evident for billet A, and these were covered with pitch matrix with a minimum of discernible CVD. In contrast, the billet B tensile sample did not display fiber pullout and the fracture tended to occur nearer the CVD-matrix interface than at the CVD-fiber interface, which indicates good bonding and the ability of a fiber to share its load with the surrounding fibers. Consequently, fiber pullout was reduced and fracture was more uniform between the fibers, resulting in higher strength.

Mechanical Properties and Fracture Behavior. Until recently, little was known about the sequence of fracture behavior that occurs in the 2D composites that are used to fabricate large C-C structures, such as the exit cones of rocket engines. In such composites, the yarns in the cloth are held together by a carbon matrix derived from a phenolic resin filled with carbon particles.

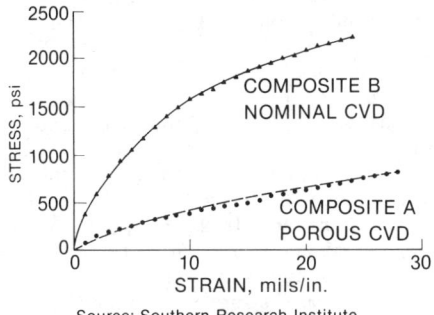

Source: Southern Research Institute

Figure 5. Torsional shear properties of two 3D composites subjected to similar processing.

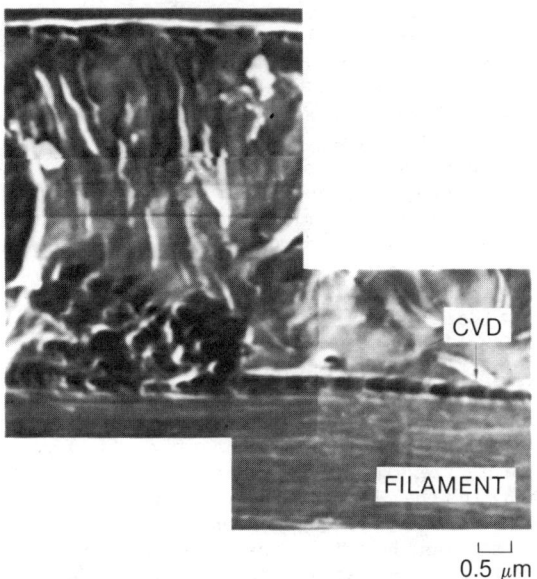

Figure 6. Composite A interfilament matrix with porous CVD coating the filament.

After pyrolysis, the carbon body is further densified by liquid pitch, resin impregnation, or CVD. Studies were undertaken to understand the role of the matrix, pores, and cracks, and fiber-matrix interactions in this material as it is stressed. The intent was to determine whether processing differences could alter the fracture behavior and possibly improve the physical characteristics.

Rather than the catastrophic failure normally observed in brittle materials, carbon-carbon composites frequently show, after reaching their maximum strength values, a gradual reduction in strength as the strain is increased. Local regions appear to fracture incrementally in a sequence that lowers the strength with increasing strain until total failure occurs (see Figure 7).

Small samples were stressed in three-point flexure while being viewed with scanning electron or optical microscopes. The fracture behavior was recorded by video and correlated with acoustic emission measurements. The samples were oriented in the flexure-test fixture with the plane of the cloth plies perpendicular to the force direction; therefore, the tensile and compressive stresses are primarily in the plane of the cloth plies where the fibers resist deformation.

Crack propagation through the matrix and between plies and fibers appeared to be a primary mode of failure. Sequential photographs taken on the side surface of the flexure sample with the tensile forces parallel to the cloth warp direction showed cracking of the matrix progressing initially between the plies from the tensile side of the specimen toward the compressive side. The cracked and weakened regions appeared to delaminate and form gaps between the yarn bundles (see Figure 8). Ultimately, the samples failed on the compressive side in a tensile manner. Thus, it appears that the weakest component of this particular 2D material in the warp direction is the matrix that bonds the fibers, yarns, and plies, because the failed samples contained a minimum of broken fibers. When similar tests were conducted with the tensile forces parallel to the cloth fill direction, fracture also initiated in the matrix, but more fibers were observed to have fractured than for the warp direction (2). This is due to the lower fiber volume fraction in the fill direction in these 2D composites.

Samples of the 2D materials were further heat-treated for 1.5 hr at 2480°C to determine whether the crystalline state of the matrix could be altered, thereby changing the fracture behavior under three-point flexure load. For typical specimens, changes were manifested in the load-deflection curves, such as those in Figures 7 and 9. Apparently, further HT reduced the ultimate strengths and modulus but increased the strain to failure. In addition, there was a difference in the amplitude of the acoustic noise emitted as the load was applied (see Figure 9) between the as-received sample and the HT sample. This difference indicates that more microcracking occurs at lower stress levels after HT. SEM observations revealed more microcracking of the HT matrix. Furthermore, there appeared to be preferential microcracking along lamellae, which is characteristic of a more crystalline graphite matrix.

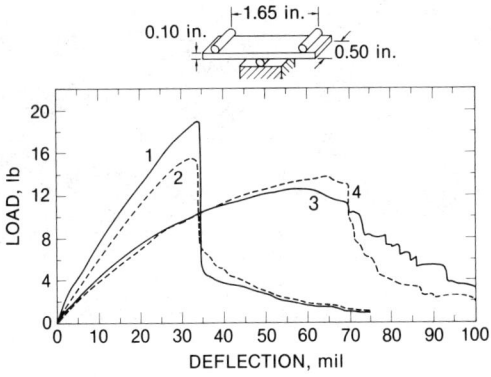

Figure 7. Three-point-bending tests of 2D C-C composite specimens: 1 and 2 in as-received condition; 3 and 4 after heat treatment at 2480°C for 1 hr.

Figure 8. Montage of photomicrographs showing crack propagation through the matrix and between plies and fibers in a flexure test of a 2D composite.

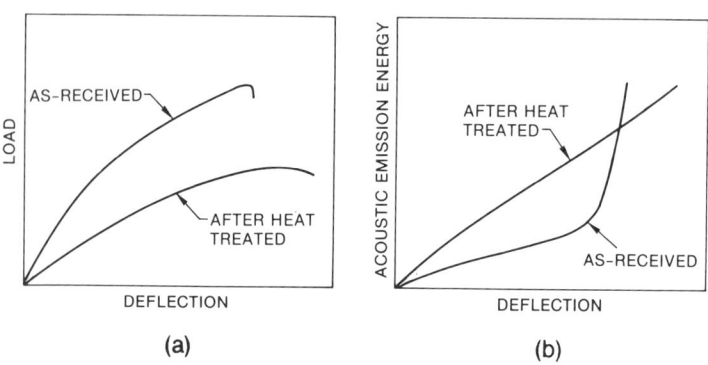

Figure 9. (a) Load-deflection and (b) acoustic emission response in three-point-bending test of some 2D C-C composites from a rocket engine exit cone.

The apparent increase of microcracking after heat treating at graphitization temperatures suggests that the matrix is more crystalline even though the resin precursor is normally considered ungraphitizable. Hishiyama et al. have shown, however, that nongraphitizing carbons can develop structure in regions that are stressed under HT at graphitization temperatures (11). Further, we have found heat-treatment effects with the same resin precursor in measurements of the dynamic modulus of unidirectional samples (13). An increase of modulus for HT > 2100°C to 2500°C occurs, which is attributed to crystallization of the matrix into more graphitic structures whose orientation is parallel to the unidirectional fibers and thereby increases their effective stiffness or dynamic modulus.

If the matrix can become more graphitic, as the above examples indicate, more shear planes become available; hence more microcracking can occur, resulting in greater strain at lower stress levels. Thus the apparent or effective modulus of the 2D composite materials is reduced, and more energy is required to cause failure--an outcome indicated by the difference of area under the as-received and heat-treated load-deflection curves in Figures 7 and 9.

Summation

Examples have been presented that suggest the fibers and matrix have an interrelationship that appears to have influenced the thermal expansion and mechanical properties of several types of C-C composites. The effects can be attributed to the densification methods, precursors, and subsequent heat treatments, which alter the crystallinity of the resulting matrix and its contained voidage, including pores, gaps, and cracks. The extent and distribution of crystallinity of the matrix appear to be important because of the very anisotropic properties of the graphite crystal. Therefore, if the degree and distribution of crystallinity can be controlled, the thermal expansivity and mechanical characteristics of the matrix can be changed. Such changes will, in turn, interact with the yarns by stressing or initiating and propagating matrix cracking around them.

The mechanical response of composites, as shown in these exploratory studies, indicates dependence on the ease with which fracture can occur between fibers, yarns, and plies. Poorly crystallized matrices result in composites that are strong and stiff but with little yield so that failure occurs catastrophically. In contrast, more crystalline matrices seem to be not quite as strong and to have a lower effective modulus, but their increased strain capability ensures that failure is not catastrophic; the composite's strength decays gradually as further strain is applied. Thus, the energy required for total failure is increased, and the composite with more crystalline matrix is more tolerant of defects or stress risers.

Finally, we note that voids, cracks, and gaps can absorb thermal expansion and provide sites at which fracture may be initiated or means by which the propagation of cracks may be interrupted.

Processing conditions have been implicated as contributors to some variations in physical properties. Possible explanations have been proposed in order to stimulate further research. More detailed information must be obtained before the cause-effect relationships can be more fully understood.

Acknowledgment

We express appreciation to our colleagues, M. Buechler, J. S. Evangelides, and J. L. White, for their helpful contributions to and discussions about this paper.

Literature Cited

1. Fitzer, E.; Gkogkidis, A. This volume.
2. Buechler, M.; Meyer, R. A. Unpublished results.
3. Evangelides, J. S. "Microstructure and Fracture of Carbon-Carbon Composites"; TOR-0075(5626)-2; The Aerospace Corp.: El Segundo, California, 1974.
4. Reynolds, W. N. In "Physical Properties of Graphite"; Elsevier Publishing Co.: New York, New York, 1968; p. 1.
5. Ibid.; p. 33.
6. Meyer, R. A.; Zimmer, J. E. "Final Report, Failure Criteria in Graphite"; ATR-74(7425)-3; The Aerospace Corp.: El Segundo, California, 1974.
7. Dubois, J.; Agace, C.; White, J. L. Metallography 1970, 3, 337.
8. Zimmer, J. E.; White, J. L. Carbon 1983, 21, 323.
9. Meyer, R. A.; Zimmer, J. E.; Almon, M. C. "Micromechanics of Failure in Carbon System"; ATR-74(7408)-2; The Aerospace Corp.: El Segundo, California, 1974; p. 21.
10. Meyer, R. A.; Gyetvay, S. R.; Chase, A. B. Ext. Abstr., 16th Conf. Carbon, 1983, p. 505.
11. Hishiyama, Y.; Inagaki, M.; Kimura, S.; Yamada, S. Carbon 1974, 12, 249.
12. Feldman, L. A. Ext. Abstr., 16th Conf. Carbon, 1983, p. 499.
13. Feldman, L. A.; Gyetvay, S. R.; Meyer, R. A. Ext. Abstr., 17th Conf. Carbon, 1985, p. 385.

RECEIVED December 12, 1985

Author Index

Bonnamy, S., 85
Bourrat, X., 85
Brandt, H.H., 172,193
Buechler, M., 62
Carney, P.R., 200
DeBiase, Robert, 155
Deutschle, F.J., 200
Dickakian, G., 118,126,134
Ehrburger, Pierre, 310
Elliott, John D., 155
Fitzer, Erich, 346
Gkogkidis, Antonios, 346
Goldberger, W.M., 200
Grimes, Gary W., 144
Gyetvay, S.R., 380
Hartnett, Thomas E., 155
Hettinger, W.P., Jr., 99
Jones, Samuel S., 234
Kakuta, M., 179
Kapner, R.S., 193
Korai, Yozo, 29
Lahaye, Jacques, 310
Latham, Carolyn S., 1
Lewis, James E., 269
Markel, R.F., 200

Marsh, Harry, 1
Matson, John A., 144
Matsubara, Kenji, 251
Meyer, R.A., 380
Miyazu, Takashi, 251
Mochida, Isao, 29
Monthioux, M., 85
Morotomi, Hidetoshi, 251
Noguchi, K., 179
O'Grady, T.M., 302
Oberlin, A., 85
Otani, Sugio, 323
Oya, Asao, 323
Rand, B., 45
Rouzaud, J.N., 85
Sato, J., 179
Tanaka, H., 179
Tibbetts, G.G., 335
Tillmanns, Harald, 215
Waller, James H., 144
Wennerberg, A.N., 302
Wesley, D.P., 99
White, J.L., 62
Wombles, R.H., 99
Yamasaki, H., 179

Subject Index

A

Acenaphthylene, carbonization scheme, 31f
Acetylene, carbon black formation role, 280
Activation energy
  carbonization, 228
  desulfurization reaction of coke, 197
Alicyclic structure, pitch chemistry importance, 326
Aluminum, dependence on anode carbon, 234
Aluminum industry, coke shortage, 193
Anode binder coke, source, 239
Anode binder pitch
  characterization, 238-239
  composition, 238-239
  properties, 238-239
  QI fraction, 238

Anode carbon
  aluminum dependence, 234
  aluminum industry use, 234-249
  baked apparent density, 237
  characteristics, 236-238
  compacted composite baking, 246
  consumption
    airburn of prebake anode tops, 236
    carbon dioxide reduction, 236
    electrolytic reactions, 235
    mechanical carbon loss, 236
  electrolytic cell described, 235
  fabrication factors, 244-246
  filler aggregate
    selection, 245
    sizing, 244-245
  graphite crystallites, 237
  paste compaction, 245
  thermal conductivity, 237
  thermal stress resistance, 237
  Young's modulus, 237

API gravity
  saturates relationship in slurry
    oil, 110
  slurry oil, 108
Ariane, rocket nozzle, 349f
Aromatic molecules, coke formation
    chemistry, 30
Aromatic pitch
  production from CCB distillate, 119
  thermal characteristics, 119
  See also Polar aromatics
Ash, coal properties, 256
Asphaltene
  catalytic cracking, 105
  examples, 96f
  LMO sizes, 94f

B

Baked apparent density, anode carbon
    property, 237
Basal carbon atoms, carbon
    supports, 311
Basic structural unit (BSU)
  concept, 88
  distorted column model, 91f
Blowdown system, process flow
    diagram, 159f

C

Calcination, process, 243
Calciner
  heat balance, 191
  photograph of model, 190f
Calcining
  coke technology
    new, 179-191
    old, 172-178
  comparison of new and traditional
    methods, 180-183
  definition, 172
  electric, 176
  kiln fuel, 189-191
  optimum process system, 189
  outline of new technology, 180
  profile, 181f
  process
    proprietary rotary hearth, 169
    rotary kiln, 167-169
  quality, 172
  schematic diagram of two-stage, 191f
  theory of new method, 184-189
  variables, 172

Carbon
  active
    adsorptive properties, 306t
    applications tested, 309t
    catalyst-carbon support
      interaction, 319
    drug overdose control for, 308
    high surface area, 302-309
    physical properties, 306t
    pilot plant, 304f
    pilot plant process
      development, 303
    potassium hydroxide used, 303-305
    process description for
      HSAA, 303-305
    structure, 305
    uses, 308
  catalyst dispersion, preparation, 311
  fiber
    development status, 65f
    graphitic fiber difference, 362
    growth, 336-341
    growth apparatus, 337f
    growth in tubes by natural gas
      pyrolysis, 335-345
    history, 335-336
    Japan, 328-332
    light intensity in the center of
      growth tube, 338f
    micrographs of growth tubes, 338f
    micrographs of PAN based, 361f
    morphology and properties, 341-345
    orientation of graphitic
      layers, 359-362
    pitch-based discovery, 323
    produced from natural gas, 337f
    scanning electron micrograph, 343f
    surface properties changed, 362
    tensile behavior, 363f
    time required to grow, 340f
  graphitic
    analysis of Desulco ash, 205t
    chemical analysis of Desulco, 205t
    density, 206
    Desulco process, 201
    Desulco process features, 204
    electrical resistivity, 207-208t
    factors influencing structure, 210
    granular, 200-213
    industrial use, 213
    nitrogen and hydrogen content, 206t
    particle size, 206
    pore size distribution, 211f
    production, 204
    properties, 205
    sintering mechanism, 318
    structure, 208,212f
    surface composition, 311
  industries relationship, 218
  raisers, carbon equivalency increase
      vs. solution time, 213f

INDEX

Carbon--Continued
  supports
    catalysts influenced, 311
    metallic derivates
      dispersion, 310-321
    surface properties, sintering
      behavior, 319-321
    See also Anode carbon
    See also Graphitizing carbons
    See also General Performance Carbon
      Fibers
Carbon black
  catalyst-carbon support
    interactions, 319
  classification and properties in
    rubber, 290
  DBP absorption, 291f
  extrusion rate, 297f
  feedstock conversion, 278
  formation
    mechanism, 269-299
    studied, 273
  gas sampling from reactor, 278
  general particle sizes, 289
  grades listed, 273
  hardness, 293f,294f
  materials distribution in
    experiment, 278-279
  mechanism, 288
  NR modulus, 292f
  particle size
    controlled, 286
    distribution, 291f
    formation, 286
  products, 288-289
  properties measured, 271
  quality, 289
  radioactive tracer used in
    experiment, 279
  reaction from oil, 286
  reactor
    cross sectional profiles at port
      locations, 281f-288f
    plotting longitudinal
      profiles, 275f
    sampling ports and quench
      locations, 274f
    sampling probe and collection
      bomb, 277f
  SBR modulus, 292f
  structure controlled, 286
  systems, 271
  tendency to smoke, 270
  tensile strength, 292f,293f
  viscosity, 297f
Carbon-carbon billets, properties, 385t
Carbon-carbon composites
  aircraft brake performance, 349f
  aircraft disc brake, 348f
  applications, 347-353

Carbon-carbon composites--Continued
  bulk strength, 372
  carbon fiber influence, 359-369
  comparison of production
    process, 356f
  fabrication
    by liquid impregnation, 346-378
    processes, 353-357
  fatigue behavior, 371f
  final properties, 369
  flammabilities, 372-378
  flexural strength
    discussed, 365f
    unidirectional, 375f
  fracture surface, 374f
  high-temperature strength of
    unidirectional, 373f
  hip joint, 352f
  hybrid matrices, 372-378
  limiting oxygen index of flammability
    resistance, 377f
  liquid impregnation process, 353-357
  load deflection and acoustic
    emission, 392f
  matrix microstructure
    influence, 380-394
  matrix pockets, 382
  mechanical properties
    compared, 354f
    discussed, 375-376f
    fracture behavior, 388-390
  medicinal use, 347-353
  micrograph of polished surface, 4f
  newest advances, 380
  polymers used as matrix
    precursors, 360f
  primary and secondary carbons
    described, 357
  process parameters, 369
  properties determined by
    matrix, 381-382
  strength, 369
  strength decrease, 390
  tool for superplastic forging of
    titanium, 351f
  toughness, 374f
  turbine rotor, 350f
  weight loss of unidirectional, 373f
  Young's modulus influenced, 369
Carbon dioxide
  carbon black formation role, 280
  reduction, anode carbon, 236
Carbon monoxide, carbon black formation
  role, 280
Carbonization
  acenaphthylene scheme, 31f
  catalytic, 34
  coke characterization, 215-232
  compatibility, 32
  composition of volatile products, 219

Carbonization—Continued
  control to mesophase, 32-36
  distillation, 220f
  electron microscopic
    observations, 85-97
  first stages, 88-95
  industries producing carbon
    products, 216
  kinetics, 225-228
  mechanisms, 30-32
  model, 221-222f
  model compounds used, 18f
  naphthalene reaction scheme, 32
  parent pitch composition
    effect, 14-15
  partial hydrogenation effect, 34
  physical and chemical processes in
    pyrolysis, 217
  physical or chemical properties, 216
  pitch quality prediction, 15
  polyvinyl chloride, 324
  pressure effect, 14
  range of carbons, 215-216
  reaction schemes, 30
  reactions, 220f
  shrinkage
    function of carbon yield, 367f
    function of heating rate, 370f
  structural stages, 90f
  testing methods, 216
  theoretical amount of mesophase, 219
  viscosity influence on mesophase, 32
  weight losses, 358f
Carbonization systems, compounds
  used, 13f
Catalyst
  carbon as a support, 310
  carbon-support interaction, 319
  efficiency of a metal, 310
  particle formation by surface
    fragmentation, 341
Catalytic cracker bottoms (CCB)
  definition, 118
  distillate
    anisotropic content, 120
    aromatic pitch production, 131t
    aromatic pitch transformation, 119
    aromatic ring distribution
      effect, 127,129t
    boiling characteristics
      discussed, 126-127
      effects, 129t
    characteristics, 120t,128t
    chemical structure, 119,129t
    contents, 118-119
    distillation characteristics, 128t
    DSC data, 122t
    DTG thermogram, 124f
    fractionated, 126
    micrograph, 124f,125f
    reaction temperature effect, 130t

Catalytic cracker bottoms (CCB)—
    Continued
    reaction time effect,
      discussion, 119,130t
    pitch composition, 121t
    residue comparison, 131t
    temperature effect on pitch
      composition, 121t
    thermogravimetric analysis, 122t
  residue
    comparison to the distillate, 127
    thermal treatment, 127-128
Catalytic cracking
  asphaltenes, 105
  definition, 100
  diaromatics, 105
  feedstock quality, 102
  major products, 102
  monoaromatics, 105
  polar aromatics, 105
  products, 102
  saturates, 105
Coal, maximum fluidity, 254
Coal-tar pitch
  characteristics, 139t
  petroleum pitch comparison, 238
Coals
  blending design, 254
  coking, evaluation, 254-256
Cocarbonization susceptibility,
  definition, 34
Coke
  addition test, 259,260t
  anisotropic, properties, 240
  anode use, 167
  blending limit, 266
  calcined
    composition, 239
    dimensional changes, 188f
    hydrogen content, 228
    pore size distribution, 187f
    properties, 179,182-183
    property range, 241t
    scanning electron
      micrographs, 185-186f
    size distribution change, 204t
    technology, 172-178
    X-ray parameters, 189t
  characterization, 228-230
  coalesced mesophase relationship, 49
  comparison to physical properties of
    Desulco, 207t
  composition, 239
  conversion from pitch, 4-5
  crushing strength, 230
  crystallite sizes and interlayer
    spacing, 209t
  desulfurized, physical
    properties, 197t
  disclinations, 68
  economic evaluation, 266

INDEX

Coke—Continued
  electrode use, 167
  factor for saturates in RCC process, 112t
  filler
    formation, 239-240
    properties, 239-241
    source, 239
  formation, 218f
  fracture surface from Chinese Shuang Ya coal, 9f
  fuel uses, 166-167
  gasification uses, 167
  high-surface-area active carbon formed, 303
  impurities, 193-194
  isotropic properties, 240
  Japanese steel companies, use, 253t
  markets
    characteristics, 149-151
    consuming industries, 149
    consuming world regions, 149
    demand, 149-151
    hierarchy, 152
    history, 144
    pricing, 151-154
    supply, 145-149
  metallurgical coke making, 251-267
  operation yields, 164-165
  optical texture
    dependence, 11
    examined, 16
    growth characteristics, 16
    increase, 12
    influenced by QI material, 22
    nomenclature, 8t
    quantified, 5-8
  overview, 144-154
  physical properties of thermally treated, 210t
  pit-type handling system, 161f
  polarized light optical microscope used, 5
  prediction
    equation in RCC unit, 110-114
    quantity and quality forecasting, 145
    RCC process, 99-117
  pricing
    conclusions, 154
    determination, 151
    European fuel market, 149
    fuel grade, 152
    variables, 154
  production
    feedstock composition relationship, 108
    forecast, 146
    methodology of forecasting, 147f
    United States, 150f

Coke
  production—Continued
    usage in 1980, 202t
    yields and product properties, 165t
  properties, 258t
  puffing behavior, 230
  quality
    calcination, 243-244
    influence of coking variables, 240-243
  refined product demand, 147f
  runs to crude stills, 147f
  South African coal comparison, 154
  strength, 251
  structure relationship to mesophase, 5-11
  sulfur content pricing, 152
  types, 160-163
  United States production, 144,193
  uses, 166-167
  utility power consumption, 149
  viscosity variation with temperature of pyrolysis, 13f
  X-ray parameters, 184
  yield factors, 108
  yield predictions for reduced crude, 111t
  See also Coking
  See also Delayed coke
  See also Fluid coke
  See also Green coke
  See also Needle coke
  See also Shot coke
  See also Sponge coke
Coke making, raw material consumption in Japan, 252f
Coker blowdown system, description, 158
Coker heater, design, 169
Coking
  buildup influences, 145
  section, flow diagram, 156
Compact composite baking, anode carbon, 246
Compounded rubber
  Firestone rebound, 296f
  Firestone running temperature, 294f
  Goodyear rebound, 296f
  particle size correlation with treadwear, 295
  properties, 269-299
  See also Rubber
Crude oil
  future origin, 146
  gravity
    future predictions, 146
    sulfur content, 146
  quality, 148f
  refining process, 242
  United States supply, 148f

D

Dark-field imaging, defined, 86
Delayed coker
  coke drum pressure, 164
  direct rail car loading, 161f
  gravity flow dewatering bin
    system, 162f
  heater outlet temperature, 164
  Pad-type dewatering system, 161f
  process
    flow diagram, 157f
    variables, 164
  recycle ratio effect on
    production, 164
  revamps and retrofits, 170
  rotary hearth furnace, 168f
  rotary kiln calciner, 168f
  slurry dewatering bin system, 162f
Delayed coking
  coker blowdown system, 158
  coker heater design, 169
  definition and uses, 155
  description, 155-156
  design features, 169-170
  drum size, 169
  energy efficiency, 169-170
  formation, 240
  fractional section, 156-158
  process
    mechanism, 156
    update, 155-170
  unit description, 156
Densification, methods for carbon
  fiber, 381
Density, real vs. bulk, 197
Desulco
  granular graphitic carbon, 200-213
  structure, 208
Desulfurization
  activation energy for coke
    reaction, 197
  annealing process for coke, 197
  calcined coke, 203f
  chemistry of new method, 196-197
  electrothermal heating, 202
  green coke, 196
  kinetics for thermal method of
    coke, 202
  thermal-chemical process,
    new, 195-197
  petroleum coke, 193-198
  process alternatives for coke, 202
  technology, 194-195
  thermal process, new, 195-197
  thermal process, old, 196t,203f
Diaromatics
  catalytic cracking, 105
  molecular parameters from slurry
    oil, 109t

Diaromatics--Continued
  slurry oil and coke factor in RCC
    process equation, 113t
Disclinations
  geometry for mesophase, 74
  pinch-off reaction, 73f
  reaction of mesophase, 73f,75f
Dispersion
  iron-phthalocyanine on carbon
    supports, 312-314
  metals on carbon supports, 310-311
Dominant Partner Effect
  coke quality upgraded, 19
  definition, 34

E

Edge carbon atoms
  carbon supports, 311
  dispersion and sintering of
    iron-phthalocyanine, 319
Effective fluidity
  petroleum coke, 264t
  residual oil, 261t,269t
Effective reflectance, residual
  oil, 261t,264t
Electrochemistry, applications for
  carbon supports, 310
Electrodes
  calcined coke used, 179
  properties of graphite, 183t
Electron diffraction pattern, ray path
  in converging lens, 87f
Electron microscopic observations,
  graphitization, 95-97
Electron spin resonance (ESR), free
  radicals in mesophase-pitch
  system, 12
Extrusion rate, carbon black, 297f

F

Fabrication processes, carbon-carbon
  composites, 353-357
FCC operation, schematic diagram, 101f
Feedstock
  carbon black formation, 271
  characterization in RCC process, 102
  coker, green coke quality, 242
  composition for slurry oil, 108
  conversion, carbon black, 278
  physical properties, 163
  quality during catalytic
    cracking, 102
  reduced crude, 107t

INDEX

Filler aggregate selection, anode carbon, 245
Filler aggregate sizing, anode carbon, 244-245
Flexural strength, composites subjected to graphitizing heat treatments, 370f
Fluid coke, formation, 240
Fogging, carbon fiber growth in stainless steel tubes, 339

G

Gas phase impregnation, difficulties, 353
General Performance Carbon Fiber, applications, 324,328-329
General Performance Carbon Fibers, description, 324
Graphite
  diffraction peak examinations, 208t
  resiliency of various materials, 209t
  rods, properties, 183t
Graphitization
  electron microscopic observations, 85-97
  structural stages, 90f
  sulfur influence, 95
Graphitizing carbons, mechanisms of formation, 2
Green coke
  binderless carbons, 50
  composition, 239
  desulfurization, 196
  dimensional changes during calcination, 184
  Gulf Coast prices, 153f
  markets for U.S. coke, 153f
  price vs. South African coal, 153f
  properties, 180t
  properties determined for new calcining method, 182
  sulfur effect, 230
  tested for new calcining method, 180
  U.S. supply and quality, 150f
  world markets for U.S. coke, 150f
  binderless carbons, 50
Gundai method
  high performance carbon fiber, 329
  preparation, 330f

H

High-modulus type fibers
  definition, 362
  surface properties, 362

High Performance Carbon Fiber
  hydrogenated pitch, 329-332
  nonhydrogenated pitch, 329
High tensile strength fibers
  definition, 362
  surface properties, 362
Hot stage microscopy
  microscope designed for quenching, 72f
  viscosity and optical texture relationship, 11
Hydrocarbons, tendency to smoke, 270
Hydrogen
  carbon black formation role, 280
  jacket, role in fiber growth, 341
Hydrogenated pyrenes, structures, 35f

I

Interlaminar shear stress, translation of the carbon-carbon composites, 368f
Iron
  carbons added, 200
  graphite added, 201
Iron-phthalocyanine
  chemical structure, 312
  crystallite size, 314
  dispersion on carbon support, 312-321
  growth, 317
  interaction with an oxygen complex, 320f
  mean particle size and mean crystallite size comparison, 316f
  particles per unit area of carbon support, 315f
  sintering, 317-318
  structure, 313f
  surface area, 312-314,320f
  surface area as a function of loading, 313f
  X-ray diffraction results, 314t
Isotropic pitch, viscosity-temperature curves, 51f

J

Japan
  imported coking coals, 251
  pitch-based carbon fiber, 328,323-332

K

Kinetics, carbonization, 225-228

Kyukoshi method
  mechanical properties of carbon
    fibers, 333f
  preparation, 330f
  viscosity of mesophase pitch, 333f

L

Lamallae, frequency of high
  temperature, 383
Lamellae, progressive crumpling, 91f
Lattice fringe imaging, principles, 86
Lignin
  molten, carbon fiber prepared, 325f
  whisker-like carbon, 325f
Liquefaction, hydrogen role, 20
Liquid crystal, wedge
  disclinations, 69f
Liquid impregnation
  pore filling, 353
  process in carbon-carbon
    composites, 353-357
  systematic studies, 357-369
Local molecular orientation (LMO)
  asphaltene sizes, 94f
  concept, 88-95
  dark-field images, 93f
  decrease with oxygen content, 95
  graphitizability and atomic ratio
    relationship, 96f
  increase with hydrogen content, 95
  lamellae similarity, 89
  morphology in bright-field, 93f
  size classification, 92t

M

Matrix
  bulk, 384f
  carbonization, 381
  crack propagation, 392f
  degree of crystallinity of
    carbon, 382
  fracturing, 381
  large-grained and fine-grained, 387f
  micrograph of sheath, 384f
  preferred orientation for carbon
    fibers, 383
Matrix microstructure
  carbon fiber, 381
    382-383
  physical properties of carbon-carbon
    composites, 383-384
  thermal expansivity influence
    explained, 386

Matrix precursor, carbon-carbon
  composites, 357-359
Maximum fluidity
  coal, 254
  coal blends, 255f
  relation to drum strength of
    coke, 255f
Medicine, carbon-carbon composites
  used, 347-353
Mesophase
  anisotropic, model structure, 37,39f
  carbonization
    control, 32-36
    rate affecting optical texture
      size, 12
  characteristics and
    preparation, 29-42
  chemistry, 1-25
  coalescence, 72f
  coke structure relationship, 5-11
  composition, 2
  crystallization, 23
  current research, 24
  definition, 1,38
  definition of bulk, 89
  deformed microstructure, 75f
  disclination structures, 64
  extrusion apparatus, 77f
  flow behavior, effect, 68
  formation
    chemical aspects, 11-20
    transferable hydrogen role, 19
  free radicals in mesophase-pitch
    system, 12
  geometry of disclination
    reactions, 74
  growth units
    anthracene-phenanthrene, 10f
    detected, 8
  hot-stage microscopy, 68-76
  incorporated oxygen, 42
  lamellae, 88
  lyotropic, 24
  molecular and strain models, 67f
  optical texture detected, 38
  pitch carbon fiber etched with
    chromic acid, 3f
  precipitation, 23-24
  preparation, 23,37
  pressure effect on formation, 8-11
  products, microstructure
    formation, 80-81
  pyrolysis, 71
  reactivity, 42
  rod
    annealing behavior, 76
    deformation and extrusion, 76-80
    microstructures, 78f
    orientation produced by light
      draw, 77f
    transverse microstructures, 79f

Mesophase--Continued
  schematic diagram of disclination
    loop, 70f
  schematic models, 69f
  schematic phase diagram, 38
  size, 8
  spider web model for constituent
    molecules, 327f
  starting materials, 37
  structure and phase transition, 37-41
  temperature effect on formation, 14
  theoretical amount during
    carbonization, 219
  time effect on temperature, 14
  uniaxial deformation of disclination
    loop, 81f
  viscosity measurements, 11
Mesophase carbon fiber
  fracture surface, 64
  microstructure formation, 62-82
  production, 63
  radial structure and open-wedge
    shape, 65f
  random-core structure and round
    shape, 66f
  recent advances, 63
  rheological behavior, 54
  rheology, 48
  schematic phase diagram, 39f
  tensile fracture surfaces, 66f,67f
Mesophase pitch, coalescence, 49
Mesophase pitch, formation, role of
    inerts, 20-22
Mesophase pitch
  pyrolyzed at constant heating, 68
  rheological properties, 49
  structural models, 67f
  transferred hydrogen in dormant and
    ordinary, 331t
Mesophase spheres, models for
    constituent molecules, 39f
Metallic derivates
  dispersion
    on carbon supports, 310-321
    on carbons, 311t
Metals, content, feedstocks
    importance, 163
Molten salt media,
    carbonization, 326-328
Monoaromatics
  catalytic cracking, 105
  slurry oil and coke factor, 113t

N

Naphthalene, condensation
  reactions, 33f

Needle coke
  blended fractions origin, 19
  characteristics, 160
  mesophase flow behavior effect, 68
  operation yield, 166t
Nenatic liquid crystals, disclination
    interactions, 74t
Nitrogen, carbon black formation
    role, 280

O

Oil--See Crude oil
Oligomerization, mechanisms of
    hydrogenated pyrene, 36f
Optical anisotropy
  cocarbonization effect, 35f
  development, 31f
  pyrolysis mechanism, 30
  solubility relationship, 38-40
  steps involved, 30
Optical microscopy, resolution
    limitation, 5-8
Optical temperature, 21f
Optical texture
  coarse-flow anisotropy, 7f
  domains, 7f
  examined for coke, 16
  fine-grained mosaics, 6f
  index (OTI), formula, 5
  medium- and coarse-grained
    mosaics, 6f
  size
    influenced by QI material, 22
    sulfur effect, 17-19
    examined for coke, 16
Organic compounds, radical chemistry of
    pyrolysis, 12
Ostwald ripening, description, 317
Oxygen, carbon black formation
    role, 280

P

Pad-type dewatering, description, 158
Paste compaction, anode carbon, 245
Pit dewatering, description, 158
Pitch
  alicyclic structure importance, 326
  blending studies, 17
  blending to produce needle coke, 19
  characterization, 49
  characterization technique for pitch
    materials, 328

Pitch--Continued
  chemical modifications, 17-19
  chemistry, 326-328
  coke conversion, 4-5
  complexity of composition, 11
  components, 15
  dispersed phase effect on the
    apparent viscosity-temperature
    curve, 52
  kinetic analysis of nonisothermal
    gravimetry, 46-48
  pyrolysis chemistry, 45
  rheological studies during
    pyrolysis, 50
  rheology, 48
  spin-lattice relaxation time
    variation, 18f
  thermogram, 226f
  thermogravimetric data for pyrolysis
    products, 47f
  thermogravimetry, 46-48
  upgrading commercial value, 16-17
  volatile content and insoluble
    contents relationship, 47f
  See also isotropic pitch
Pitch-mesophase-coke transformation
  analytical and rheological
    techniques, 45-59
  diagram, 54,56f
Pitches
  basic structural units (BSU), 88
  pretreatment processes, 324
Polar aromatics
  catalytic cracking, 105
  molecular parameters for slurry
    oil, 109t
  slurry oil and coke factor in RCC
    process equation, 115t
Polyacrylonitrile (PAN)
  carbon fibers
    derived, 324
    produced, 359
Polygranular graphites, two-phase
  structures, 358f
Polyphenyleneacetylene
  carbonization shrinkage and
    translation of fiber
    strength, 366t
  shrinkage observations for
    composites, 364
Polyvinyl chloride, mechanisms of
  carbonization, 324
Pore blocking
  graphite body, 354f
  schematic mechanism, 356f
Pore filling
  graphite body, 354f
  schematic mechanisms, 356f
Porosity, Desulco graphitic carbon, 210
Potassium hydroxide, used in active
  carbon formation, 303-305

Prebake cells
  advantages, 247
  anode performance, 246-247
Pyrolysis
  carbon fiber growth in steel
    tubes, 339-341
  effluent gas
    fiber growth reactor, 339-341
    obtained in carbon fiber growth
      tube, 342f
  hydrocarbons described, 223
  processes
    carbonization, 217
    contribution, 219
  reaction path of hydrocarbons, 223
  rheological studies of pitch, 50

Q

Quinoline insolubles (QI)
  carbonization influenced, 223-225
    226f
  coke structure, 223
  formation
    rate, 224f
    reaction time effect, 123f
    temperature effect, 123f,127
    weight loss during
      carbonization, 224f
  fraction, anode binder pitch, 238
  material
    definition and
      classification, 20-22
    function during carbonization
      process, 22
  pitch size reduced, 383
  types, 223

R

Radical chemistry, mesophase
  formation, 12
RCC feedstock, slurry oil and coke
  prediction, 115t,116t
RCC operation, schematic diagram, 103f
Recarburiser, materials analysis, 201t
Reduced crude conversion (RCC) process
  feedstock characterization, 102
  slurry oil and coke yield
    prediction, 99-117
Reduced crude oil, analysis, 106-108
Residual oil
  addition test, 259
  effective reflectance and effective
    fluidity, 261t

Residual oil—Continued
  equations for effective reflectance and fluidity, 264t
  properties, 256t
  slurry oil and coke yield prediction, 99-117
Rheology, mesophase pitch, 49
Rotary hearth calciner
  advantages, 176
  cross section, 177f
  description, 176,243-244
  top view, 177f
Rotary hearth furnace, delayed coker, 168f
Rotary hearth method, calcining coke, 169
Rotary kiln calciner
  coke feed rates, 243
  delayed coker, 168f
  described, 173-176,243-244
  history of coke calcining, 172-173
  modern, 175f
  original, 174f
Rotary kiln process, calcining, 167-169
Rubber
  carbon black properties, 290
  See also Compounded rubber

S

Saturates
  catalytic cracking, 105
  slurry oil and coke factor for RCC process, 112t
Selected area diffraction (SAD), unscattered beam displacement, 87f
Shear deformation, bent flake of natural graphite, 363f
Shot coke, characteristics, 160
Sintering
  graphitized carbon mechanism, 318
  iron-phthalocyanine, 317-318
Slurry oil
  analysis, 106-108
  API gravity, 108
  factor for diaromatics in RCC process equation, 113t
  factor for monoaromatics in RCC process, 113t
  HPLC analysis, 107t
  HPLC scheme, 107f
  models of molecular types, 104f
  molecular parameters
    diaromatics, 109t
    polar aromatics, 109t
  percent diaromatic molecules, 110
  prediction equation in RCC unit, 110-114

Slurry oil—Continued
  saturates
    API gravity relationship, 110
    content vs. API gravity, 111f
Soderberg cells, anode performance, 246-247
Space shuttle, carbon composite parts, 348f
Sponge coke, characteristics, 160
Steam cracker tar
  aromatic pitch conversion, 134
  characteristics, 134,136t
  definition, 134
  pitches
    aromatic oil determined, 135
    characteristics, 137t,139t
    coking yields, 134
    molecular weight distribution, 135
    thermal process temperature effect, 138t
    thermogravimetric analysis, 135
    viscosity, 135
Sulfur
  coal properties, 256
  coke removal methods, 194
  content, feedstocks importance, 163
  graphitization influenced, 95
  reduced by calcination, 244
  thermal removal from coke, 194
  thermal-chemical removal from coke, 195
  See also Desulfurization
Synthetic aromatic pitch
  CCB distillate and residue fractions, 126-133
  production using steam cracker tar, 134-143

T

Tetrabenzo (a,c,h,j) phenazine
  carbonization, 324
  pitch, 327f
  structure, 325f
Thermal conductivity, anode carbon, 237
Thermal expansion
  carbon fiber, 382
  relationship to direction of composite, 387f
Thermal expansion coefficient
  calcining coke, 183,184,189
  crushing strength, 230,232f
Thermal expansivity, influence on microstructure, 385-388
Thermal stress resistance, anode carbon, 237

Thermogravimetric analysis
  CCB distillate fraction, 122t
  kinetics of carbonization, 225
  steam cracker tar pitches, 135
THF-insoluble pyridine soluble
    fractions
  optical micrographs, 41f
  structural indices, 40t
THF-soluble fractions, structural
    indices, 40t
Torsional shear
  bonding characteristics of carbon
    matrix, 388
  properties, composites, 391f
Transformation diagram
  definition of coke, 59
  mesophase-pitch properties
    predicted, 59
  pitch, 58f
Transmission electron microscopy
  basic structural units (BSU), 86
  LMO classification, 92
  LMO extent, 92
  mesophase growth units detected, 8
  techniques, 86-87

V

Viscosity
  carbon black, 297f
  mesophase influenced during
    carbonization, 32
  mesophase pitch pyrolyzed, 68,70f
Viscosity-temperature curves, isotropic
    pitch, 51f

W

Williams-Landel-Ferry equation,
    viscosity of pitch systems, 48-49

X

X-ray diffraction,
    iron-phthalocyanine, 314t

Y

Yield predictions, slurry oil
    development, 108
Young's modulus
  anode carbon, 237
  carbon-carbon composites effect, 369

Production and indexing by Keith B. Belton
Jacket design by Pamela Lewis

Elements typeset by Hot Type Ltd., Washington, DC
Printed and bound by Maple Press Co., York, PA

DATE DUE